HEAVY OIL AND TAR SANDS RECOVERY AND UPGRADING

HEAVY OIL AND TAR SANDS RECOVERY AND UPGRADING

International Technology

Edited by
M.M. Schumacher

Based on research by:

Roebuck Associates
Booz-Allen & Hamilton Inc.

NOYES DATA CORPORATION
Park Ridge, New Jersey, U.S.A.
1982

Library of Congress Catalog Card Number: 82-2229
ISBN: 0-8155-0893-X
ISSN: 0270-9155
Printed in the United States

Published in the United States of America by
Noyes Data Corporation
Mill Road, Park Ridge, New Jersey 07656

10 9 8 7 6 5 4 3 2 1

Library of Congress Cataloging in Publication Data
Main entry under title:

Heavy oil and tar sands recovery and upgrading.

(Energy technology review, ISSN 0270-9155 ;
no. 78)
Includes bibliographies and index.
1. Petroleum engineering. 2. Oil sands.
I. Schumacher, M. M. II. Roebuck Associates.
III. Booz, Allen & Hamilton, Inc. IV. Series.
TN871.H38 622'.3382 82-2229
ISBN 0-8155-0893-X AACR2

Foreword

This book, based on research by Roebuck Associates and Booz-Allen & Hamilton Inc., provides an in-depth assessment of international technology for the recovery and upgrading of heavy crude oil and tar sands. The technologies included are currently in use, under development, or planned; emphasis is placed on post-1978 activities.

Processes for converting heavy crudes, and bitumen from tar sands, to clean liquid products are becoming more important to the refiner. At present, an imbalance exists between petroleum supply and demand because the industrialized countries with high petroleum demand either have minimal oil production or are rapidly depleting their oil reserves. Until 1981 imported oil prices had increased substantially, and unstable political situations in some producing areas have previously interrupted oil supplies and could do so again. Thus, the oil-consuming countries are now faced with economic instability and possible future threats to their national security. These forces have combined to give increased urgency to the consuming countries' efforts to develop alternative sources of liquid fossil fuels that are based on indigenous resources. Most of the easily developed on-shore deposits of higher quality oil have already been developed; thus, it may be necessary in the future to exploit lower grade oil resources, such as heavy oil reserves and the very heavy oil in tar sands.

The heavy oil technologies and processes considered in Part I include methods relating to the exploitation of heavy oil reservoirs, such as production from underground workings, all types of improved or enhanced recovery, subsurface extraction, and well rate stimulation. Furthermore, even though heavy crudes are understood to include only those liquid or semiliquid hydrocarbons with a gravity of 20°API or less, technology applied to lighter crude oils with in situ viscosities of the same order of magnitude as some U.S. heavy oils is also included.

The scope of the tar sands section (Part II) includes sizing the resource base and reviewing and evaluating past, present, and planned research and field developments on processes for mining, producing, extracting, and upgrading very heavy oils recovered from tar sands, e.g., bitumen recovery from tar sands where primary production was impossible because of the oil's high viscosity. On the production side, very heavy oil is defined as having a gravity less than 10° to 12°API and greater than 100,000-centipoise viscosity at 50°F. On the upgrading side, hydrocarbons whose characteristics dictated additional processing prior to conventional refining into salable products (1050+° material) were included, regardless of origin, in order to encompass all pertinent upgrading technologies.

The information in the book is from the following documents:

> *A Study of Foreign Technologies Relating to the Exploitation of Heavy Oil Reserves,* prepared by Roebuck Associates, Dallas, Texas, for International Energy Technology Assessment Program, Energy and Research Planning Group, under Glenn Werth, Lawrence Livermore National Laboratory, Livermore, California, under the auspices of the U.S. Department of Energy, September 1980.
>
> *An Assessment of International Tar Sands Recovery and Upgrading Processes,* prepared by Booz-Allen & Hamilton Inc., for International Energy Technology Assessment Program, Energy and Resource Planning Group, under Glenn Werth, Lawrence Livermore National Laboratory, Livermore, California, under the auspices of the U.S. Department of Energy, December 1980.

The expanded table of contents is organized in such a way as to serve as a subject index and provides easy access to the information contained in the book.

Advanced composition and production methods developed by Noyes Data are employed to bring this durably bound book to you in a minimum of time. Special techniques are used to close the gap between "manuscript" and "completed book." In order to keep the price of this book to a reasonable level, it has been partially reproduced by photo-offset directly from the original reports and the cost savings passed on to the reader. Due to this method of publishing, certain portions of the book may be less legible than desired.

Notice

CONTENTS AND SUBJECT INDEX

PART I: HEAVY OIL

PART II: TAR SANDS

Part I

Heavy Oil

The information in Part I is from *A Study of Foreign Technologies Relating to the Exploitation of Heavy Oil Reserves* (UCRL Report 15311), prepared by Roebuck Associates, Dallas, Texas for International Energy Technology Assessment Program, Energy and Research Planning Group, under Glenn Werth, Lawrence Livermore National Laboratory, Livermore, California, under the auspices of the U.S. Department of Energy, September 1980.

1. Introduction and Executive Summary

1.1 INTRODUCTION

The purpose of this study and report is to provide an in-depth assessment of foreign technology relating to the recovery of heavy crude oil in order to recognize and appraise processes and techniques that could influence or aid the development and exploitation of comparable resources in the U.S.A. To accomplish this purpose, it was assumed and stipulated that this should include technologies in use, under development, and planned in the context of U.S. technology, with emphasis on post-1978 activities. Furthermore, even though heavy crude oil is understood to include only those liquid or semi-liquid hydrocarbons with a gravity of 20°API or less, technology applied to lighter crude oils with in situ viscosities of the same order of magnitude as some U.S. heavy oils are also included. By the same token, technologies primarily useful for the recovery of tar sand deposits and shale oils have been included if also applicable to heavy oil recovery.

The technologies and processes considered include all means relating to the exploitation of heavy oil reservoirs, excluding any type of direct mining, but including production from underground workings, all types of improved or enhanced recovery, subsurface extraction, and well rate stimulation.

The mechanics of the study included a comprehensive search and evaluation of available published materials relating to the recovery of heavy oils outside of the U.S.A., with emphasis on activities by foreign companies, individuals, universities, and governmental entities. Also included were attendance at and information from technical meetings and conferences, as well as direct conversations and correspondence with scientists and other experts engaged directly or indirectly in oil recovery research, planning, and operations outside of the U.S.A. However, no foreign travel was permitted according to the specifications of the assignment, so considerable use was made of telephone communications, and foreign persons traveling in the U.S.A. were contacted whenever possible for personal conversations.

The principal problem relating to the exploitation of heavy oil deposits is the high to ultrahigh viscosity of the subsurface fluids. Such viscosities result in low or nonexistent primary recoveries and well producing rates, as well as unacceptable displacement efficiencies by any conventional means of fluid injection. Therefore, most past, current, and planned recovery and well stimulation processes, both domestic and foreign, involve techniques for in situ viscosity reduction or dilution, chiefly by thermal means. Other technologies include those developed initially for lighter, relatively viscous oils which can by applied directly or modified in some manner for exploitation of some range of heavy oils.

Therefore, Section 2.0 of this report deals with the basic, relatively well-known recovery processes for heavy and other viscous oils. Then, Section 3.0 describes the modified and innovative processes related to heavy oil recovery being developed and in operation outside of the U.S.A., and Section 4.0 describes new developments in surface and downhole equipment. Some U.S. technology is also included in these sections to assist in the overall evaluation and where such processes are related to foreign developments. Finally, Section 5.0 includes a survey and discussion of the location, by country, of known heavy oil reserves and past and present research, tests, and field operations related to heavy oil recovery in each country.

Where possible and when pertinent, the physical properties of the reservoir rocks and fluids are related in the text. Also, specific economics or economic examples are given in a few cases, but no claim for any detailed economic evaluation of any process or application is made. Generally, the true economics of any method of oil recovery can be determined only after several years of field-scale operations. Such history and data are just not available for any innovative foreign technology, with few exceptions, and in those exceptional cases the economic standards and environment do not permit comparisons with applications in the U.S.A. Also, the rapidly changing world-wide situation with respect to oil supply and pricing makes the economics of any process very site-specific and makes comparisons of economics country-to-country very difficult, if not impossible.

1.2 EXECUTIVE SUMMARY

Significant and important deposits of heavy crude oils are known to exist in a relatively (or totally) unexploited state in many countries of the world. This section presents major conclusions and observations regarding those deposits and significant past and current foreign efforts in research, pilot testing, and field applications of heavy oil recovery processes.

1.2.1 Definition and Problem Statement

For purposes of this study, heavy oils have been defined as those with a gravity range of 10 to 20°API; however, lower gravity deposits have also been so-classified where the hydrocarbons exist in the subsurface in a fluid or semi-fluid state.

Recovery processes have been included and described where such are being considered or applied to heavy oil deposits. Other processes, intended for application to either lighter oils or heavier "tars," have also been considered if some applications to heavy oil recovery can be identified or inferred.

1.2.2 Basic Recovery Processes

During the past 30 years, certain recovery processes have become relatively well-defined and of almost routine application to some type of heavy oil reserves, at least on a technical, if not economic, basis. Most of these involve viscosity reduction of the heavy oil by thermal means, but viscosity reduction and displacement by physical and chemical means are also used.

In situ combustion has been applied for many years, but the current trend for heavy oil deposits is toward the "wet" forward combustion process. Reverse combustion essentially has been abandoned.

The majority of thermal projects consist of the injection of surface-generated steam, initially on a cyclical basis for well rate stimulation, followed by continuous steam injection and displacement. Various gaseous and liquid steam-additives are being tried to enhance the displacement efficiency.

Carbon dioxide injection for viscosity reduction and viscous displacement by injected polymer and microemulsion solutions can be applicable to some heavy oil deposits with viscosities of no more than 100 cp.

Chemical flooding for in situ wettability reversal, emulsification, or both can also be applicable to less viscous heavy crude oils, especially when combined with some means of thermal or viscous displacement.

1.2.3 Modified and Innovative Processes for Heavy Oils

Considerable research and testing is being conducted for the improvement of basic thermal recovery processes and for the development and application of other methods of reservoir heating.

Electrical heating by resistance, induction, and magnetic fields is receiving considerable attention, especially in Canada. These methods would serve primarily as well rate stimulation processes, but are also being considered and/or tested for reservoir preheating to improve the thermal efficiency of a steam injection process to follow.

The Soviets have successfully applied cyclic steam injection through boreholes drilled from shafts and chambers in drainage mines at Yarega. They are installing other thermal mines of this type elsewhere, but labor costs would prohibit the direct application of this technique in the U.S.A. and other Western Hemisphere countries under current economic environments.

Research and field testing of a modified "thermal mining" process are progressing in the U.S.A. and Canada under the labels of HOP and MAISP processes. These consist of steam injection and oil production, with lateral boreholes drilled from central shafts or tunnels, combining experience with steam injection processes and older gravity drainage techniques and with impetus from Soviet thermal mining activities.

Other promising techniques for improving the efficiency of steam injection processes include the addition of viscosity-reducing additives, gases such as carbon dioxide and methane, and caustic emulsifiers.

Miscellaneous other processes for application to heavy oil recovery include the injection of direct solvents and in situ oil modification or generation of biopolymers or surfactants by microbiological means. The Soviets have also reported laboratory results from the application of sonic vibrations and displacement with "magnetized" water. None of these processes or methods are sufficiently developed to be of near-term interest, and the sonic and "magnetized" water experiments are of doubtful value for heavy oil recovery.

1.2.4 Surface and Downhole Equipment

Considerable, valuable work is being done on the development and application of equipment and materials to assist in the production of heavy oils and to improve the thermal efficiency, economics, and pollution aspects of steam injection processes. However, most of this technology is being developed in the U.S.A or has sprung directly from such efforts, even though major applications are under way in foreign countries.

The most significant foreign activities are in Venezuela where more emphasis has been placed on the application of means of thermal insulation in steam injection wells. However, this work is based primarily on U.S. technology.

1.2.5 National Efforts in Heavy Oil Recovery

Heavy oil and other viscous oil deposits are found in most of the major depositional basins of the world, with the largest known deposits being in Canada, Malagasy, the U.S.A., the U.S.S.R., and Venezuela. Not surprisingly, the major research and field tests and applications with respect to heavy oil recovery processes are in Canada, the U.S.A., the U.S.S.R., and Venezuela. Elsewhere, little or no original work is being done.

Canada has large deposits of heavy oil and tar sands in Alberta and Saskatchewan and is heavily engaged in the development of processes to exploit these major reserves. Processes in operation or under field test in Canada at this time include conventional steam stimulation and displacement, "dry" and "wet" forward combustion, caustic flooding, and steam injection with preheating by combustion and by electrical resistance heating. Active research projects for heavy oils include the use of steam additives, mine-assisted steam injection, electromagnetic and electrical induction heating, in situ emulsification, improved hydraulic fracturing, and microbiological processes.

France has only minor amounts of domestic heavy oil reserves but is actively engaged in research and some field testing of improved and enhanced oil recovery processes, some of which are applicable to heavy oils, because of a heavy dependence on foreign oil sources. Currently, the first priority for production methods research at the French Petroleum Institute (IFP) is related to heavy oil recovery processes. The Association for Research on the Techniques of Petroleum Exploitation (ARTEP) is guiding all research, with most of the fundamental research being done at IFP and supplemented by applied testing in the laboratories of the producing companies. Several universities are also involved in these activities. Much work is being done on steam injection, microemulsions, and polymer flooding.

Romania has significant deposits of heavy crude oil and has been engaged for many years in recovery-improvement activities, including conventional waterflooding, in situ combustion, steam injection, polymer flooding, micellar solutions, carbon dioxide injection, microbiological processes, and drainage mining.

The U.S.S.R. has large, scattered deposits of heavy and other viscous oils and, through the years, has at least experimented with essentially all the various oil recovery improvement techniques, in addition to developing and applying at least one unique and original method. Conventional waterflooding is, by far, the most extensively used improved recovery technique, with steam injection and some in situ combustion being used in heavy and otherwise viscous oil reservoirs. Thermal mining with steam is a large operation at Yarega and is under developement in the Baku area. Other methods being considered or in use for heavy oils include alternating steam-water injection, electrical downhole heating, thermochemical heating, chemical and micellar flooding, carbon dioxide injection, microbiological processes, and sonic vibrations.

Due to a variety of reasons, Soviet oil production technology, logistical support, and operations lag behind that of the West, except for the thermal mining techniques which would be difficult to apply elsewhere under current economic conditions. However, a growing emphasis on the development of oil reserves within the U.S.S.R. will likely see this situation changing.

Venezuela has large heavy oil deposits, with in-place amounts estimated to range up to two trillion barrels or more. As a result, thermal processes for heavy oil recovery, including in situ combustion and steam injection, have been in operation for many years. These and other activities in the development of thermal insulation and heavy-crude lift systems, are either the direct application of U.S. technology, or a continuation of technology first applied by U.S. oil companies operating in Venezuela. More recently, cooperative arrangements with groups in Canada and the U.S. have been made in an attempt to introduce new technology as it becomes available and to direct some efforts toward the specific properties of Venezuela reservoirs rocks and fluids.

Various other countries have applied relatively conventional means of heavy oil recovery or are turning their attention in that direction. No innovative, unique technology has been developed. It is of interest to note that Hungary has had some success in caustic flooding with ammonium hydroxide and that no major developments have yet occurred in the area of microbiological processes for improving heavy oil recovery.

1.2.6 Survey and Evaluation Results

It appears that the most likely sources for new heavy oil recovery technology applicable to U.S. reserves will be from the activities in Canada, where most of the original research and technology development outside the U.S.A. is occurring. This is not to presume that some unique and valuable development has not and will not be made elsewhere, but at this time the most important contributions from other countries are the technical data that can be acquired to reveal the performance characteristics of newer technology which is also being applied in the U.S.A.

2. Basic Recovery Processes for Heavy Crude Oils

2.1 DEFINITION AND PROBLEM STATEMENT

Heavy crude oils have been defined, somewhat arbitrarily, as those crude oils with an in situ gravity of 10 to 20°API. The lower gravity limit is not well established and includes those hydrocarbons resident in porous media in a fluid or semi-fluid state and, therefore, may include oils with a gravity as low as 6 to 7°API in some cases. This total gravity range corresponds to a range of specific gravity of from 0.934 to approximately 1.029. Such crude oils generally exist in relatively young, often poorly consolidated beds, and at relatively shallow depths. Heavy oils are usually young oils with respect to the overall petroleum maturation process, thereby normally containing large percentages of non-paraffinic hydrocarbons. They are more often found in clastic rocks, as opposed to carbonates, and are found in many of the world's oil producing provinces.

The physical and chemical properties of heavy crude oils and the circumstances of their occurrence establish immediately the low or even non-existent recovery potential by primary means and the difficulty of obtaining substantial recovery by existing means of improved or enhanced recovery. With relatively low reservoir pressures and minimal gas evolution potential, there is only a low level of internal expulsive energy. High viscosity and density require more energy for moving such oils through the reservoir and up the wellbores. In addition, many if not most, heavy oil reservoirs tend to be more oil-wet (less water-wet) than do reservoirs containing older, more mature, low sulfur, lighter crude oils. This characteristic results in a larger residual wetting volume and greater displacement pressures in pore throat passages under displacement by any non-wetting fluid phase.

As a result, primary recovery of heavy crude oils by natural energy mechanisms ranges from essentially zero to perhaps as much as eight to 10 percent of the original oil-in-place. Standard commercial waterflooding may increase recovery of the lighter and less viscous heavy crude oils to only as much as 20 to 25 percent of the original oil-in-place, still an entirely inadequate recovery of a large potential energy source.

In some areas, the very unconsolidated nature of the heavy oil bearing reservoirs constitutes another problem: that of excluding formation sand from the producing well-bores -- sand that can be tightly entrained in the dense, viscous crude oil and difficult if not impossible to exclude by usual downhole screens, slotted liners, and gravel-packing techniques.

In one locale or another, one or all of these problems must be attacked and solved to obtain commercial production from heavy oil reservoirs, to maximize recovery efficiency, or both. Methods used, experimentally or commercially, have been and are many, but most deal with the problem of high viscosity either by reducing the oil viscosity by thermal or other means, or by displacing the crude with a fluid of equal or similarly high viscosity. The former are at least theoretically feasible for crude of any measurable viscosity, but the latter are generally limited to crude oils with a viscosity of 60 to 80 cp or less, thereby eliminating all but the higher gravity, deeper (higher temperature) heavy crude oils. Following is a description of basic and generally well known techniques applicable to the recovery of heavy crude oils.

2.2 THERMAL RECOVERY METHODS

All methods of so-called thermal recovery are designed to heat the reservoir rock. This heat energy is then transferred to the crude oil by conduction and convection. Heat energy can be generated directly in the reservoir itself, such as in the in situ combustion process, or it can be generated externally and transferred to the reservoir by a carrier fluid such as steam.

2.2.1 In Situ Combustion

The in situ combustion process consists of the injection of oxygen (air) into the reservoir to create and propagate a combustion front through the reservoir. The rock is heated directly by and at the combustion zone and indirectly by the movement of combustion gases, unburned air, volatile hydrocarbons, and steam through the reservoir ahead of the combustion front. Most of this secondary preheating occurs as steam, resulting from heating of interstitial water and as a product of combustion, condenses to liquid water ahead of the front, thereby giving up its latent heat of vaporization.

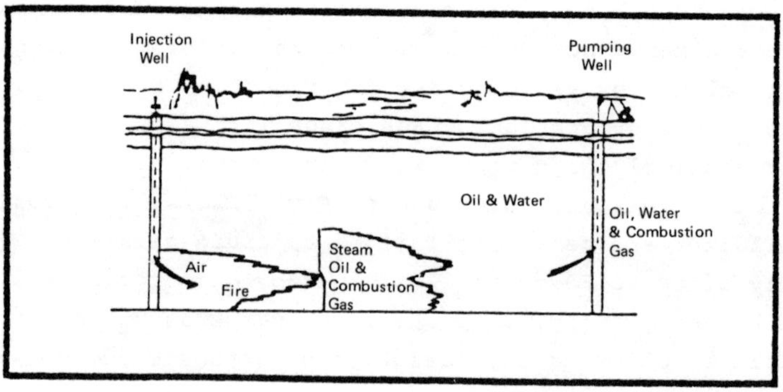

Figure 2-1. Dry Forward Combustion

Crude oil itself serves as the fuel for the combustion process but not in the liquid state. Rather, as the heating proceeds, volatile components are driven off, leaving behind a solid coke-like residue which serves as the actual fuel to be burned. This phenomenon serves to limit the classic forward combustion process in several ways. Heavy oils can form so much coke the reservoir is actually plugged to the extent that gas flow is restricted to the point of extinguishing the combustion. Also, air injection requirements increase with the quantity of coke formed, decreasing and even negating the economic attractiveness of the process in many heavier oils.

For these reasons, the standard forward combustion process is technically and economically most feasible for generally higher gravity oils in the 28 to 34°API range, oils which are succeptible to recovery by other, more efficient processes.

Other problems with the combustion process include formation plugging by swelling clays contacted by fresh, condensed water ahead of the combustion front, by overburning of the reservoir as the injected air tends to move toward the top of the formation, and by preferential up-dip burning in the tilted patterns of steeply dipping beds. Mechanical problems include severe corrosion in producing wells and the formation of oil-water emulsions which may require extensive and expensive treating.

The basic recovery mechanism of the in situ combustion process is water displacement of a thinned, heated oil augmented by a minor amount of gas drive and internal displacement. Other benefits are derived from some solvent extraction by the lighter materials boiled from the oil in the coking process, CO_2 solution with some attendant viscosity reduction, and the lighter gravity of the produced oil due to the removal and burning of the heaviest ends. Some thermal cracking of heavy ends can also be obtained.

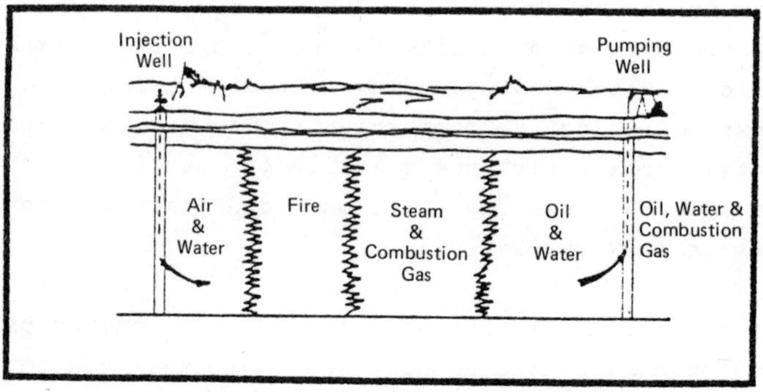

Figure 2-2. Wet Forward Combustion

The disadvantages of the process are being attacked mostly by the injection of a mist of water with the air and by the alternate injection of air and water. This so-called wet or partially quenched combustion is primarily an attempt to extend the applicability of the process to a broader range of lower gravity crude oils by cooling the combustion process. This can result in less coke formation in some oils, thereby reducing air requirements and inhibiting formation plugging, or in incomplete burning of the coke formed, thereby reducing air requirements. In some cases, this method can also reduce overburning tendencies and assist in uniform burnout of dipping, non-horizontal patterns.

Research and field trials of a so-called reverse combustion process have been abandoned for all practical purposes. This consists of establishing a burning front in one injection well, moving it out some distance, and then injecting air into surrounding wells while producing from the initial injection well. In this manner, the front continues to move toward the air source (surrounding injection wells) with the oil moving through the combustion zone and the preheated rock to the central producing well. This can result in a higher oil producing rate and, in some cases, less coking of the crude oil, but adequate control of the rate and direction of movement of the burning front has been very poor or impossible.

Because of the thermal degradation of carbonate rocks, the combustion process has been restricted to clastic, non-calcareous formations.

2.2.2 Steam Injection

External heat generation and injection is accomplished by means of steam stimulation and drive processes. Again, the ultimate recovery mechanism is by water displacement of a heated, lower viscosity oil, but several other mechanisms and effects are involved in these methods. Recoveries beyond that obtainable solely by the effects of viscosity reduction and mobility ratio improvement include steam distillation of lighter ends with a resulting solvent extraction effect, thermal expansion which yields a lower-mass residual oil, and to some extent a developed internal gas drive.

Basically, steam serves only as the medium for heat transport into the reservoir, heating the rock primarily by the latent heat of vaporization released upon condensation to liquid water. Problems related to this method are inherent to both the process itself and to the reservoir rock and fluid characteristics. The presence of significant amounts of hydratable clays can prevent steam transport and displacing water phase movement for any significant distance away from the injection well. The presence of formations with a high water content within, above, and below the reservoir rock, including thick shale sections, can act as heat sinks, thereby robbing heat energy to the point of disrupting the technical or economic feasibility of the process, or both.

Applicability of the process is restricted to those oils where the thermally reduced viscosity is low enough to mobilize the oil and result in displacement efficiencies of a reasonable magnitude. Likewise, it is restricted to depths where downhole injection pressures do not exceed the critical pressure of the injected steam, and surface and downhole pipe and equipment must be such as to minimize heat losses to the point that steam actually enters the formation at a reasonable cost and with practical fuel requirements. Finally, the steam process is limited to areas with an economical fuel supply and with adequate quantities of water suitable for conversion to steam. This last restriction may prove to be the most serious problem of all in the years to come in many areas of the world.

As generally employed, the steam process is begun as a well rate stimulation treatment. Steam is injected for periods ranging from several days to several weeks to heat the reservoir rock surrounding the wellbore. This is followed by a period of weeks to months (or years) duration with the injection well on production. As the heavy oil moves toward the wellbore, it passes through an increasing temperature, absorbing heat and undergoing a reduction in viscosity with an accompanying increase in mobility. Since the oil carries heat from the reservoir, the process has a diminishing return, and ultimately steam must be injected into the well again to reheat the rock. Then the well is put back on production. After three to five of these six to 18-month cycles (hence the names: cyclic·steam injection and "huff and puff" steam injection), steam is injected continuously in alternate wells, and the method becomes one of straightforward steam/water displacement, commonly referred to as steam drive.

In heavy oils, cyclic steam injection is usually effective over an area of only two or three acres around a well. Therefore, most steam projects are restricted to reservoirs that are developed or can be developed economically to no more than five acres per well. Actually, current average spacing is less than that, but many of the older fields were already developed to two or three-acre spacing before the advent of steam injection operations, so current project spacing patterns can be misleading.

Other problems can be associated with steam injection operations. Emulsification of the oil may result which can be disadvantageous and costly. Steam, as a gas, will tend to rise and move through the top of the reservoir, creating steam and water layers in the interwell zone. With the same gravity forces, steam will tend to move preferentially updip in non-horizontal beds, disrupting areal coverage of the reservoir. The low reservoir pressures associated with heavy oils and with depleted reservoirs provide only a low level of internal energy to move oil to producing wells during stimulation cycles.

Various of these problems are being attacked, and considerable experimentation is under way to improve the efficiency and the efficacy of the steam process. Lateral boreholes are being drilled for significant distances -- adapting well known water well and early petroleum technology to obtain deeper heat penetration and distribution and to serve as drainage sumps in low pressure reservoirs. Alkalis, surfactants, nitrogen, and CO_2 are being injected with steam for additional viscosity reduction, emulsification and entrainment, wettability alteration, and overall improvement in the displacement efficiency. Downhole steam generation units are being perfected for more concentrated application of generated heat to the reservoir. Solid fuel steam boilers are being developed to ease the restrictions on current fuel requirements. Finally, means of insulating downhole injection/production tubing for heat conservation are being tested with as-yet little economic benefit.

2.3 OTHER VISCOSITY REDUCTION METHODS

Methods other than thermal have been used for viscosity reduction of heavy oils. The first of these was the so-called Orco Process, or carbonated waterflooding.

2.3.1 The Orco Process

The solution of CO_2 into viscous oils can cause dramatic decreases in viscosity, thereby increasing overall water displacement efficiency. The Orco and similar processes consisted of the injection of pressurized, carbonated water into "viscous oil" reservoirs. As pressure decreased with distance from the injection well, CO_2 evolved from the water and was absorbed by oil with an attendent reduction in viscosity. Water was therefore used both as a carrier fluid for the CO_2 and as the displacing fluid. This approach was developed because of relatively short supplies of CO_2 at the time and the expense and difficulty of injecting gaseous CO_2. More areal contact between crude oil and CO_2 within the pore system was also achieved by this technique.

2.3.2 Flue Gas Injection

Flue gas (actually stack gas) injection grew out of carbonated waterflooding as a means of obtaining larger quantities of CO_2. In this process, hydrocarbons are burned to generate CO_2, and all the gaseous products of the combustion are injected. Ultimately, in heavy and viscous crudes, the flue gas slug is followed by displacing water. Although viscosity reduction by CO_2 solution was and, in many cases, remains the primary aim, additional benefits can be obtained from the other gaseous elements, such as nitrogen, carbon monoxide, steam and water, and acidic products of sulfur radicals in the hydrocarbon fuel. Such benefits result from additional viscosity reduction, some vaporization of lighter ends from the crude oil, and the formation of surface-active (surfactant) compounds which can result in wettability alterations, emulsification, and perhaps some miscibility effects.

2.3.3 Carbon Dioxide Injection

With increasing supplies of relatively pure, naturally produced carbon dioxide, these processes have evolved into the direct injection of gaseous or liquid CO_2 to the point of achieving a miscible-type displacement with some oils existing in a relatively narrow range of reservoir temperatures and pressures. However, all these processes require the absorption of gaseous CO_2 by the crude oil, since liquid CO_2 is not miscible with liquid hydrocarbons.

Flue gas and pure CO_2 injection methods are generally confined to crude oils of intermediate gravities, with only the higher gravity range of heavy crude oils being susceptible to the viscosity reduction processes of these methods. Achievement of miscibility in heavy crude oils is generally not practical or possible. Too, many of the heavy crude oils will precipitate solids in the presence of CO_2 and (sometimes) nitrogen. Unburned oxygen in untreated flue gases can actually increase the viscosity of some heavy oils by oxidation or can even initiate the in situ combustion process.

2.4 VISCOUS DISPLACEMENT METHODS

For mobility ratio and fluid displacement improvement, an alternative to a decrease in oil viscosity is an increase in the displacing fluid viscosity. One means of accomplishing this is by the thickening of injection water by the addition of high molecular weight polymers.

2.4.1 Polymer-Water Injection

The use of polymers, chiefly polysaccharides and polyacrylamides, was pioneered and patented by Dow Chemical Company as the Dow Pusher process. Various polymer products from several commercial sources are now available.

The addition of such polymers to injection water, when tailored to the formation and water chemistry, can be used to increase water viscosity to perhaps 40 cp or more, but most practical use has been in the range of four to 20 cp. This can reduce the displacing water-oil mobility ratio to near unity for more viscous intermediate gravity crude oils and can extend waterflooding, as an acceptably efficient process, into the upper gravity range of those heavy crude oils with viscosities of 40 to 100 cp.

Because of the large sizes of the polymer molecules and the difficulty in moving viscous fluid systems through porous media, this process is restricted to formations with relatively large pore sizes and attendant high porosity and permeability, at least in the case of heavy crude oils. Polymer concentration requirements in terms of costs and mechanical problems associated with stabilization and pumping of the thickened water are also significant. Therefore, this process can be considered as generally feasible only for those heavy crude oils with gravities in excess of 17°API resident in high permeability formations where hydrocarbon compositions and temperatures yield in situ viscosities of 100 cp or less.

It is interesting to note that polysaccharide polymers achieve mobility control entirely by increased water viscosity; whereas, the polyacrylamides obtain mobility control by viscosity and by adsorption with an attendent reduction in water permeability within the reservoir.

2.4.2 Micellar Injection

Micellar or microemulsion flooding may ultimately offer another means of improved recovery for certain heavy crude oil reservoirs, although at this time such processes are being considered for commercial application only for lower viscosity, intermediate gravity crudes. Generally considered as a chemical flooding technique and for tertiary recovery applications, the chief benefit in non-depleted, high oil saturation, heavy oil reservoirs would come from mobility control rather than from wettability and miscible displacement effects.

In this process, either a water-external surfactant solution or oil-external microemulsion is injected and then pushed by a viscous polymer-water mobility buffer. The polymer concentration is tapered so that the ultimate injection fluid is untreated water. In tertiary applications, oil-external emulsions displace the reservoir oil miscibly; whereas, the water external aqueous surfactant solutions depend on the detergent effect of the surfactants.

For use in suitably permeable rocks, these materials can be constructed so as to exhibit effective reservoir viscosities of up to 20 to 40 cp. They must then be displaced by polymer-water buffers of equal or slightly lower viscosity. It is in this range of viscosities that microemulsion flooding is technically feasible for the higher gravity heavy crude oils. With or without the solvent or miscibility effects, the improved mobility ratio of such floods leads to increased oil recovery. At the present, however, costs of the chemicals required make this a marginal process even for lower viscosity crude oils. In addition, the necessary surfactants (petroleum sulfonates) are in short supply and cannot tolerate temperatures above 200°F, salinities greater than 150,000 ppm, or hardness greater than 1000 ppm.

2.5 CHEMICAL FLOODING METHODS

The injection of water containing various acids, bases, and some salts can result in improved waterflood recoveries of crude oils by one of four distinct effects: emulsification and entrainment, emulsification and entrapment, oil-wet to water-wet wettability reversal, and water-wet to oil-wet wettability reversal. Only those processes entailing the injection of caustic solutions have been applied to any extent for the exploitation of heavy crude oil reserves.

2.5.1 Emulsification and Entrainment

Caustic solutions, such as sodium hydroxide or potassium hydroxide, and alkaline salts, such as sodium carbonate and sodium silicate, will react with organic acids in oils to form soaps and other surfactants, thereby reducing oil-water interfacial tension and creating emulsions or altering the wetting conditions of the oil-water-rock system, or both. The chemical reactions that occur and the phenomena that result depend on the chemical nature of the oil, the formation water, and the solutions injected and, hence, are very complex and even unpredictable at times. Emulsifying soaps can create an oil-in-water emulsion, thereby entraining oil in a continuous flowing alkaline water phase and resulting in an increased displacement efficiency.

2.5.2 Emulsification and Entrapment

Larger caustic concentrations can result in water-in-oil emulsions (or no emulsion at all) due to lower interfacial tensions. In this manner oil droplets can be mobilized to flow with the caustic solution only to be entrapped again in pore throats too small to permit passage under normal pressure gradients. Although not recoverable, these entrapped droplets can result in a reduced water mobility and an improvement in both vertical conformance and areal sweep efficiency, effects very important to the recovery of viscous crude oils.

2.5.3 Oil-Wet to Water-Wet Reversal

Sodium hydroxide will react with some oils to create surfactants that result in a wettability reversal from oil-wet to water-wet. The chief benefit of this effect is to decrease the water relative permeability which results in an improved mobility ratio and increased conformance and displacement efficiencies.

2.5.4 Water-Wet to Oil-Wet Reversal

On the other hand, under certain conditions of temperature and natural or controlled pH and salinity, some acidic crude oils and reservoir rocks react with sodium hydroxide to reverse the system wettability from water-wet to oil-wet, while at the same time reducing the oil-water interfacial tension to a low value. The overall displacement mechanism at the front, in the best cases, then becomes a sequence of events: the oil-water interfacial tension is drastically reduced; the rock matrix is wet by oil; water droplets are dispersed (emulsified) within the now-continous oil phase and oil drains from between the alkaline water droplets leaving a residual emulsion containing only a relatively small amount of oil. An additional, final effect has also been noted. Since oil is left as a continuous, wetting phase within the alkaline flushed zone, extensive gravity segregation of oil can occur, thereby further reducing the final residual oil saturation to a very low amount.

A severe limitation to alkaline flooding processes is the intolerence to the presence of calcium and magnesium salts in any significant concentrations. In sands and sandstones, this can be overcome by a saline preflush or with a sodium carbonate buffer, but this limitation precludes the use of alkaline processes in carbonate reservoirs.

2.6 SUMMARY OF METHODS

All of the recovery methods described have been in existence in the United States and in at least the field research stage for 18 years or longer. All of them have undergone some field testing, and several now have widespread and growing applications. All of them offer at least some degree of technical feasibility for the recovery of portions of the world's heavy crude oil reserves. At this time, however, by far the largest number of routine applications in heavy oil reservoirs involve thermal-type operations. Still, much research into the other basic methods, into improved thermal methods, and into development of new, innovative methods continues. These activities, as they exist and proceed outside of the United States, are described in the following sections of this report.

3. Modified and Innovative Heavy Oil Processes

3.1 THERMAL RECOVERY METHODS

The majority of efforts directed toward the development of improved recovery techniques for heavy oil deposits involve mobilization of the oil through viscosity reduction by thermal means. Research and field tests involve both new and innovative technology and modifications and adaptations of previously used methods or of techniques proposed in years past but not economically feasible under pricing schedules of that time. Also, in several cases, combinations

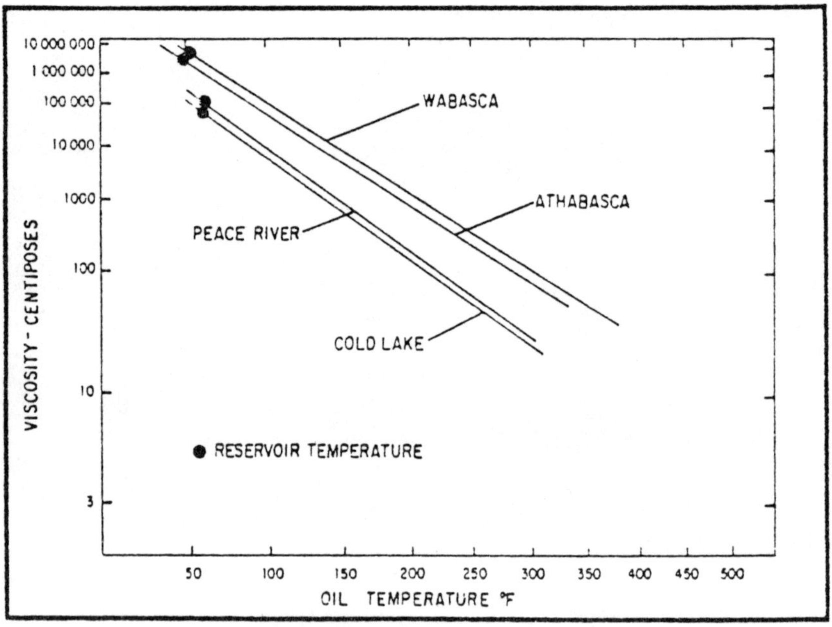

Figure 3-1. Viscosity-Temperature of Some Heavy Canadian Oils

of methods are under consideration and testing in attempts to improve upon an otherwise less-than-desirable application. Advancing technology and methods basically involving the application of thermal energy to the reservoir are discussed here.

3.1.1 Electrical Resistance Heating

Research into the application of alternating currents for resistive heating of in situ reservoir fluids was begun in the early 1970's in both Canada and the U.S.A. In Canada, work was initiated by ARCO Canada and is being continued today by Petro-Canada with the support and cooperation of Cities Service, Esso Resources, and Japan Canada Oil Sands. Installation of a field pilot test is nearing completion and is expected to require three years for evaluation.

This method consists of the resistive heating of the conductive reservoir fluids (primarily interstitial or injected water) by the imposition of a high-voltage A.C. potential between electrodes in adjacent wells. The construction

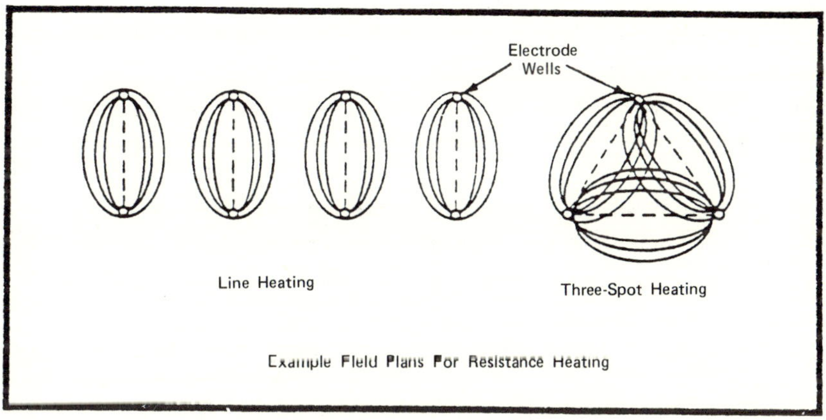

Figure 3-2. Electrical Resistance Heating

and placement of the electrodes and the voltage imposed can be designed to concentrate the current flow in a desired configuration and minimize energy losses to the rocks above and below the productive zone.

Much of the detailed design of this technology is yet to be established, but the electrodes will be some type of insulated, electrically conductive tubing on non-conductive centralizers designed so as to affect the open face of the productive zone. Annular and vertical insulation is necessary to minimize power loss and to protect equipment and workers at the surface. In field applications, two example arrangements of wells would be alternating lines of opposed-terminal wells and a triangular arrangement using three wells in each pattern and a three-phase alternating current. In either case, the wells must serve as both electrodes and as production or fluid injection wells.

The most obvious problem with this method is the generally high cost of electrical power where the generating fuel must be gas, oil, or coal. Therefore, resistance heating is being field tested primarily as a means of preheating the reservoir prior to the application of conventional steam injection or in situ combustion operations. Also, attempts will be made to recover and utilize "waste" heat energy; i.e., heat energy contained within the produced fluids.

It has been estimated by Petro-Canada that this method will require a minimum of nine years of additional trial design, testing, and experimentation before commercial development can begin. Problems that must still be solved include the flashing to steam of water near the wellbores (electrodes) due to high current density in those regions. This effect would disrupt the path of current flow and prevent any further heating. The solution will involve wellbore cooling, large diameter electrodes, and careful selection and control of input power. Other problems relate to the design and manufacture of optimum downhole and surface equipment, integration of electrical heating and production facilities, reservoir monitoring methods and equipment, and fool-proof hazard prevention.

The first field pilot consists of four electrode wells and eight observation wells in the Fort McMurray area of the Athabasca Oil Sands, Alberta, Canada. This pilot will test the method as a preheating process prior to conventional steam displacement operations.

A second proposed application of electrical resistance heating is the result of research conducted at the University of Missouri-Rolla. The theory of resistance heating is the same as in the preheating technique, but it has been proposed as a method to mobilize oil in lower permeability zones or in pores bypassed by fluid displacement. This is accomplished by injecting water of a selected resistivity before and/or during the electrical heating to create selective heating of the lower resistivity pore water.

In the case of multizones, due either to actual stratification or adverse permeability distribution, a high resistivity water can be injected to pass through and occupy the more permeable pores. The electric current would then pass preferentially through and heat the tighter pores still containing low resistivity formation water.

In the case of oil trapped in the reservoir outside of the area swept by fluid displacement patterns, water of an appropriate resistivity can be injected through the swept portion of the pattern to create a zone of low-conductivity. The maximum current flow and heating would then be in the unswept areas of high conductivity. This method has been shown to be analogous to electrical resistors in series; whereas, the first technique is analogous to resistors in parallel.

As yet, no field tests have been reported for these applications, but data from laboratory studies and a mathamatical model are available. It has been shown that the imposition of 1000 to 2000 volts of 60 Hz alternating current will result in a current flow of up to 12,000 amperes and an increase in average formation temperature of up to 160°F in 42 days. These results were derived from a model of a five-spot injection pattern encompassing 4.65 acres and with a formation thickness of 100 feet.

3.1.2 Electrical Induction Heating

As an alternative to resistance and dielectric heating, a process of induction heating primarily by eddy currents has been proposed and patented by Sidney T. Fisher of Montreal, Canada, for the in situ recovery of coal, shale oil, tars, and heavy crude oils. Various versions and aspects of the process have been tested in the laboratory at atmospheric conditions, and a field pilot test has been proposed.

For full-scale application, a vertical shaft 10 feet in diameter would be driven to the base of the heavy oil reservoir. From this shaft, a series of paired, horizontal tunnels two feet in diameter would be drilled at the top and base of the reservoir to form a radial or rectangular pattern over the project area for a distance of up to 1650 feet from the central shaft. Ten-inch diameter verticals would then be drilled from the surface to connect the ends of each pair of horizontal tunnels and to intersect each pair near

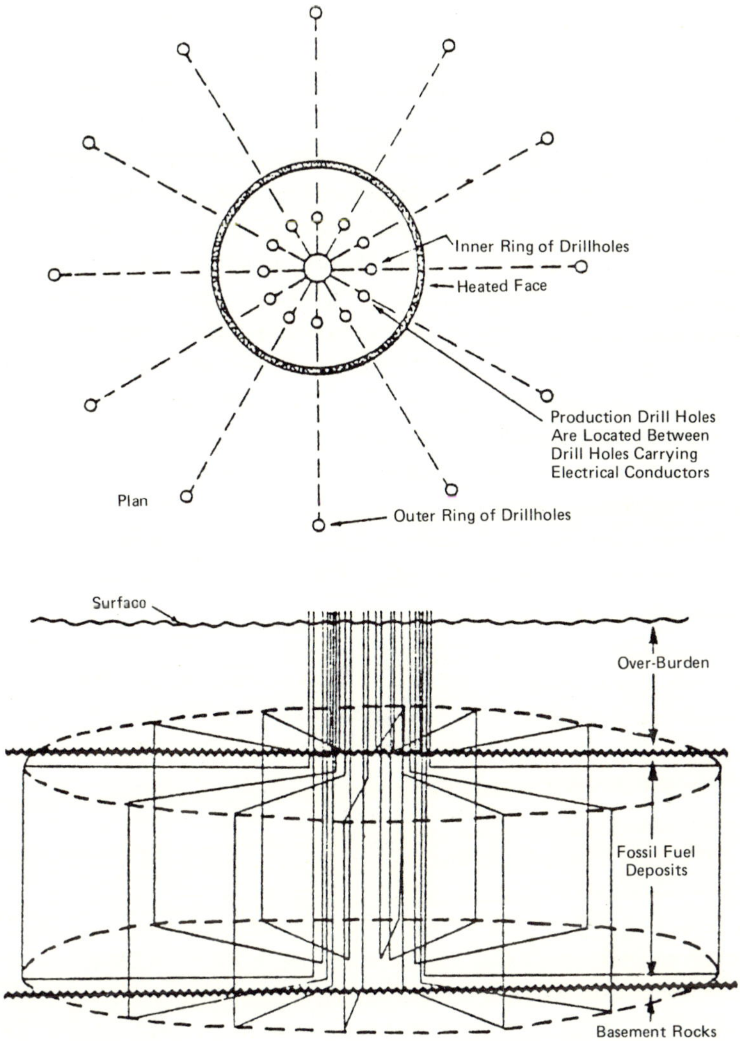

Figure 3-3. Induction Heating of Heavy Oil Reservoirs - I (After Fisher)

the central shaft. Production wells would also be drilled within the network.

The tunnels and connecting wells would be cased with electrically conducting copper pipe with an inner steel liner, or they would be lined with silica cement to serve as conduits for electrical conductor cable. In either case, the horizontals and the vertical connectors would thus form a toroidal coil capable of setting up a strong inductance field within the outer ring of verticals. When current is applied from the surface, a heating zone with temperatures of 200 to 300°C for heavy oils would begin as a thin cylindrical shell a few centimeters thick around the inner ring of conductors and would move slowly outward to the outer ring of conductors. These temperatures are sufficient to mobilize or volatize heavy crude oils and to assure that pyrolysis begins before vaporization of the interstitial water. Cooling water would be circulated through the system to protect the coils.

For a typical heavy oil, it has been estimated that, for ideal conditions, 154 Kw-hr/ton (25 Kw-hr/bbl) of oil produced would be required with a thermal/electrical net energy ratio of 8.8. However, it is expected that a net energy ratio of only about one-half of that value could actually be achieved.

An alternating current of 60 to 300 Hz would be used to maximize eddy-current heating. The lower frequencies can be used with increasing reservoir temperature and decreasing interstitial water resistivities. Therefore, it has been considered that a hot saturated solution of a metallic salt such as calcium chloride could be injected to obtain minimum resistivities, thereby forcing the induction heating operation to an entire dependency on low-frequency eddy currents. This would be especially applicable to coal and shale oil applications and less necessary for heavy oil projects.

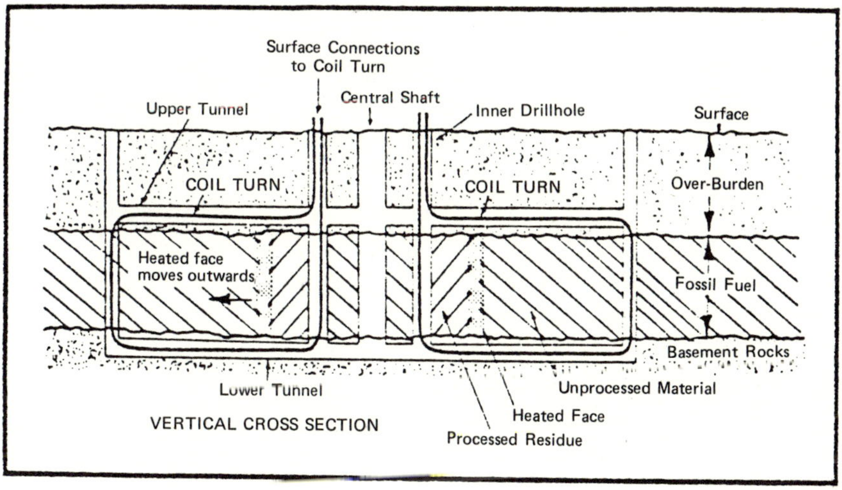

Figure 3-4. Induction Heating of Heavy Oil Reservoirs - 2 (After Fisher)

In the published articles on this process there are several estimates of oil recovery and economics for a "typical" application. From these an example project can be constructed in terms of 1980 economics:

Project Area = 550 acres
Porosity = 30 percent
Sand Thickness = 150 feet
Oil Saturation = 70 percent
Original Oil-in-Place = 134,410,000 barrels
Recovery Factor = 80 percent
Royalty Interest = 20 percent

	Total	Per Gross Bbl
Gross Oil Produced @ 80%, bbls	107,528,000	
W.I. Oil @ 80%, bbls	86,022,000	
W.I. Income @ $30.00	$2,580,672,000	$24.00
Operating Costs, 7 years	(1,020,000,000)	(9.49)
NET OPERATING INCOME	$1,560,672,000	$14.51
Investment	$ (860,000,000)	$(8.00)
FINAL NET REVENUE	$ 700,672,000	$ 6.51

Return of Investment = 0.81

Of course, these results are very approximate and must await a field pilot test for refinement, but they can serve for rough comparisons with other methods of heavy oil recovery.

3.1.3 Thermal Mining

The use of mining methods and technology to drive shafts and boreholes into and through oil-bearing sands to improve drainage, when combined with cyclic injection of steam to heat the rock and reduce oil viscosity, is referred to as thermal mining.

The concept and practice of subterranean drainage of oil-bearing sands through mine workings driven directly into and through the sands have been in existance for many years. As early as 1866, some 31 separate drainage tunnels, one of which had a length of 1900 feet, were driven to intersect oil sands in Ventura County, California. Prior to 1940, similar efforts were made in Kentucky, Pennsylvania, and

Texas, with other projects in the Pechelbronn field in the Alsace region of France, in the Wietze field in Germany, in the Higaskiyama field in Japan, in the Sarata-Monteoru field in Romania, in the Ebel field in Czechoslovakia, and in the Grozny field in the U.S.S.R.

This methodology has been applied to shallow, low pressure reservoirs containing relatively heavy oils or lighter crudes with little or no natural or remaining internal gas energy. The best-known and most extensive of these projects, that of the Pechelbronn field in Alsace, was begun in 1916 by the Germans in pressure depleted sands less than 800 feet deep containing a paraffinic 29°API crude oil.

In 1925, Leo Ranney patented and published the "Ranney Oil-Mining System" employing boreholes draining to mine shafts situated directly above or below the oil sands and driven from one or more shafts communicating to the surface. Ranney proposed that gas or air could be injected through one series of boreholes to displace oil to a parallel, producing series. He also proposed that steam could be applied through boreholes in alternate period of heating and drainage, elevating rock temperatures to reduce oil viscosity and improve drainage rates and recoveries. This basic technique is what is referred to today as thermal mining.

The currently most extensive thermal mining project is being conducted by the Soviets in the Yarega field complex near Ukhta in the KOMI A.S.S.R. as a joint effort between the All-Union Scientific Research Institute and the Yaregsk Petroleum Mining Department of "Komineft" Union. At Yarega cyclic steam injection is utilized with a system of mine shafts, galleries, and boreholes to recover up to 60 percent of a 17.5°API crude oil. This sulfurous, non-paraffinic oil has a viscosity of from 15,000 to 20,000 cp at a reservoir temperature of approximately 43°F, which decreases to 30 cp at 200°F, and is found in a consolidated, fine to medium grained quartzitic sandstone of middle Devonian age at a depth of 180 to 200 meters (590 to 655 feet).

The Yarega reservoir is a stratigraphic trap, folded, with some fracturing and minor faulting, containing some conglomeritic zones and argillitic intrusions, and completely underlain by water; but no active water influx mechanism has been detected. Net pay thickness averages 14 to 27 meters (45 to 90 feet); average porosity ranges from 20 to 25 percent; and average permeability is 3.2 darcies, with a range from 0.5 to 5.0 darcies. Initial oil saturation was reported to be 97 percent, and the original pressure was about 1500 kPa (218 psi).

The Yarega oil fields were first developed in 1940 by conventional wells with sucker-rod pumps, but even with well spacing distances of 70 to 100 meters (1.2 to 2.4 acres/well) only one to two percent of the oil-in-place was recovered. This was followed by conventional drainage-mining methods. These included two different systems of shafts and drainage wells.

The Uhktinsk system used mine shafts and galleries in the reservoir cap rock with wells 12 to 20 meters (39 to 65 feet) apart drilled from the workings to depths of 50 meters (164 feet). The second system used horizontal and inclined holes drilled from production galleries in the reservoir rocks. In this manner, 200 to 300 wells up to 300 meters (984 feet) long could drain an area of 8 to 12 hectares (20 to 30 acres). Since these methods increased the oil production to only as much as six percent of the oil-in-place, tests were begun to develop a method of thermal stimulation by the cyclic injection of surface-generated steam.

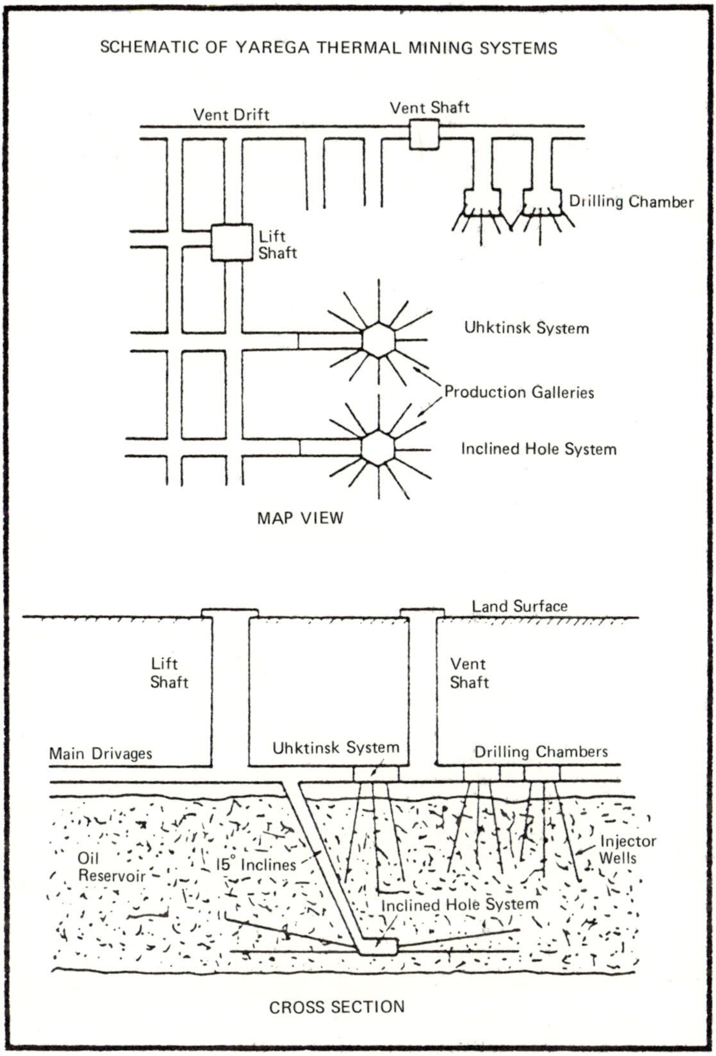

Figure 3-5. The Yarega Thermal-Mining System

Borehole steam injection in the Yarega reservoir was begun in 1968 and expanded to a commercial size project in 1972. The so-called thermal mines are adaptations of the drainage mining systems with the addition of steam injection wells drilled downward into the oil sands from chambers at the main working levels. Two types of vertical shafts are utilized, one to move men and machinery and one which serves for ventilation purposes. With the inclined-hole system, production wells radiate outward and upward from galleries near the bottom of the reservoir for distances of 182 to 244 meters (600 to 800 feet). Production wells drilled downward from galleries above the reservoir, according to the Uhktinsk system, are also utilized.

Steam utilization for this project is reported to have ranged from 5.5 to 2.2 tons of steam per ton of oil recovered, decreasing with time and improved techniques. A minimum steam requirement of 2.0 tons per ton of oil has been projected, amounting to approximately 670 lbs of steam per barrel of oil. Annual production rate has reportedly exceeded one million tons (16,500 bbl/day) of oil.

One unique aspect of the Yarega project is that produced oil and water are moved underground through open gutters, pumping substations, and pipelines to a central underground facility. There the oil and water are separated before the oil is pumped to the surface for storage, transportation and processing.

The productive formation at Yarega is a well-consolidated sandstone, so that tunnels do not have to be lined or supported. This reduces investment costs but limits steam injection pressures and volumes, since steam must not break through into the tunnel system. This method is also complicated by increased ventilation requirements with any crude oil containing significant quantities of gaseous components.

At least three mines are in operation at Yarega. The Soviets report that Mine No. 1 cost a total of 6.5 million rubles (about 8.8 million 1976 U.S. dollars), including costs of the complete underground mine system, aboveground structures, underground equipment, and all oil handling, storage, and transfer facilities. Production costs are reported to be about $5.75/bbl of oil. Of course, there is no way to either verify these costs or to relate them to equipment, labor, and operating costs in the U.S., but the Soviets state that the initial investment amounts to about one-half of the investment cost of a coal mine yielding a comparable amount of energy.

3.1.4 Mine-Assisted In Situ Recovery Process

A modification of the thermal steam/mining system at Yarega is the Mine-Assisted In Situ Recovery Process (MAISP) which is being considered for use in Canadian heavy oil sands, such as those at Cold Lake in Alberta. A vertical shaft five meters (16 feet) in diameter would be sunk 460 meters (1500 feet) to the base of the Cold Lake sands. From this shaft, a horizontal tunnel 3.5 meters (11.5 feet) in diameter would run a distance of up to 5000 meters (16,400 feet) along the base of the sand. This tunnel would be lined with concrete or cast iron because of the unconsolidated nature of the Cold Lake (and Alberta, in general) oil sands and to permit steam injection pressures of 300 to 500 psi (2070 to 3450 kPa).

From both sides of the horizontal tunnel a total of 332 12-cm (30-in.) injection/drainage holes would be drilled out through the base of the reservoir on 30-meter (100-foot) centers to a distance of 460 meters (1500 feet). Such a system would then underly a potential drainage area of 460 hectares (1137 acres).

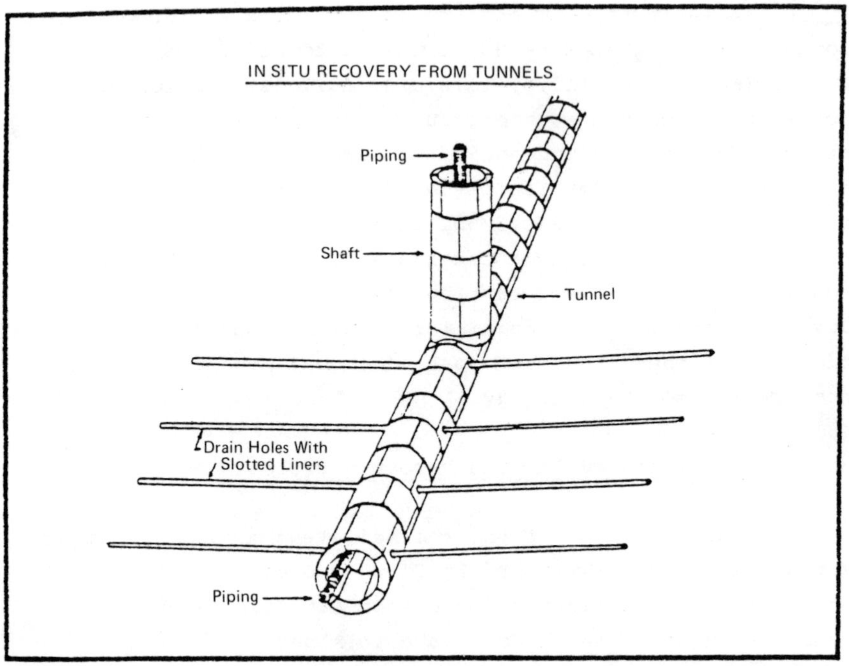

Figure 3-6. The Mine-Assisted In Situ Recovery Process (After Smith and Butler)

Steam would be injected from surface facilities through an insulated line running down the central shaft and through the horizontal tunnel. It could then be metered on a cyclic basis into the lateral holes which would be completed with slotted casing. This, in turn, would heat the reservoir and the oil, taking full advantage of gravity for the rise of steam in the reservoir and downward drainage of thinned oil into the laterals. Alternatively, recovery processes employing solvents or combustion techniques could be considered as technically feasible.

For application to the Cold Lake oil sands, it has been estimated that the MAISP system as described could be installed for a cost of approximately $80 million. Such a system could recover 50 percent of the oil-in-place, or approximately 86.8 million barrels, so this investment would amount to $0.92/bbl produced. The productive life was estimated to be five years with an average daily production of 150 barrels per tunnel well, or about twice that of a conventional cyclic steam (huff and puff) well.

The chief advantages of the MAISP method over conventional cyclic steam projects is in the recovery expected and the economics proposed. To obtain the same recovery from coventional wells, the field would have to be developed on very close spacing as is now the case in some areas of California where 50 percent recovery is achieved from some steam projects. For new development, though, conventional wells drilled on four-acre spacing would cost about the same as the tunnel well system described but would recover only 40 percent as much oil at one-half the average daily rate. Also, the tunnel well system can be used under terrain which is difficult or impractical to develop with conventional wells.

Disadvantages of this method include potential problems in the disposal of mine tailings, the large initial investment requirements, and the primitive status of current technology.

3.1.5 The HOP Process

A third method involving the use of thermally-assisted drain holes is the patented Heavy Oil Process (HOP) currently under installation for test purposes in the Kern River field of California by Barber Oil Corporation of New York. Although this cannot be considered foreign technology, it is under consideration for use in Canada and Venezuela and is included here for completeness.

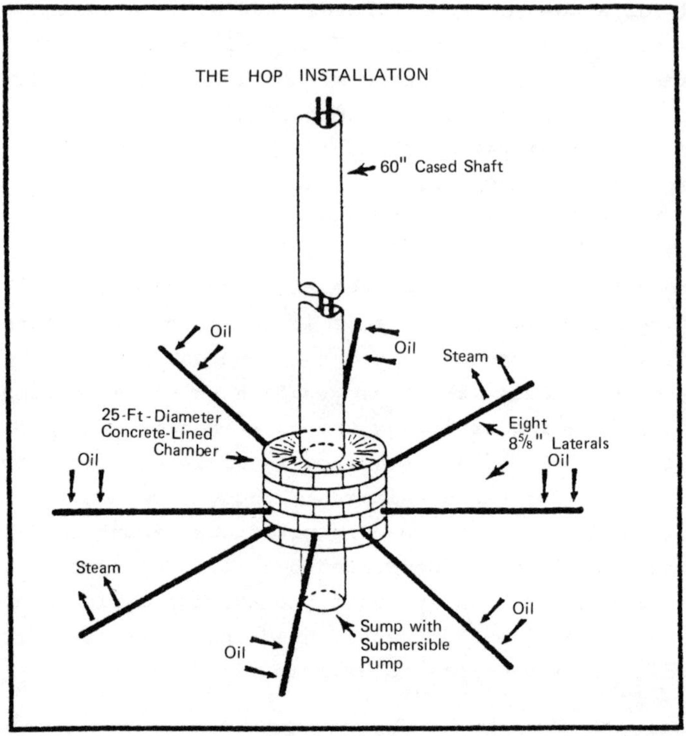

Figure 3-7. Heavy Oil Process (HOP), Kern River Field, California

This method is similar in concept to the MAISP method but utilizes a series of horizontal injection/production wells drilled radially from a subterranean chamber rather than from linear tunnels. For this reason it is also referred to as the "Hot Plate Method".

At Kern River, the installation consists of a central shaft 84 in. in diameter which is drilled in the center of a rectangular 25-acre block to the base of the oil sand at a depth of 700 feet. This shaft is cased with 60-in. diameter pipe, and a concrete-lined chamber is constructed at the bottom. From this 25-feet diameter and 20-feet tall chamber, eight lateral holes 8-5/8 in. in diameter are drilled out horizontally 400 to 700 feet to underly the 25-acre block.

Surface-generated steam is injected on a cyclic basis into opposite pairs of perforated laterals in a sequence that rotates around the area. Heated oil will drain downward into idle laterals and be lifted to the surface from a sump in the circular chamber.

For the initial installation, it has been estimated that approximately 68 percent of the 14°API oil-in-place, or 1.6 million barrels, will be recovered in a seven-year period. The estimated construction and installation costs are $7 million, or $4.38/bbl of produced oil. Steam injection requirements should be about 1050 lbs/bbl of oil, or approximately three tons per ton of oil. This would be competitive with standard steam stimulation projects in the Kern River field.

3.1.6 Steam with Preheating

Low permeability in many heavy oil reservoirs is due to pore blockage by the hydrocarbon presence itself. This occurs in reservoirs containing heavy viscous crudes at low reservoir temperatures of around 50 to 60°F. Results of conventional steam displacement projects can be severely hampered if not destroyed by channeling, low well injectivities, and general lack of mobility control in both the condensed water and heated oil phases. Methods are being applied, mostly in Canada, to attack the problem by warming the oil zone prior to starting steam injection operations.

In situ combustion is one such method being applied as a preheat process. A downhole fire is generated and kept burning to create a heat wave throughout the oil zone. This is done in the same way as conventional combustion projects. Selection of the air injection interval and location is critical to best utilize the heat generated to contact as much of the oil zone as possible.

Electric resistance heating, as previously described, is a recovery process in itself. When applied as a preheating operation its purpose is to mobilize oil-in-place which can then be displaced to the producing wells by subsequent injection.

Many heavy oil deposits are underlain by water. Where such water zones exist and are thin, 10 to 25 feet, steam is injected into the water zone to take advantage of the high mobility. The heat migrates upward to heat the oil zone mobilizing oil which can then be displaced by follow-up steam injection into the oil zone.

3.1.7 Combination Methods

Effects of additives injected simultaneously with steam are being tested in laboratory and field projects for the purpose of increasing steam recovery efficiency. The physical phenomena created by various additives in conjunction with heating by steam are:

(1.) Initial viscosity reduction of hydrocarbons near the injection wellbore for higher early injectivity.

(2.) Viscosity reduction out in the reservoir supplementing increasing temperature to increase the area contacted by steam.

(3.) Supplemental energy behind the steam displacement front to aid in driving out mobilized oil.

Solvents, when mixed and injected with steam supplement the effects of heat in viscosity reduction in the oil zone to improve oil mobility.

Gases such as CO_2 and natural gas (methane) can be miscible with the mobilized oil phase and serve to reduce the viscosity while also providing additional driving force for the steam displacement mechanism. Nitrogen has also been used as a steam additive and merely provides an available replacement volume for filling pore space in the reservoir as it is vacated by the heat-mobilized oil.

Caustic emulsifiers, namely NaOH, are being tested in conjunction with steam. The principal mechanism involved in steam-NaOH flooding is the combined effect of temperature-viscosity reduction and emulsification and entrainment.

3.2 MISCELLANEOUS OTHER METHODS

3.2.1 Viscosity-Reducing Surfactants and Solvents

Certain polymers and polyisobutylenes act to reduce the resistance to flow of oils through pipe. The mechanism is not well understood but has been postulated as a result of a surface active phenomena. These relatively low molecular weight compounds evidently form very long chain molecules, adhering to the surface which causes an apparent reduction in viscosity. This effect is noted only for non-laminar flow regimes and consequently is of no specific benefit in increasing flow within the reservoir rock. It could be of benefit for particular applications involving gathering and transport of heavy oils.

Solvents may be used to aid in the displacement of heavy oils from the reservoir, as well as in stimulation of individual wells. Laboratory tests have resulted in oil recoveries of 50 percent or more with solvents such as carbon disulphide, alcohol, methyl chloride, naptha, and heavier condensates; however, no field wide solvent floods for heavy oils are currently operative. Solvents act to reduce the viscosity; however, precipitation of asphaltenes may occur to reduce the effective permeability to flow.

3.2.2 Microbiological Methods

At the present time, there are a number of research groups, both U.S. and foreign, actively studying ways in which microorganisms can be used to improve oil recovery. These approaches can be divided into two general areas which overlap to some extent. One is the development of microbiological products (biosurfactants and biopolymers) which can be produced by normal fermentation methods, purified, and injected into reservoirs to improve recovery by various displacement mechanisms. The second approach is aimed at determining the limiting parameters involved in the injection of microorganisms directly into oil bearing formations where their activities will improve oil recovery, again by various mechanisms.

In Canada, several groups have been interested in working on microbiological methods for recovery improvement; however, none are actively engaged at this time due to lack of funding. One group at the University of Ontario in London, Ontario, has been working on the production of biosurfactants with the objective of developing in situ methods for using biosurfactant microorganisms in the tar and heavy oil sands of Alberta. However, the leader of this group, Dr. J. R. Zajic, has left to become Dean of Science at The University of Texas at El Paso, and funding problems may turn the group effort to more conventional fermentation studies.

The prime worker in Australia appears to be Dr. B. Bubela at Baas Becking Geobiological Laboratory. His work involves in situ production of biosurfactants. The only scientist known to be active in England is Dr. Vivian Moses at Queen Mary College in London. One segment of his work is reportedly a study of North Sea oils to develop enhanced recovery methods to be used after reservoirs have been depleted by primary means.

One group in France is reported to be working on microbiological processes. A group in Braunsweig, Germany, is working on biosurfactants with in situ use of the microorganisms as the ultimate objective.

Some of the most interesting work abroad has been done in Eastern Europe. Groups in Czechoslovakia, Hungary, Poland, Romania, and the U.S.S.R. have published extensively, and some have apparently carried out field testing of oil recovery by in situ use of microorganisms. There was still activity in this area as late as 1976. Some positive results were reported, so there is reason to believe that there is still some activity. However, it has been reported that these efforts have been abandoned in Poland, and no recent reports are available from the U.S.S.R.

More information on this subject will be forthcoming as papers are received for presentation at the International Conference on Microbial Enhancement of Oil Recovery to be held early in 1982. In the meantime, only the work in Canada has been definitely applied to heavy oils, but any research and field tests of these methods can be of interest in that regard.

3.2.3 Hydraulically Induced Vibrations

The concept of using compression waves to disrupt the natural capillary forces in fluid-containing porous media is not new. Sonic devices have been lowered into wellbores in the United States for the past few decades to "shake loose" heavy asphaltenes from the wellbore and adjacent formation. This action affects not only crude oil retained by capillary forces, but those heavy ends precipitated due to production processes and retained by mechanical bonding.

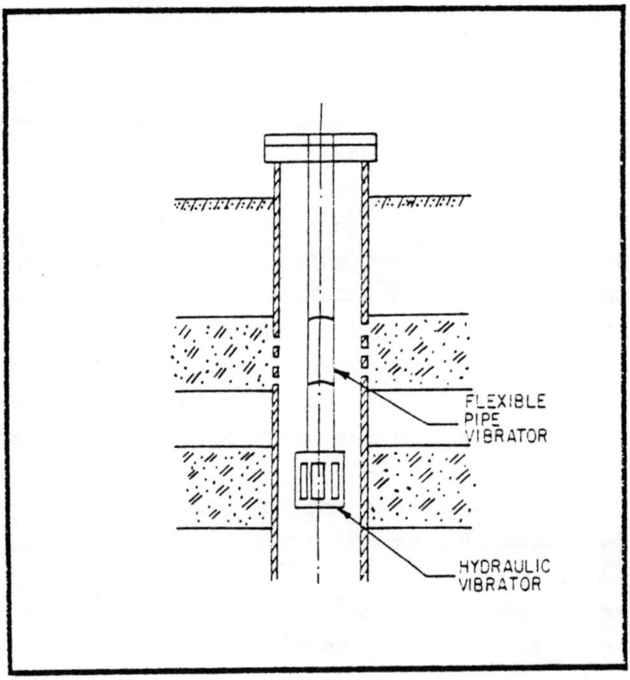

Figure 3-8. Hydraulically Induced Vibrations Equipment

The use of hydraulically induced vibrations as an oil recovery process would be a novel extension of this technology, and the effect of hydraulically generated, high frequency compression waves has been laboratory tested in the U.S.S.R.

In these laboratory tests, a rational vibrator was used in conjunction with a high pressure pump to generate compression waves over a frequency range of 10 to 100 Hz. Subsequently, modification of the vibrator rotor (increasing the number of openings) permitted the test to be conducted with variable frequency up to 500 Hz. A special membrane separated the wave generating equipment and the oil containing core. Although many important aspects of the laboratory tests are not explained, reported results were a 10 to 12 percent increase in recovery of one to 14 cp oil.

Laboratory results obtained at the Gubkin-Moscow Petrochemical and Gas Industry Institute were sufficiently encouraging to justify the design and development of downhole equipment. The two essential elements of this apparatus are a hydraulic vibrator and a flexible pipe vibrator. This permits an enhanced hydraulic impact on the primary zone of interest and simultaneously subjects a secondary zone, in multipay wells, to the same compression wave.

The amplitude, and subsequently the impact, imparted to the fluid bearing stratum or strata is dependent upon the characteristics of the pipe vibrator and the energy transfer fluid between the hydraulic vibrator and the pipe vibrator. The more important pipe characteristics include the elasticity, radius, wall thickness, and length. The principal fluid property affecting energy transfer is ultimately the compressibility of the fluid; therefore, the actual design of the total downhole apparatus must be governed by the fluid saturation(s) within the stratum or strata being stimulated.

Similarly, the pipe vibrator is used in conjunction with the hydraulic vibrator to enhance the emitted compression wave to the oil bearing stratum. The actual design of the pipe vibrator depends on fluid saturation within the productive stratum; i.e., as the liquid saturation in the stratum decreases, the flexibility of the pipe vibrator increases. Conversely, if the porous stratum is 100 percent fluid filled, maximum vibration effects will be realized with a pipe vibrator designed for minimum flexibility.

No information was found to indicate that any field testing of this apparatus has been done. Furthermore, there was no indication of the variability of results as a function of oil viscosity or gravity.

3.2.4 Magnetized Water and Electromagnetic Heating

For several years magnetic water treatment devices that prevent or inhibit calcium scale formation in water heaters and boilers have been available in Europe. Although the scientific community in general has been skeptical of any such effect, recent research at Johns Hopkins University in Baltimore, Maryland, has confirmed a beneficial magnetic effect in calcium-hard water. Still, the actual mechanism of the magnetic influence is unexplained.

In 1977, the Ivanov-Frankov Institute of Oil and Gas in the U.S.S.R. reported the results of a series of laboratory waterflood tests of an "aleurite" core saturated with an apparently viscous oil from the Bitkov formation. Displacement tests were made with magnetized water and with untreated water. Hardness of the water used was reported to be a comparatively low value of 8.6 mg/liter (ppm). An average recovery of an additional 4.5 percent of the oil-in-place was observed with the magnetized water. No explanation of this effect was reported.

There are discrepancies and oddities in the data reported, and sufficient information is lacking for a proper evaluation of the experiment. It was reported that approximately 351 cm^3 of oil was used to saturate the test core, but a cylindrical core of the dimensions and porosity given would have a pore volume of only 314 cm^3. Likewise, the oil recovery of 39.0 to 43.4 percent reported is very minimal for laboratory flooding tests of any sort. If the core were 100 percent saturated with 314 cm^3 of oil, then the resulting recoveries of 43.5 to 48.5 percent are still inordinately low. Also, there is no reason to believe or infer that an irreducible, connate water saturation was established in the test core prior to flooding.

If it is accepted that magnetization of water has a "softening" effect, then it follows that the surface tension could be altered. This, in turn, could have some effect on the unit displacement characteristics of an oil-water system. However, the degree of hardness reported for the Soviet experiment would have an insignificant effect on the water surface tension, so that possibility is remote for this case. Therefore, the significance of this experiment must remain in doubt pending further information and experimentation.

A different approach to the use of magnetic fields is that of in situ electromagnetic heating of formation fluids. Research is currently in progress at the University of Alberta, Canada, on the potential of electromagnetic heating for use as a heavy oil recovery process. This research is concentrating on developing scaled laboratory physical models which will simulate the thermal and electromagnetic events simultaneously. It has been shown that a synthetic porous medium and analog fluids exhibiting particular thermal and electrical properties, must be used to give the proper scaling criteria.

These same researchers are also measuring the electrical conductivities and dielectric constants of sand samples saturated with heavy oil to develop better correlations for estimating these properties in the reservoir. These data are used primarily in calculating oil contents of hydrocarbon deposits by wireline logging methods.

4. Surface and Downhole Equipment

4.1 THERMAL INSULATION

One of the most significant problems associated with steam injection is heat loss through the tubing and casing into the adjacent strata. The quantity of heat lost to these strata is dependent on the length of the injection string and the thermal conductivity of the metal, annular fluid, cement, and the adjacent strata and their interstitial fluids. Insertion of a low conductive medium in the tubing-casing annulus, consequently, has obvious advantages.

In attempts to deliver maximum heat to the injection zone, two main elements have received considerable efforts and attention. These efforts have led to the design of a number of high temperature packers and annular insulating materials.

High temperature packers should fulfill two main functions; i.e., prevent elongation of the tubing downward below the perforations and prevent steam from entering the annulus. Insulating media have only one function; i.e., to decrease heat loss from the steam thus delivering more thermal energy to the hydrocarbon bearing formation.

High temperature packers differ from normal packers mainly through the material used in construction of the sealing rings. Most high temperature packers utilize a non-elastometric material composed of asbestos, stainless steel, and carbon. When equipment failure does occur it is almost always a breakdown of the seals. A number of the packer manufacturers build these tools for high temperature instalations.

A wide variety of materials, gas, liquid, and solid, have been used and are marketed to reduce heat loss through tubing. During the early 1960's in the San Joaquin Valley of Southern California, many cyclic steam injection projects were initiated. After the tubing and packer had been run into the hole, nitrogen was injected into the annulus and allowed to return to the surface via the tubing; at this point the packer was set and the well was readied for steam injection.

More recently, and prior to nationalization of the petroleum industry in Venezuela, Amoco used there a thixotropic grease mixed with diesel oil and lime. The resulting gel was pumped into the annular volume. This substance is now marketed under the name Ken Pak and reportedly is in wide use in Venezuela.

Numerous low conductive solid materials are used to reduce heat loss in steam injection projects. Obviously, solids must be prefabricated on the tubing before running into the wellbore. In recent field tests, mineral wool, calcium silicate, and a multi-layered insulation were used very effectively to yield up to a 400°F temperature difference between steam and casing. One consistent problem; however, with the prefabricated tubing exists at the couplings. The accessory couplings insulation, which must be installed while the tubing is run into the hole, permits "hot spots." Despite the design deficiency of the coupling insulators, these hot spots are substantially lower in temperature than that of the steam.

Research continues on methods to effectively insulate steam heat from adjacent strata. However, with present equipment it is possible to inject steam substantially deeper than the 2500 feet which a few short years ago was considered maximum depth for steam injection.

4.2 OTIS HEAVY-CRUDE LIFT SYSTEM

Artificial lift equipment for heavy oil reservoirs routinely has included sucker-rod pumps, plunger lift, and Kobe power-oil systems. In eastern Venezuela, the Mene Grande Oil Company successfully used Kobe systems for their deeper heavy oil wells, which led to the practice of mixing a light blend oil with heavy crude at the bottom of wells equipped with sucker-rod pumps. From these ideas and efforts, a developing system for artificial lift of 10°API and lighter oils is being marketed by Otis Engineering Corporation.

This system, the Otis Heavy-Crude Lift System, was developed originally by Bill Waters, now deceased, an American service company representative and consultant in Venzuela. It was tested and used on ARCO and Mene Grande wells near Lake Maracaibo and in eastern Venezuela.

For the current version, a production tubing string with a perforated nipple is run on a single packer to the production zone, and a second power-oil/gas-lift string is run on a dual packer to a shallower depth, forming a production chamber between the two packers. Surface equipment includes a power-oil reservoir, pump, and separator and, if needed, a gas compressor. In operation, crude oil rises in the production string and fills the production chamber. Power fluid (blend crude, diesel oil, or water) is then pumped down the power-oil string to empty the chamber of crude oil and displace it to the surface. The power fluid is then gas-lifted out of the system. If necessary, a small diameter spaghetti-string can be run from the lowest gas-lift valve to the bottom of the production chamber to assist in discharging the power fluid.

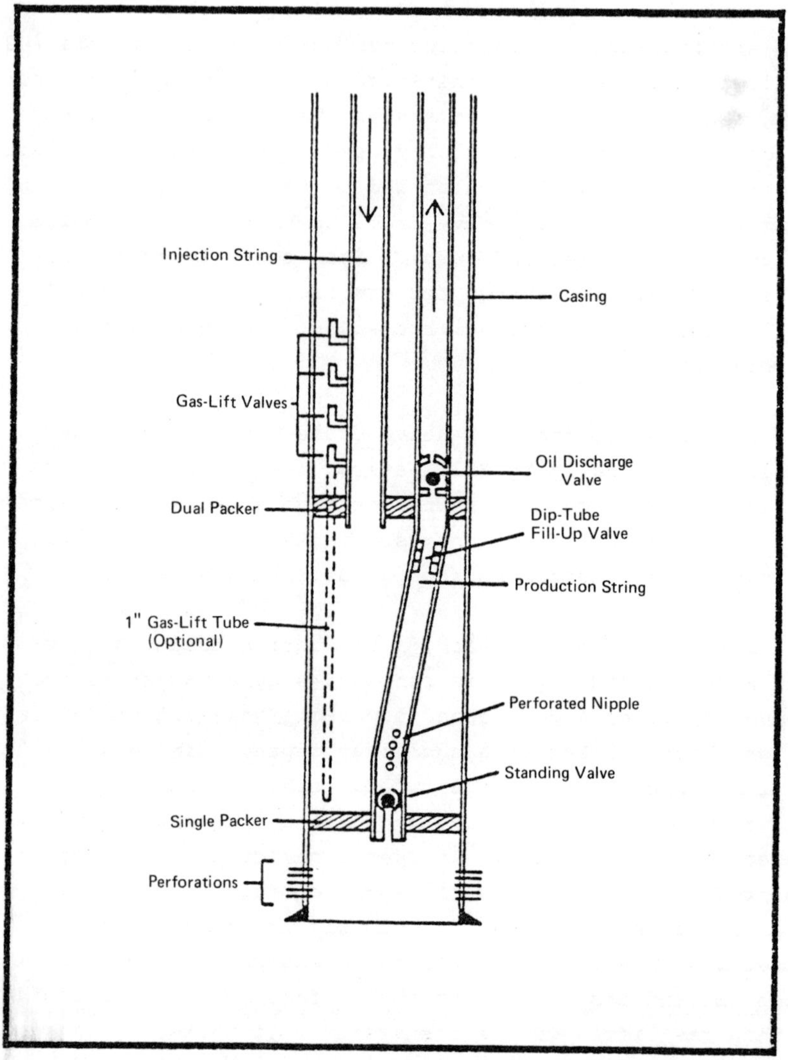

Figure 4-1. Otis Heavy-Crude Lift System

The original system, as tested by Waters et al in Venezuela, used a system of concentric strings, such as a 1-1/4 in. power-oil string inside 3 in. tubing inside 5-1/2 in. casing. Rates of up to 1600 bbl/day were obtained with this early version.

4.3 STEAM GENERATION EQUIPMENT

4.3.1 Downhole Generators

Two problems associated with the generation and injection of steam in both steam stimulation and displacement projects are the polution potential of combustion gases and surface and downhole heat losses. Bottomhole generation of steam would treat both these problems. Sandia Laboratories of the U.S. Department of Energy has reportedly developed and is testing equipment for downhole steam generation. Although not foreign technology, this brief description is included here for completeness and because of an expressed interest in such equipment for use in deep wells in Venezuela and elsewhere.

The equipment under test is a modified version of a propane-fired steam generator developed by Vapor Energy, Inc., of Grand Prairie, Texas. It has been used by Sandia to generate steam at a rate of 800 cu ft/min from feed water and a 400 psi mixture of propane and air. Steam and combustion products together are injected directly into the oil-bearing formation, so the recovery process becomes one of a combination steam and flue gas (stack gas) injection.

Field tests to date have been with a California reservoir 800 feet deep. Further testing is planned for a reservoir 2500 feet deep, and the equipment is expected to be applicable to a depth of 4500 feet.

In addition to the Sandia generator, Sullair Petro-steam of California is developing a diesel-fueled downhole steam generator which uses a heat exchanger and vents low-pressure, treated exhaust gases at the surface well below current air quality standards for admissable pollutants. This unit has been subjected to a seven-day field test at 1200 feet and is projected to be operational to 6000 feet some time in 1981.

4.3.2 Solid Fuel Boilers

The use of solid fuels to supply the heat required for steam generation in heavy oil recovery may result in an increase in the economically recoverable reserves. Conventional oil fired boilers may consume as much as one-third of the oil produced as a result of the steaming operations. Conoco and Struthers Thermo-Flood Corporation plan a pilot plant operation to generate five million Btu/hr using high sulfur petroleum coke and low Btu lignite as fuels.

4.3.3 Solar Preheaters and Boilers

Solar energy plants could possibly provide a substantial portion of the input heat required for steam generation. Systems are now available to generate 850 psi steam and to provide as much as twenty million Btu/hr from 70 acres of land area.

4.3.4 Atomic Generators

It is of interest to note that the Canadian government has expressed an interest in the use of atomic reactors for steam generation with regard to heavy oil and tar sand exploitation in Alberta. A feasibility study is reportedly in progress.

5. National Efforts in Heavy Oil Recovery

Heavy crude oil deposits have been found in most of the major depositional basins of the world. Although many of these potential reserves are small or even insignificant in a global context, some of them, such as those of Canada, the U.S.A., the U.S.S.R., and Venezuela, represent large and important, relatively unexploited hydrocarbon resources. As would be expected, those nations with major reserves are the most active in the development and application of heavy oil recovery processes. A few nations, such as France and Japan, are also active in this pursuit, even though they have relatively insignificant domestic deposits, because of a heavy reliance on foreign oil. Others, such as Venezuela and Brazil, rely primarily or entirely on foreign technology; and others are not yet engaged in any significant efforts directed toward exploitation of heavy oil reserves.

Following is a worldwide survey of known heavy oil deposits outside of the U.S.A. and discussions of national efforts to develop and apply more efficient recovery processes for the exploitation of these important resources.

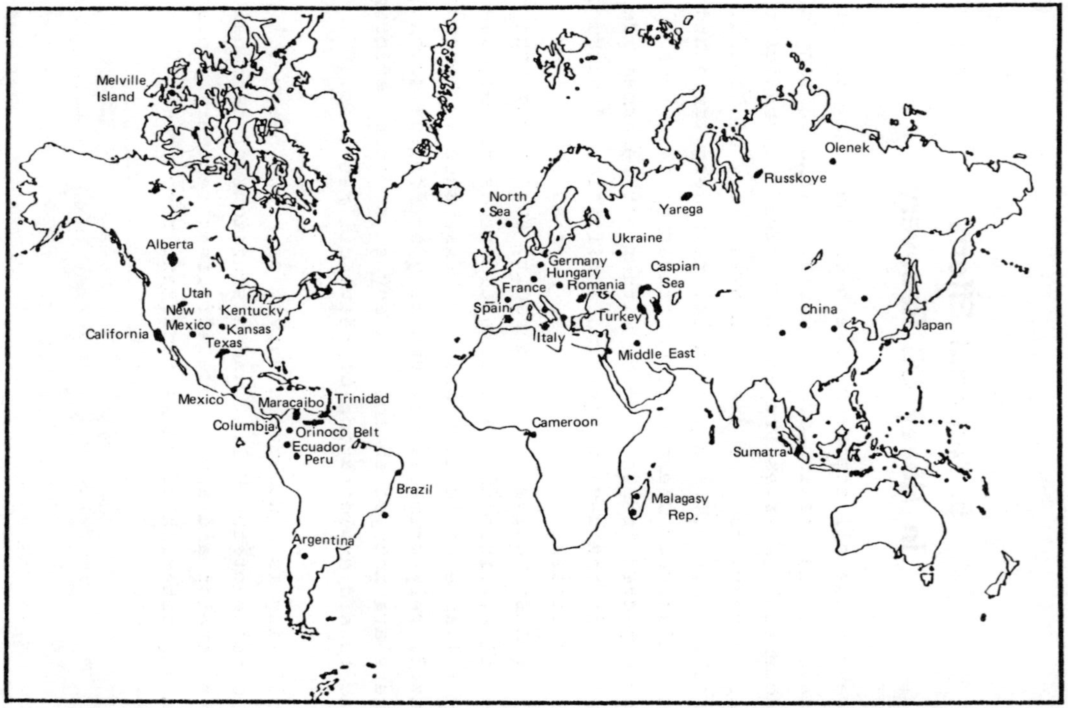

Figure 5-1. Worldwide Heavy Oil Occurrences

5.1 BRAZIL

For such a large land area, the country of Brazil has very few known deposits of heavy oil. Until recently the only reserves of any significance were in some of the productive zones of the Carmopolis field in Sergipe-Alagoas in the northeast. Here oil was found over twenty years ago with gravity ranging from 14°API to the high twenties. In

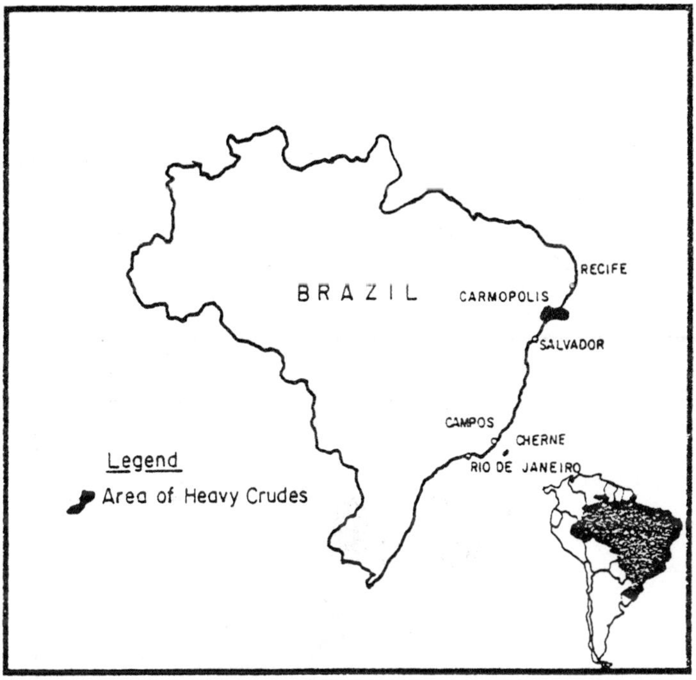

Figure 5-2. Heavy Oil Occurrences in Brazil

1962 feasibility studies were begun, and ultimately a forward dry combustion project was initiated. Later, steam stimulation and steam displacement was tried. These projects have had only limited success and were all accomplished with the aid of consultants from the U. S. A.

At this time, Petrobras, the Brazilian national oil comany, is considering the use of polymer flooding and micellar slugs, again working with American consultants. No original research as been done in Brazil with respect to heavy oil recovery.

Recently, another heavy oil reservoir has been discovered offshore of Campos near Rio de Janerio. This has been designated as the Cherne field and has an oil gravity of 20°API.

5.2 CANADA

The occurrence of heavy oil deposits in Canada trend eastward from the Peace River deposit in northwestern Alberta through Wabasca to the Athabasca deposit and then south to the Cold Lake, Lloydminster, and Kindersley area Coleville

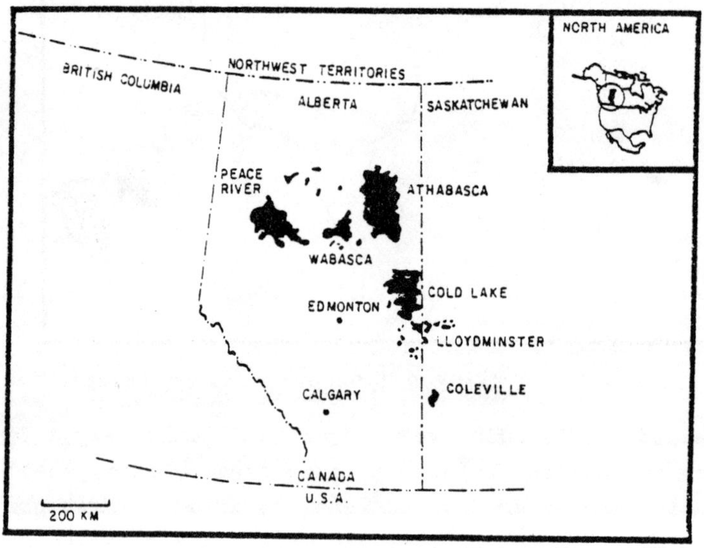

Figure 5-3. Heavy Oil Occurrences in Canada

deposits. An estimated 965 billion barrels of oil-in-place exists in the largely Lower Cretaceous sandstone deposits. Table 5-I presents a breakdown of oil-in-place in each of the major deposit areas as well as some average reservoir parameters.

Oil gravities are extremely low, ranging from 6°API at Wabasca to as high as 19°API in the Lloydminster area pools with the largest percentage being in the 7 to 8°API range. The critical factor in recovering these heavy oils is the extremely high viscosities of the oil at reservoir conditions; therefore, thermal-based methods are the primary means being researched and tested in Canada today. Not surprisingly, current emphasis is centered on ways of applying heat to the oil deposits as efficiently and economically as possible.

Table 5-I. Canadian Oil Sands Formation Properties

	Athabasca	Cold Lake	Wabasca	Peace River	Lloydminster
Oil in place, 10^9 bbl	626	164	85	75	15
Depth, ft	0-200	1,000-2,000	250-1,000	1,000-2,500	1,000-3,000
Average pay, ft	70	42	28	37	30
Average oil in place, bbl/acre-ft	1,500	1,240	1,100	1,150	1,700
Porosity, %	30-40	30-35	30-35	25-30	23-35
Gravity, °API	7-8	10-12	6	8.5-9.0	11-19

To accomplish this goal, steam and variations of steam-based methods is by far the leading approach. Research and field testing of steam injection in the Athabasca oil sands was begun by Shell Canada in the period 1957 to 1962. The study involved injecting steam into hydraulically created horizontal fracture channels for increased reservoir communication and better distribution of heat. Also studied at that time was production of oil by emulsification with alkaline synthetic detergents and caustic solutions. It is reported that best recovery efficiencies were obtained by fracturing with an alkaline solution followed by a hot caustic solution and steam injection.

The year 1957 also marked the start of Amoco Canada's field testing of forward and reverse in situ combustion in the same Athabasca oil sands. Promising laboratory results suggested that reverse combustion would be a successful recovery method, but it was a failure in field trials. However, positive results with the forward combustion process culminated in the later development of the Combination of Forward Combustion and Water Process (COFCAW) in 1968 after additional laboratory and field testing.

During the 1960's, numerous field pilot operations were constructed to investigate thermal heavy oil recovery processes. Among the processes tested were many forerunners of current active investigations, including cyclic steam injection (huff and puff), steam drive, caustic, and surfactant enhancements to reservoir communication, horizontal fracturing, and in situ combustion. The majority of the field pilots were located in the Athabasca and Cold Lake deposits, with the exception of Shell Canada's Peace River steam injection and combustion projects. Three wet combustion projects were started in the late 1960's by Mobil Oil and Husky Oil in the Lloydminster heavy oil area. All three, the Aberfeldy, Golden Lake South, and Battrum projects, remain active in 1980.

Current heavy oil recovery activities in Canada remain heavily weighted toward conventional steam methods. Active projects are summarized in Table 5-II which identifies 15 out of 37 as being either conventional cyclic steam injection, steam drive, or a combination of the two in which steam drive follows cyclic injection operations. Field trials of these methods are being studied in all five of the major deposit areas. Esso's May and Leming cyclic steam pilot projects in the Cold Lake deposits over the past 16 years have shown positive results, leading to a recent decision by Esso to prepare for a planned commercial size operation by the mid 1980's.

Table 5-II. Active Field Pilot Projects in Heavy Oil Deposits, Alberta and Saskatchewan, Canada

Operator	Area (Formation)	Oil Gravity °API	Recovery Methods Investigated	Test Pattern and Number of Wells	Estimated Initial Phase Cost (10^6 $)
		Cold Lake Oil Sands Deposit			
B P Exploration	Marguerite Lake (Mannville "C")	10–12	Combination of steam drive and combustion	4–5 spots plus 3" out of pattern wells 17 wells total	18.4
Chevron-Standard	Beaver Crossing (L. Grand Rapids "B")	10–12	Cyclic steam injection	7 spots 7 wells total	?
Esso Resources	May (Clearwater)	10–12	Cyclic steam injection	5 spots 30 wells total	40.0
	Leming/Borgue (Clearwater)	10–12	Cyclic steam injection	7 spots 80 wells total	included above
Gulf Canada	Cold Lake (Clearwater)	10–12	Steam drive	multiwell 6 wells total	15.0
Murphy Oil	Lindbergh (Grand Rapids)	10–12	Steam drive	7 spots 7 wells total	?
Norcen Energy	Primrose (Clearwater)	10–12	Steam drive	9 wells diamond 9 wells total	?
Petro-Canada	Muriel Lake	10–12	?	?	?
Sun Oil Co. (formerly WECO)	Fort Kent (Colony "C")	10–12	Cyclic steam injection	multiwell 20 wells total	9.1
Union Texas of Canada	Ardmore (Clearwater)	10–12	Cyclic steam injection	multiwell 15 wells total	?

(continued)

Table 5-II. (continued)

Operator	Area (Formation)	Oil Gravity °API	Recovery Methods Investigated	Test Pattern and Number of Wells	Estimated Initial Phase Cost (10^6 $)
Athabasca Oil Sands Deposit					
Amoco Canada	Gregoire Lake (McMurray)	7-8	COFCAW	9-5 spots 25 wells total	46.0
Gulf Canada	Surmont (McMurray)	7-8	Horizontal fracturing and cyclic steam injection	multiwell 6 wells total	1.3
Petro-Canada	Mildred Lake	7-8	Mine assisted steam injected	Radial horizontal well shafts	?
	PCEJ Project (McMurray)	7-8	Electric preheating steam drive	– 16 wells total	?
	Primrose	7-8	Steam drive	– 1 well	11.8
Texaco Canada	McMurray	7-8	Steam drive with additives	5 spots, 9 spots 7 spots 30 wells total	?
Athabasca Grosmont Carbonates					
Union Oil	Buffalo Creek (Grosmont)	7-8	Cyclic steam injection and wet combustion	multiwell 5 wells total	?
	Chipewyan River (Grosmont)	7-8	Cyclic steam injection	multiwell 3 wells total	?

(continued)

Table 5-II. (continued)

Operator	Area (Formation)	Oil Gravity °API	Recovery Methods Investigated	Test Pattern and Number of Wells	Estimated Initial Phase Cost (10^6 $)
Wabasca Oil Sands Deposit					
Gulf Canada	Pelican	6	Cyclic steam injection	– 1 well	?
	Wood River (Grand Rapids "A")	6	Cyclic steam injection, steam-solvent, steam drive and combustion	5 spots 16 wells total	?
Peace River Oil Sands Deposit					
Shell Canada Resources	Cadotte Lake (U. Bullhead)	8.5–9	Steam drive	7 spots 31 wells total	58.0
Lloydminster Oil Sands Deposit					
Canadian Reserve	Epping (Sparky)	15	Caustic waterflood	5 spots 5 wells total	?
Dome Petroleum	Hughenden	19	Dual completion in aquifer and oil zone	–	?
General Crude	Kitscotty	11–15	Combustion	?	?
	Silverdale	16	Wet combustion (combined air-water injection)	4–7 spots 33 wells total	?
Norcen Energy Resources Ltd.	Provost	12–13	Combustion	?	?
	Provost	12–13	Cyclic steam injection	– 1 well	?

(continued)

Table 5-II. (continued)

Operator	Area (Formation)	Oil Gravity °API	Recovery Methods Investigated	Test Pattern and Number of Wells	Estimated Initial Phase Cost (10^6 $)
Husky Oil	Aberfeldy (Sparky)	11-15	Wet combustion; and cyclic steam injection followed by steam drive	? 38 wells total	?
	Blackfoot	11-15	Steam drive	– 3 wells total	?
	Golden Lake	11-15	Wet combustion	? 19 wells total	?
Mobil Oil	Battrum	11-15	Wet combustion	?	?
	Fosterton N.W.	11-15	Wet combustion	?	?
	Celtic	11-15	Wet combustion; with cyclic steam and solvent injection	? 25 wells total	15.0
Murphy Oil	Eyehill-Senlac	11-15	Dry combustion; with cyclic steam injection	? 25 wells total	?
	Silverdale (Sparky)	14	Wet combustion	Inverted 9 spots 9 wells total	10.0
Petro-Canada	Viking-Kinsella (Wainwright)	15	Combustion and steam drive	5 spots 10 wells total	?
Texasgulf	Meota (Lloydminster)	11-15	Cyclic steam injection with steam drive	? 9 wells total	10.0

Second in number of active projects are those involving in situ combustion. Amoco Canada's COFCAW process being applied at Gregoire Lake in the Athabasca area is among those best known. Other projects labeled "wet combustion" in operation at several Lloydminster area test sites are similar, making use of water injected with or behind air to increase recovery efficiency.

Active projects, many of which are in very early development stages, are as follows:

(1.) Combustion and Steam - Several companies, including BP Exploration at Marguerite Lake, Petro-Canada in the Viking-Kinsella area, Murphy Oil at Eyehill-Senlac, Mobil Oil at Celtic, Gulf Canada at Wood River, and Union Oil at Buffalo Creek, are combining tests of steam stimulation and in situ combustion. When combined in this manner, one process is normally used as a preheating phase to improve reservoir communication and recovery efficiency of the process to follow.

(2.) Steam Additives - A number of companies have carried out single well testing of various steam additives. The additives mostly being tested are solvents, surfactants, nitrogen, and CO_2. Turbo Chemicals is involved in laboratory and field tests involving high temperature stable surfactants for steam flooding. Field tests were carried out in a single-well cyclic injection program in a reservoir containing 13°API oil. The oil production rate and cumulative production is reported to have been doubled in each stimulation cycle.

(3.) Mine Assisted Steam Injection - Several companies, Esso and Petro-Canada being the most active ones, are drilling shafts into the oil bearing zone with usually four to six holes radiating from each well. Petro-Canada is testing this method at their Mildred Lake site as operator for a group of companies.

(4.) Electric Preheating - Petro-Canada, as operator, is testing this process as a preparatory step to steam displacement operations at their PCEJ project site in the Athabasca area.

(5.) Caustic Flooding - There is intense interest and recent activity in Canada concerning the use of NaOH solutions alone and in conjunction with steam. Some field tests have been undertaken, and extensive laboratory tests and evaluations are underway. Companies most active in NaOH flooding of heavy oils are Petro-Canada, Canadian Reserve, Husky Oil, Dome Petroleum, Hudson's Bay Oil and Gas Company, Union Oil, and Chevron-Standard. Since the principal recovery mechanism involved in NaOH flooding is emulsification and entrainment, as opposed to thermal viscosity reduction, its testing and use is being centered in the Lloydminster area pools where heavy oil gravities of no lower than 15 to 20°API are found. Canadian Reserve Oil and Gas is currently undertaking their inverted five-spot pilot test in the Epping area.

In addition to the industry projects just described, there is considerable research activity taking place at universities and laboratories throughout Canada, as follows:

(1.) In Situ Electromagnetic and Induction Heating - Research is being carried out at the Department of Electrical Engineering of the University of Alberta, Edmonton, by Professors F. S. Chute, F. E. Vermullen, and M. R. Cervenan on the potential of electromagnetic heating for heavy oil recovery.

Considerable research and design of electrical induction heating methods, as described previously in this report, have been done by Sidney Fisher of F. T. Fisher's Sons, Ltd., of Montreal. Application of the induction heating method is yet to be applied in field tests of heavy oil recovery.

(2.) <u>In Situ Emulsification of Oil-Water System During Steam Injection</u> - This work is being carried out on a contract basis by the Petroleum Recovery Institute for the Alberta Research Council. The study deals with the formation and flow properties of oil-water emulsions which form during steam injection to displace heavy crude oil.

(3.) <u>Hydraulic Fracturing</u> - The most difficult problem in heavy oil displacement is how to increase contact between the injected fluids and the oil. This project deals with creating and keeping open fractures in heavy oil bearing sand. The project is carried out at the Civil Engineering Department of the University of Alberta, Edmonton, by M. B. Dusseault, S. M. Maloney, and D. J. Barnes.

(4.) <u>Theoretical and Experimental Aspects of Laboratory In Situ Combustion Tests</u> - The Department of Chemical Engineering of the University of Calgary, Alberta, has an adiabatic combustion tube to study dry and wet combustion processes for the recovery of heavy oil. The research is directed by Professors R. G. Moore, J. K. Donnely, L. D. L. Vorndran, and D. W. Bennion. The specific phenomena covered in this project are:

- Low temperature reactions, especially in wet combustion
- Carbonate decomposition
- Wettability changes
- Plugging
- Channeling and bypassing
- Stalling of the combustion front

(5.) Bacteria - The use of bacteria in heavy oil recovery may be one forthcoming process. The bacteria of prime interest are the ones which will produce exopolysaccharides in situ and thicken the displacing water.

(6.) Chemical Reaction of Water with Heavy Oils and Sands at Elevated Temperatures - This project is carried out at the Department of Chemistry of the University of Calgary, Alberta, by Professors J. B. Hyne and J. W. Freidanus. The project focuses mostly on reactions with steam in the form of chemical hydrylosis as distinct from thermal degradation. The study covers a temperature range of 350 to 500°C.

5.3 CHINA

There are several major oil fields in China. Only five of these, however, are presumed to contain heavy, near heavy, or viscous crude oils. The Sheng-li and Ta-Kang fields are located near Peking and produce 20 to 24.6°API gravity and "some very thick" oil, respectively. The Fu-Yu field, located near Harbin, is of unknown gravity, but reports indicate that extraction is very difficult, and presumably the oil is of a heavy type.

In western China, near the cities of Chital and Wulumuchi, lies the K'o-la-ma-i field. Oil seeps and asphalt mounds are visible in the immediate area, and the first well to be drilled in this field was completed in 1955. Highest reported production was 10,000 bbl/day in 1965, although estimated reserves enabled the Chinese to set a target production of 60,000 bbl/day from K'o-la-ma-i as early as 1962. Water injection started in this field in 1959.

The Yu-Men field in west central China was discovered in 1939, peaked in production in 1957, and accounted for 52 percent of the national production in that year. There are four independently productive intervals in this field, and the crude characteristics vary over a range of properties. Reports indicate that some of the crude produced is of high viscosity and has a high paraffin content.

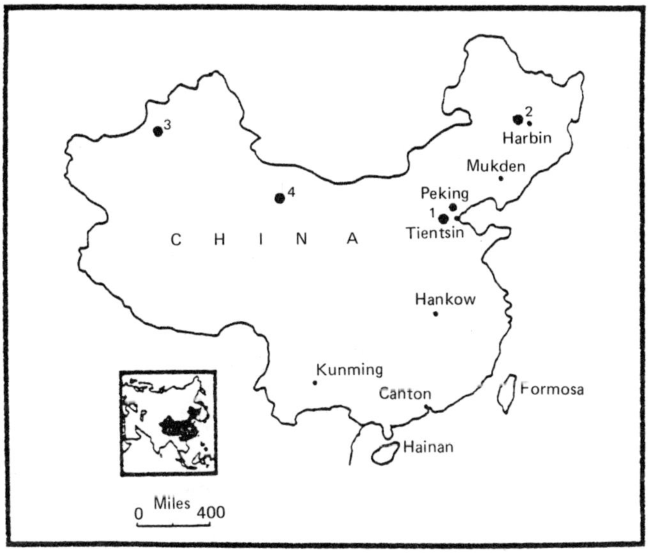

Figure 5-4. Probable Heavy Oil Occurrences in China

The above fields are the only known producing fields, which, at this time, can be classified as probable heavy oils. However, it must be noted that much of China, terrestrial and offshore, remains to be explored. No heavy oil recovery processes are known to be under test or under serious consideration at this time.

5.4 FRANCE

Heavy oil reserves in France are essentially limited to the Pechelbronn area (Alsace) and a small deposit in the Grenade area. The Pechelbronn area deposits have been subjected to many small scale recovery experiments since early in the century, but the most notable efforts have been in the "mining" realm, with some recent activity attempting improvements utilizing thermal (steam) processes. Commercially viable exploitation is not anticipated, and efforts continue to be pilot in nature to the end of supplementing basic research into methodology.

Figure 5-5. Heavy Oil Occurrences in Western Europe

The major French involvement in improved recovery processes of all types is currently in the realm of research. The first priority of production methods research at the French Petroleum Institute (IFP) is related to heavy oil recovery. The Association for Research on the Techniques of Petroleum Exploitation (ARTEP) is guiding all research within the industry, with the majority of the fundamental research performed at IFP and supplemented by applied testing in the laboratories of the operating companies. Further basic research is conducted by several universities through the National Center for Scientific Research (CNRS). Much of the IFP effort is also supported by contracts with other producing countries (Romania, Turkey, etc.). As is the case with the ARTEP studies, the results are proprietary in nature and publication is limited.

Within France itself, pilot operations using both microemulsion and polymer flooding have yielded satisfactory results in the Chateaurenard field in the Paris basin. The crude is of a low gravity but above 20°API. While there are several other fields in the basin, reserves are insignificant. Of greater significance is the steam flood underway in the Lacq (Superior) field in the Aquitaine basin where steam injection is into a fractured limestone. Results obtained into 1980 show favorable response without any adverse breakthrough. Limited reserves prevent this project from being of great significance in itself; however, the results obtained can be of great importance due to the widespread occurrence of heavy oil in reservoirs of this type.

5.5 GERMANY

Two fields in the northern West German oil belt appear to have oil gravities of less than 20°API. The Nordhorn field in the extreme northwestern part of Germany is the larger with an estimated 300 million barrels of oil-in-place and a gravity of 14°API. Second is the Wesendorf field located approximately 250 km to the east of Nordhorn. Less information is available on the Wesendorf field except that the oil properties are very similar to that found in the Nordhorn field.

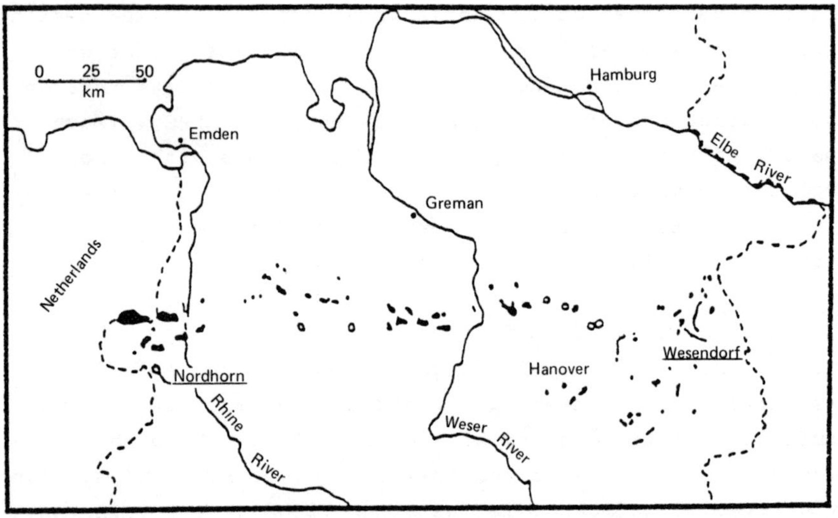

Figure 5-6. Heavy Oil Occurrences in Germany

Based on good laboratory results showing steam flooding to be a viable recovery mechanism, a single-well steam injection field test was begun in the Nordham field in November 1977. This test, conducted by Wintershall AG as operator for a consortium of four German oil companies, basically confirmed laboratory results which indicated that thermal methods such as steam stimulation would be the most suitable recovery technique.

Both fields are currently inactive although reports indicate further investigations concerning crude utilization from these fields are underway. The primary deterrents are the extremely high oil viscosity of about one million centipoises at reservoir conditons and the low market value of the oil. This latter factor is due to a lack of light hydrocarbon components in the crude oil which also is otherwise not suitable for either lube oil, asphalt, or extra heavy fuel oil.

5.6 INDONESIA

The Duri field of Central Sumatra is the largest known occurrence of heavy or viscous oil in Indonesia. Estimates from knowledgeable sources place initial oil-in-place at nearly three billion barrels. Conflicting information shows the crude gravity to be either 12 to 15° or 22°API with a viscosity of 140 to 180 cp at reservoir conditions. Formation characteristics of this 150 to 200 meters deep formation indicate a 37 percent porosity, a 100 to 5000 md permeability, 62 percent oil saturation, and a bottomhole temperature of 100°F.

Other known heavy oil fields exist in southern and offshore southeastern Sumatra and in Borneo. A southern Sumatra field is operated by Natomas and contains 18°API crude. Two offshore fields (Kitty and Utari) contain 15°API oil at 700 to 900 meters of depth with a 60 to 100 cp viscosity. Primary recovery from these two fields is estimated to reach about 12 percent at a 98 percent watercut. The smaller heavy oil fields of Borneo are operated by Tesoro and Pertamina.

The only heavy oil recovery process known to be in operation is a steam soak project operated by Caltex Pacific Indonesia in the Duri field. This project is beset with three major problems; i.e., lack of an adequate fresh water source, casing failures, and sand production. Despite these problems, the project is considered to be promising and ultimate recovery is expected to significantly surpass the less than 10 percent primary recovery. Through December 1976, production from steam injection was estimated to be about 20 million barrels, eight percent of the field's total production.

5.7 ITALY

Heavy oil deposits are located along the southern coast of Sicily, in the Adriatic Sea area offshore Pescara, and reportedly in the Pisticci field located near the Gulf of Taranto. Oil-in-place from these areas is estimated to be in excess of two billion barrels, of which approximately 450 million barrels are in the Adriatic Sea area, and 1300 million barrels are in the Gela field in Sicily.

The Gela field produces from depths ranging from 9900 to 10,900 feet under a natural water drive mechanism. Gravity ranges from two to 11°API and viscosity ranges between 140 to 240 centipoise. Due to its depth and low porosity, thermal processes are not considered to be feasible. Research by AGIP has been conducted under partial funding by the EEC (EEC Regulation 3056/73) to investigate the possibility of increasing recovery by means of high pressure nonmiscible gas injection. Both CO_2 and natural gas injection were considered. Under natural gas injection, gas imported from Algeria via the Trans-Mediterranean pipeline has been considered, but field testing of the injection scheme has not yet been done.

The Emilio structure is located in about 325 feet of water, offshore Pescara, in the Italian sector of the Adriatic

Sea. Heavy oil (11.6°API) has been found at about 9750 feet. This structure was found by AGIP and Deutsche Shell under an EEC subsidized joint venture. Original estimates of 1800 million barrels were later revised downward to 450 million barrels as the result of subsequent drilling. This venture was discontinued in 1977 due to the "unconventional nature of the oil distribution and migration."

5.8 JAPAN

There are only very small known deposits of heavy oil on the islands of Japan, all on the western side of the main island of Honshu. Early on, two of these deposits were developed with small drainage mines, and in the early 1960's an in situ combustion project was attempted. After a serious accident, further attempts were abandoned until the middle 1970's when U.S. consultants were contracted to survey and study processes that might be applicable to these reservoirs.

Today, some small amounts of research and computational activities in improved oil recovery processes are conducted by the research arm of the Japan National Oil Company (JNOC), but the major activity in heavy oil recovery is as a participant in the group of companies which will apply steam processes, with electrical preheating, in the oil sands of Alberta, Canada.

5.9 MALAGASY REPUBLIC

Heavy oil and tar sand occurrences in the Malagasy Republic lie in the west-central and southwest parts of the island. Very little has been accomplished toward recovery of the deposits at Bemolanga, Tsimiroro, or Folakara. Bemolanga reportedly contains three billion tons of heavy oil and tar and is, by far, the largest of these deposits.

It has been estimated that 780 million tons of the Bemolanga deposits are mineable.

The Bemolanga deposits are usually classified as tar sands because of the ultrahigh viscosity of the in situ hydrocarbons. On the other hand, physical data indicate that the gravity ranges from less than zero to as high as 12.6°API in this area, so at least some of these deposits may be classified as heavy oil on that basis.

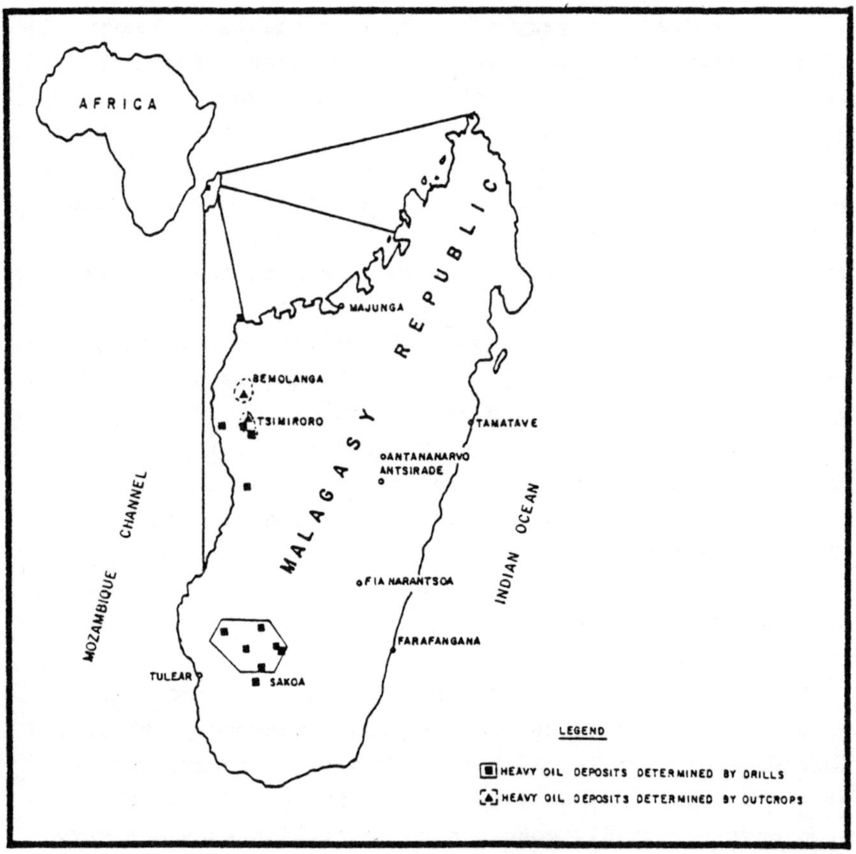

Figure 5-7. Heavy Oil Occurrences in The Malagasy Republic

Despite the large amounts of shallow, heavy deposits in Malagasy, only minimal efforts have been applied to recovering this reserve. In 1962 Societe des Petroles de Madagascar (SPM) opened two quarries and a pilot plant at Bemolanga in an effort to find a hot water recovery process such that the results could be applied to an industrial plant. In 1965, the French society ELF-RE attempted unsuccessfully to institute an in situ combustion process. Between 1972 and 1974, two thermal studies were made by the University of Tananarive and The Oil Shale Company (TOSCO); presumably these efforts were directed at problems of in situ combustion. Apparently no field or pilot test were conducted. No further heavy oil recovery processes have been studied or attempted.

5.10 MEXICO

The known heavy oil occurrences in Mexico are located around the Gulf of Mexico near Tampico and Veracruz and in the State of Tabasco. The amount of heavy oil existing in Mexico has not been defined but is believed to be less than 500 million barrels. This figure likely will increase as exploration continues in presently unexplored sedimentary basins.

It has been suggested that some of the new oil fields currently under development on the Campeche Bank contain oil of less than 20°API. This; however, is not the case; gravity in these fields ranges from 23 to 38°API according to laboratory fluid measurements.

Efforts to improve recovery of heavy oil in Mexico have been meager at best. In fact, only two attempts at cyclic steam injection have been made. A "huff and puff" pilot test was conducted in three wells near Tampico where the oil-bearing formation lies at a depth of about 700 meters and is composed entirely of limestone. These efforts proved unsuccessful and the steam generator was moved to the Samaria field in the State of Tabasco where steam was injected into a shallow (600 meter) sand. Early equipment failure occurred, and the test was aborted and inconclusive. No further attempts have been made to improve the recovery of any heavy oils in Mexico.

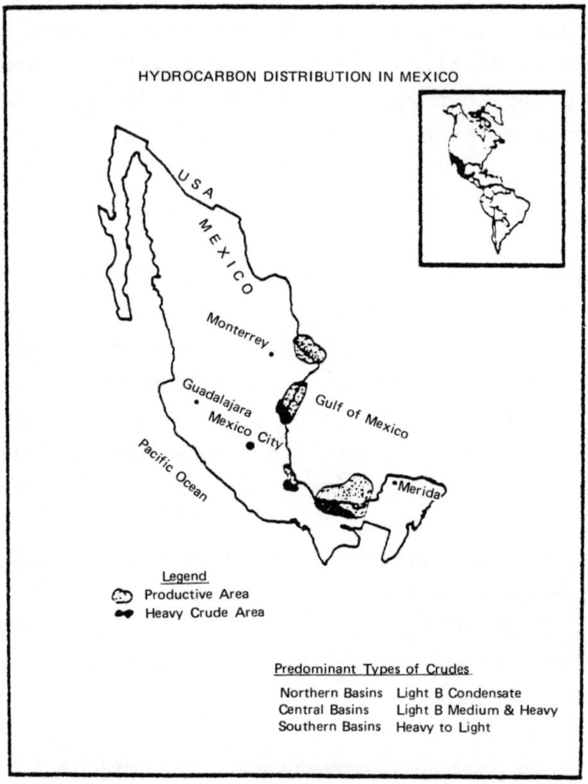

Figure 5-8. Heavy Oil Occurrences in Mexico

5.11 PERU

Heavy oil accumulations in Peru are primarily within the Marañon basin in northeastern Peru. Oil-in-place for these heavy oil fields has been estimated at 1500 million barrels. Crude oil gravities range between eight and 20°API. Occidental Petroleum Corporation has initiated pilot projects for development of the Cunambo, San Jacinto, Bartra, and Jibaro structures. These reservoirs normally produce by natural water drive where recovery and rate are primarily functions of the total fluid lifting capacity and costs.

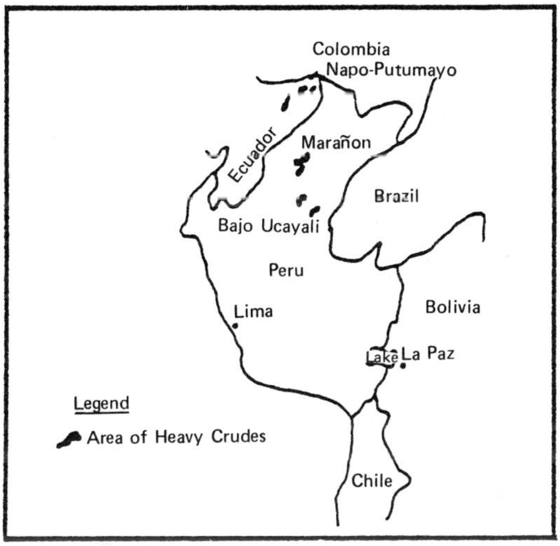

Figure 5-9. Heavy Oil Occurrences in Peru

5.12 ROMANIA

Within the southern and eastern portions of Romania lies the Pre-Carpathian Balkan basin. It is within this geological province that the known deposits of heavy oil are contained. Until 1976, Romania was self-sufficient with regard to oil production, partially due to efforts to improve the recovery of heavy oil. A wide variety of modern methods are being employed to increase recoveries of both the heavy and the lighter, less viscous oils.

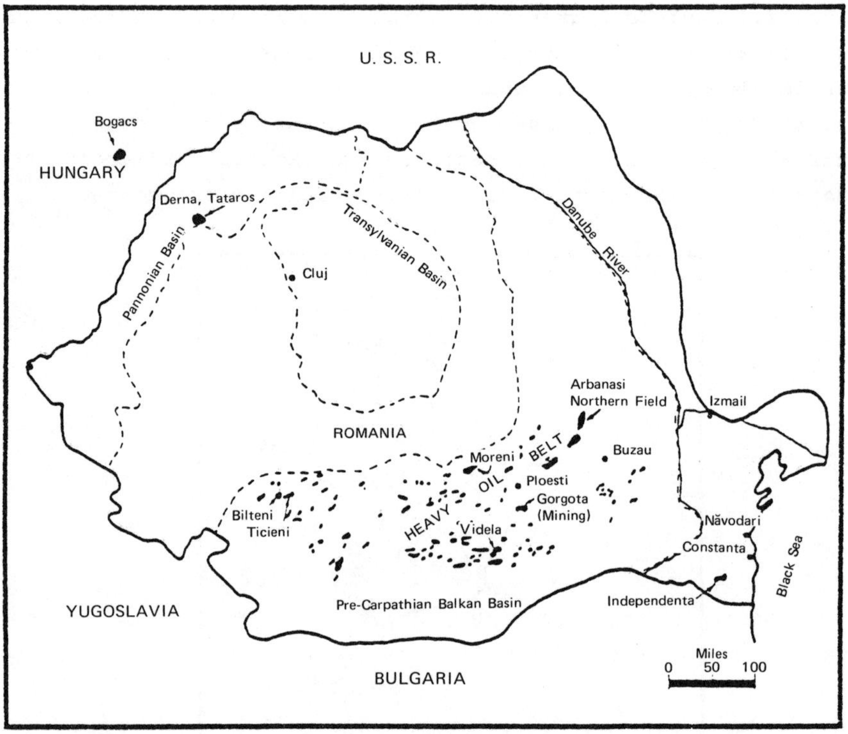

Figure 5-10. Heavy Oil Occurrences in Romania

Romanian efforts to increase recovery of oil include: (1) conventional waterflooding, (2) in situ combustion, (3) steam injection (cyclic and continuous), (4) polymer flooding, (5) micellar solution flooding, (6) CO_2 injection, (7) bacterial activity, and (8) mining. Numerous variations of these efforts are in pilot test or in active field status.

Reservoir engineering studies as a viable discipline began in 1945, and conventional waterflooding was started in 1952. Presently, there are over 100 fields under active artificial water drive. It was not until 1964, however, that Romania initiated its first in situ combustion project in the Suplacu de Barcau field. By 1978, this project had been expanded to 32 injectors and 205 producers.

The efforts applied to the Suplacu de Barcau field are an indication of Romanian interest in thermal recovery. Presently, there are four full-scale in situ combustion projects in progress; i.e., in the Suplacu de Barcau, Poseti, Gura-Ocnitci, and Balaria fields. Eight pilot projects are under way in the Ochiuri, Babeni, Slaticarele, Dealu Batrin, Stanesti, Otesti, and Moinesti fields, and eight additional tests are under design or in the early stages of implementation. Oil gravities in these planned projects are not avail able except for the Independenta (20°API) and South Moreni (21°API) fields. It is interesting to note that Romanian in situ combustion projects are usually designed to start upstructure.

Steam injection (cyclic) was the second thermal method applied in Romania. The Moreni field was considered unfavorable for exploitation due to its high oil viscosity (800 to 1500 cp and 21.5°API). However, reservoir properties made this field a candidate for cyclic steam injection and daily production rates by 1979 had reached 300 tons. Steam injection to production ratio averages 2.7 and yields an estimated ultimate recovery of 25 percent compared to the five percent estimated recovery by primary means.

Romania is also very active in polymer and micellar flooding; however, none of these efforts are believed to be directed toward heavy oil.

Table 5-III. Summary of Romanian Oil Recovery Activities

	FIELD	DEPTH (M)	VIS. (cp)	TEMP. (°C)	POR. %	PERM. (md)	SG / °API	PROJECT * TYPE	RECOVERY RATE PRIMARY ACTUAL	PRIMARY ESTIMATED	ULTIMATE ESTIMATED
1.	Suplacu de Bareau	60-200	2000	18	32.0	2000	0.960 / 16	F.C.	--	9.0	50
2.	Dealu Batrin	350	120	20	33.0	1000	0.920 / 22	F.C.	7.0	10.0	35
3.	Balaria	700	120	46	31.0	140-400	0.940 / 19	F.C./S.D.	8.2	12.0	36
4.	Otesti	900	180	35	30.0	700	0.940 / 19	F.C.	14.0	23.0	45
5.	Videle (G_5-II)	650	69-101	50	30.0	700	0.950 / 17	F.C.	5.4	12.0	37
6.	Videle (D_3-5)	760	10-33	52	30.0	1000	0.940 / 19	F.C.	14.0	25.0	51
7.	Dlejesti	950	40	42	26.5	410	0.940 / 19	F.C.	--	24.0	35
8.	Doleni	850	110	35	28.0	100-1000	0.940 / 19	F.C.	5.2	16.0	45
9.	Independenta	600-800	28-36	33	30.0	1000	0.936 / 20	F.C.	11.0	31.0	41
10.	Moreni	200-300	800-1500	18	29.0	1000	0.925 / 21	F.C.	--	2.0	35
11.	Poseti	250-480	6.4	25	27.0	800	0.870 / 31	F.C.	18.0	20.0	53
12.	Babeni	1000	2.0	42	20.0	2000	0.867 / 32	F.C.	24.5	30.6	41
13.	Gura-Ocnitci	800	5.0	38	26.0	400	0.840 / 37	F.C.	18.1	20.0	62
14.	Slaticarele	850	16	16	30.0	115	0.910 / 24	F.C.	9.7	15.0	40
15.	Monesti	500	12	27	29.0	2000	0.863 / 32	F.C.	9.4	27.0	39
16.	Stanesti	200-500	4-9	27	18.0	20	0.870 / 31	F.C.	10.2	15.7	27
17.	Ochuri	300	6	25	34.0	804	0.870 / 31	F.C./S.D.	45.0	46.0	60
18.	Moreni (Sub)	1000	4	45	33.0	1500	0.880 / 29	F.C.	36.7	45.0	54

Table 5-III. (Cont.)

FIELD	DEPTH (M)	VIS. (cp)	TEMP. (°C)	POR. %	PERM. (md)	SG / °API	PROJECT * TYPE	RECOVERY RATE PRIMARY ACTUAL	PRIMARY ESTIMATED	ULTIMATE ESTIMATED
19. Bolteana	700-1300	12	30-45	16.0	20	0.890 / 27	F.C.	12.9	21.0	--
20. Dragaesti	1600	1.8	63	27.0	1050	0.870 / 31	Polymer	10.9	40.0	50
21. Moreni (Sud)	1900	2.0	60	30.0	1500	0.850 / 35	Polymer	43.6	46.9	52
22. Bordei-Verde	1430	5.6	64	29.0	400-1660	0.910 / 24	Polymer	7.0	39.0	48
23. Tioleni	1460	1.5	62	30.0	1400	0.860 / 33	Micellar/ Polymer	60.0	60.0	70
24. Boldesti	2500	80	73	20.0	7350	0.840 / 37	Micellar/ Polymer	39.0	44.7	58
25. Surdulesti	845	35	45	26.0	1000	0.910 / 24	Micellar/ Polymer	29.6	36.2	--
26. Cascioarele	1500	120	70	27.0	400	0.840 / 37	Micellar/ Polymer	34.6	38.9	50
27. Satchinez	1300	80	85	25.0	1000	0.790 / 48	CO_2	30.0	46.0	63
28. Bradu	950	95	40	25.0	90	0.914 / 23	CO_2	8.4	26.3	45
29. Cerdec	1700	90	54	16.0	51	0.860 / 33	CO_2	8.0	20.0	27
30. Calacsa	1100	30-79	60	25.0	130	0.790 / 48	CO_2	16.8	40.0	48
31. N. Baicoi	Caustic-Surfactant Flood Planned - Drader Formation									
32. Baicoi	Bacterial Test Conducted in Field - Dscian Formation									
ACTIVE MINES										
1. Busterari										
2. Gorgotta										
3. Montecru										

* F.C. = Forward Combustion
S.D. = Steam Drive

Injection of CO_2 appears natural for Romania owing to the large CO_2 resources in that country. There are several CO_2 projects active and under design; however, none are directed toward reservoirs containing 20°API or less crude oil. The Bradu field (Meotian formation) is under active CO_2 drive; this reservoir contains oil of about 23.5°API which is intended to experience CO_2 solubility, viscosity reduction, and an increase in the formation volume factor.

Other methods are also being used to improve oil recovery. Bacterial activity for production stimulation has been and is being tested in the laboratory and field. The most notable project of this type was attempted in the Dscian formation of the Baicoi field. The gravity and other important parameters are not known for this field; however, considering the location of Romanian shallow crude, it is assumed that the gravity is at most in the high 20's. The field test using bacteria as a recovery agent in Romania did not prove effective.

Owing to the large amount of heavy oil deposits in Romania, it is only natural that mining be included as a heavy oil recovery process. Several oil mines are currently in operation, most notably in the Bustenari, Ochiuri-Gorgota, and Sarata-Monteoru fields. Both the mine gallery and excavation methods are used. For example, the Sarata-Monteoru deposit was developed in 1925 utilizing gravity drainage into mine galleries. Later, horizontally descending wells were drilled from these galleries, all of which has yielded a 55 to 60 percent recovery factor.

Reportedly, the Romanians have experimented with a unique and imaginative method for tracking combustion fronts. It has been noted that in cold weather European hares can be released into the project area where they will congregate and lie down where the soil is warmed above ambient temperatures. For relatively shallow projects, heat from the combustion process can reach the surface, and the ground temperatures will be at a maximum immediately over the front. Therefore, the front position can be detected by observing the bedding and congregating patterns of the hares. It is also reported that the Romanians have suggested the use of this method in Argentina.

5.13 TRINIDAD

It is natural to expect that Trinidad, being an extension of the Venezuelan land mass, would contain considerable heavy oil. Some estimates give Trinidad as much as 40 billion barrels of heavy oil, in addition to the famous "tar lake" on the south end of the island. However, these estimates often include deposits in the Guanoco area, which is actually within the boundaries of Venezuela.

Heavy oil recovery processes actively being pursued consist of steam displacement and are operated by Texaco and Tesoro-Trinidad, as follows:

OPERATOR	FIELD	GRAVITY (°API)	VIS. (cp)	POROSITY (%)	PERM. md
Texaco	Forest Reserve	16	95 c	33	300
Tesoro	NW Palo Seco	12	<400	33	<2000
Tesoro	Fyzabad	24	35	34	412
Tesoro	Guapo	12 - 15	1600	35	800

5.14 TURKEY

All known heavy oil deposits in Turkey lie in the eastern part of that country and appear on trend with some of the oil deposits of Iraq and Iran. Exploitation of heavy oil has been seriously attempted in only one field, Bati Raman.

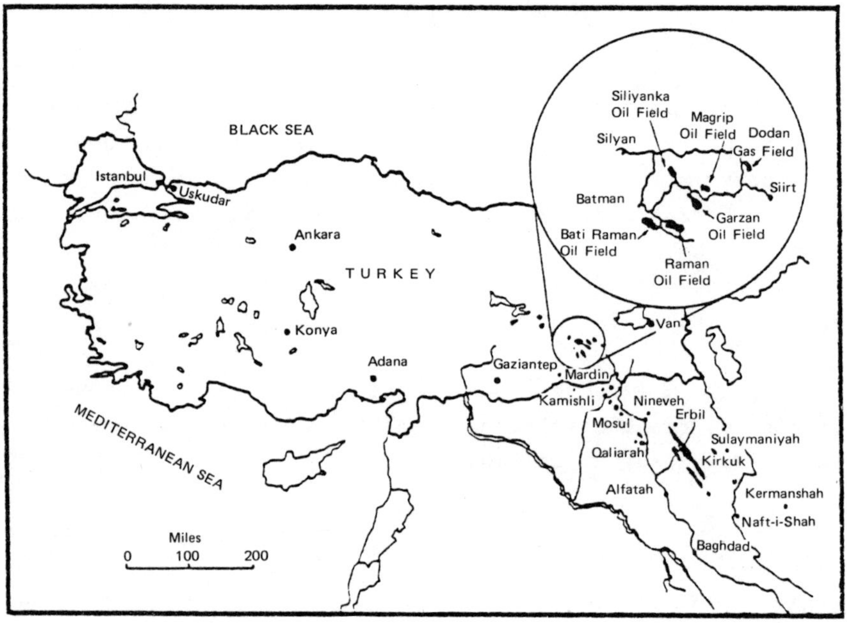

Figure 5-11. Heavy Oil Occurrences in Turkey

The Bati Raman field in eastern Turkey was discovered in 1961 at a depth of about 1300 meters (4265 feet). The productive Garzan formation is a carbonate deposit with vuggy, moldic, chalkification, and fracture porosity which averages from 14 to 19 percent depending upon location within the field. Permeability ranges from 200 to 475 md as determined from field production tests. Original oil-in-place was estimated to be two billion barrels, ranging in viscosity from 450 to 1000 cp. Initial pressure was about 1800 psig; solution gas oil ratio was 18 scf/STB; and the gravity of the crude oil varied from 9.7 to 15.1°API. Average temperature within the 200 feet of productive interval ranges from 150 to 160°F. Accumulated field data and reservoir engineering calculations project a primary recovery of 1.5 percent of oil-in-place, or only about 30 million barrels.

Since 1967, waterflooding and several other oil recovery methods have been attempted on a pilot basis. These efforts include cyclic steam injection, continuous steam injection, air injection, and several waterflood experiments.

Cyclic steam injection was initiated on a pilot basis in October 1967, in the south central section of the field. After five days of steam injection and while the well was being returned to production, the casing collapsed preventing further cycles. Fifteen days elapsed before oil production was commenced at a rate of 155 bbl/day, an increase of 65 bbl/day over pre-steam production rates. However, this test was considered inconclusive.

A pilot continuous steam injection test was initiated in the field during May 1969. Injection continued through October, during which time the entire process was monitored through surrounding wells. A total of 170,000 barrels of water equivalent steam was injected. Results in surrounding wells varied from no response to an increase from 62 bbl/day to 240 bbl/day; however, this increased oil rate was accompanied by a corresponding increase in water production.

Air injection commenced in May 1969, and continued over a total period of about 110 days. Again, as in the continuous steam injection, offset wells were used to monitor the air injection project.

After about two months of injection, air breakthrough occurred in one offset well. However, no significant increase in production of oil was observed and, in fact, the well rate began declining to below that of the pre-injection period. Other offset wells experienced this same decline, which was attributed to the creation of a gas phase and oxidation of the oil, yielding lower relative permeability and higher viscosity, respectively.

There have been a series of water injection projects attempted in the Bati Raman field, all of which showed positive responses. The single problem associated with all of these projects has been premature water breathrough. It is believed that premature breakthrough is a reponse to fractures and vuggy porosity in addition to the effect of high oil viscosity.

The two billion barrels of heavy oil trapped at Bati Raman and the very low primary recovery anticipated make this field a prime candidate for a variety of heavy oil recovery methods. In this regard, the Turkish Petroleum Company (TPAO) has contracted with the French Petroleum Institute to perform laboratory studies on the following basic aspects:

(1.) The effects of CO_2 on crude mobility; relative permeability to water, oil, and gas.

(2.) Dissolution and liberation kinetics of CO_2 in Bati Raman crude.

(3.) Effectiveness of using CO_2 after waterflooding.

(4.) Geological resolution of the fracture and vuggy porosity system.

Additionally, engineering studies are in progress by the Godsey-Barlougner Division of Williams Brothers Engineering and by Intercomp Research and Development to evaluate alternative recovery methods. In this regard, hydrocarbon miscible flooding, dry and wet in situ combustion, continuous and cyclic steam injection, and combinations with the use of CO_2 are under study.

Immediately upon completion of the above efforts, TPAO plans to initiate a pilot test to demonstrate the feasibility of the selected method. It is noteworthy that the Dodan Gas Field, containing about 0.75 trillion cubic feet of CO_2, is located just 90 km to the northeast of Bati Raman.

5.15 U.S.S.R

There are several areas of heavy oil deposits in the U.S.S.R. and a large number of reservoirs with higher gravity, viscous crude oils. Consequently, the Soviets have been engaged for several years in research and field tests of a wide variety of methods for recovery of viscous crude oils.

Petroleum occurrences in the U.S.S.R. are generally divided into five major provinces; i.e., in order of estimated reserves, Western Siberia, the Volga-Urals region, the Caspian Sea and North Caucasus, Eastern Siberia, and Central Asia. Heavy crude oils are reported in each of these areas, but essentially all of the Soviet activities related to heavy oil recovery processes have been in the Volga-Urals region and in the Caspian Sea and North Caucasus area.

In Western Siberia, two large heavy oil deposits have been located at Russkoye and Olenek. Russkoye is just north of the Arctic Circle and east of the Gulf of Obskaya. This

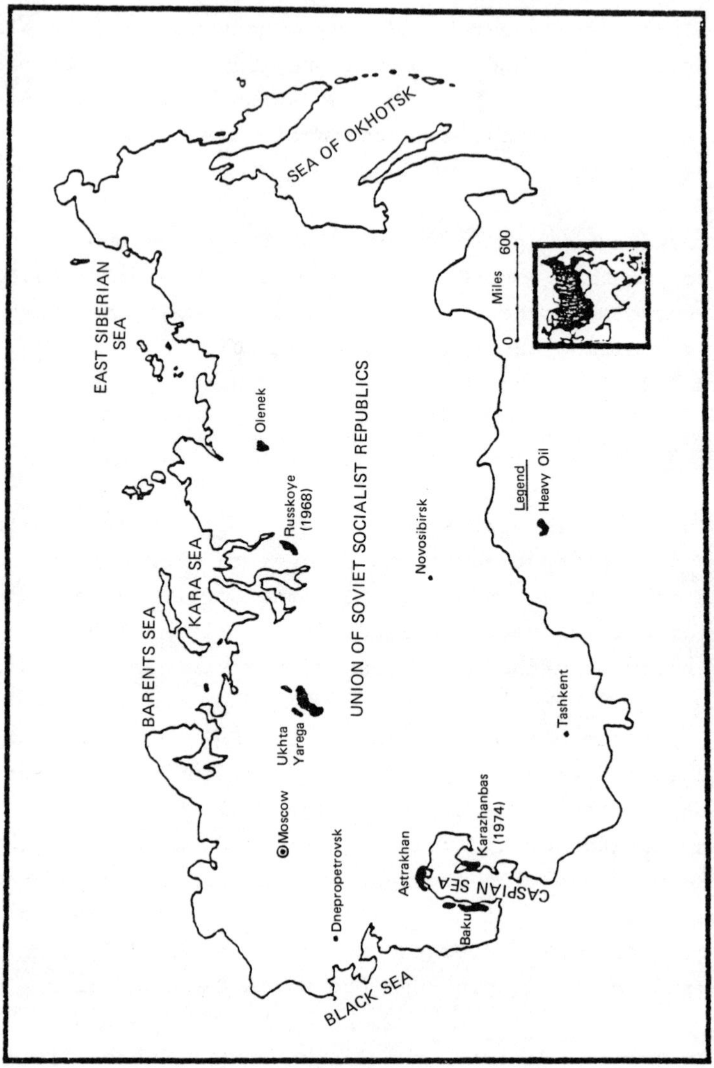

Figure 5-12. Heavy Oil Occurrences in the U.S.S.R.

field was discovered in 1968 but did not go on primary production until August of 1975. The 18°API crude oil is found just below the permafrost at a depth of approximately 800 meters. The Olenek Field was discovered more recently and lies east of Russkoye at 112° East Longitude.

In 1974, the Karazhanbas heavy oil field was discovered on the Buzachi Peninsula at the northeastern corner of the Caspian Sea. A primary recovery of seven percent of oil-in-place has been projected for this field. Both Karazhanbas and Russkoye are large fields containing at least 500 million barrels of reportedly recoverable oil.

Although the Soviets have applied or experimented with essentially all the heavy oil recovery processes investigated elsewhere, and have engaged in some innovative research, major efforts have consisted of various types of mining methods, thermal operations, and combinations of the two. Both surface and drainage mining have been used for many years. More recently, thermal stimulation by cyclic steam injection has been combined with drainage mining to evolve the "thermal mining" technique applied so successfully in the Yarega fields. U.S. patents have been issued for this method of thermal mining with Resource Sciences Corporation of Tulsa, Oklahoma, as the licensee.

Currently, the success at Yarega has resulted in vigorous activities in thermal and conventional mining processes in the Baku area on the western side of the Caspian Sea. In that area, the heavy oil reservoirs are located at shallow depths of less than 700 meters, the oil is highly viscous, and the deposits are sandstone. A thermal mine of the Yarega type is under construction near an active strip mine and an in situ combustion project. At the strip mine, the "ore" is crushed at the surface and mixed with a reagent mixture of kerosene, alkali, and water or other inexpensive solvents for extraction of the oil.

The underground thermal mine in the Balakhany field of Baku will be of the same type as those at Yarega, but the deeper sands and greater gas content have required some modifications and led to several design and operational problems. The project consists of two vertical shafts, 340 meters deep and 300 meters apart, connected by lined horizontal tunnels from which injection and drain wells are drilled into the productive sandstone. This mine was originally scheduled for completion in 1978, but a series of problems have caused a delay, including inadequate and faulty equipment and the need for additional and more positive ventilation because of the relatively gassy nature of the crude. The Soviets project that these problems will be overcome in time for an operational start in 1981.

The Soviets estimate that there are at least 25 heavy oil fields in the Baku area at a depth of 500 meters or less where various mining methods can be applied successfully. They state that recovery from these fields has been no more than 20 percent of the oil-in-place and that they still contain "several hundred million metric tons."

The increase in steam requirements in the Baku area has forced the Soviets to develop an additional source of high quality water. This has resulted in the acquisition and installation of nine reverse osmosis (RO) units from Water Services of America, Inc., of Milwaukee, Wisconsin. These units will treat Caspian Sea water to yield 80,000 bbl/day of high quality water for steam generation.

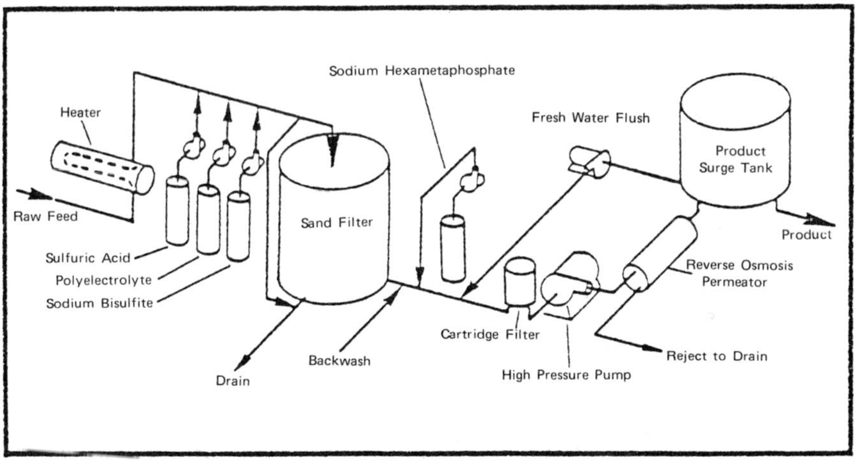

Figure 5-13. Reverse Osmosis (RO) Unit for Treatment of Caspian Sea Water

Steam injection is the most active thermal stimulation method currently used in the U.S.S.R., with a reported additional 200,000 bbl/day being produced by this method. Various Soviet literature cites D. J. Mendeleev's work of the late 1800's relating to the increase of heavy oil recovery by the introduction of steam into oil-bearing rocks. During the period of 1947 to 1950, P. T. Zakirov's proposal to mix steam and flue gas was tested in the Shodnitsa Field in the Ukraine. In 1951, a comparison test was conducted in the same field using steam and unheated water. Finally, the 1965-66 period marks the beginning of thermal stimulation on a practical basis in the U.S.S.R. when cyclic steam treatment was applied in the Krasnodar and Sakhalin areas. The steam displacement process was field tested in 1968 in the Yarega fields and in the Okha field in the Sakhalin area.

The Soviets credit steam stimulation of heavy oil reservoirs with yielding "hundreds of thousands of tons per year" and state that this method "continues to grow." They claim that the Okha and Yarega fields have produced "more than two million tons" of additional oil by steam injection. Steam injection has also been applied successfully in several other heavy oil fields, notably Katangli, Borislav, Kenkiyak, and Sibsa.

Although the steam soak and continuous steam drive methods are widely used in the U.S.S.R., it appears that the injection of a steam slug in combination with waterflooding is now gaining in popularity. This alternating steam-water-steam process was first tested in the Okha field where the heavy crude oil is found at depths of from 30 to 700 meters. The oil sands vary in thickness from 22 to 56 meters with permeabilities of from 1000 to 1500 md, and the oil has a viscosity ranging from 300 to 3000 cp within the same reservoir. It was reported that a 39 percent improvement in recovery was obtained with the injection of 1.5 pore volumes of steam. A 46 percent improvement was achieved with the injection of 1.9 pore volumes.

Fields in the Emba area are characterized by the presence of oil of high to ultrahigh viscosity. In spite of production histories encompassing from 10 to 50 years, 65 to 80 percent oil saturations still exist in these sands, which have porosities of from 21 to 36 percent. Waterflooding has proved to be generally unsuccessful due to the high oil viscosities, so both steam injection and in situ combustion are being applied as improved recovery processes.

In situ combustion is a popular and apparently effective heavy oil recovery process in several regions of the U.S.S.R. Generally, this process is applied to those reservoirs where depth limitations are believed to exclude the viable use of steam processes. However, combustion methods are reported to have experienced less than desirable results in some reservoirs.

Although both "dry" and "wet" combustion projects are currently active in the U.S.S.R., the "wet" method seems to be gaining in popularity. There are two reasons given for this. First, it is believed that the farther the combustion front moves from the injection well in the "dry" process, the farther it lags the displacement front, thereby decreasing the displacement effectiveness and limiting the allowable well spacing. Second, after years of experience and experimentations, the Soviets have confidence in their ability to control the more complex "wet" combustion process.

Apparently, the first field to be subjected to the in situ combustion process on a truly commercial basis was in the Khorosany area of the Azerbaijan S.S.R. This project was initiated in 1973 and converted to a "wet" process in 1974. Through 1977, it was reported that production rates had increased by a factor of 3.5 times the 1972 rates obtained from conventional waterflooding. This reservoir is 300 meters deep and has a net pay thickness of 35 meters, a porosity of about 24 percent, and an average permeability of 500 md. The 20.3°API crude oil has a reservoir viscosity of 140 cp.

Overall, several of the Soviet oil administrative agencies are engaged in combustion operations, including the Mangyshlakneft, Leninneft, Embaneft, Kirovneft, and Karadagneft. Generally, in situ combustion methods are employed in the same areas as are steam processes, namely the Caspian Sea and Baku areas, the Volga-Urals, and to a lessor extent the Ukraine.

In addition to steam and combustion processes, the Soviets are working on two other methods of in situ heat generation. In 1978, at least two administrative entities, Dossorneft and Kul'saryneft, were pursuing the implementation of electrical downhole heating. At the same time, G. D. Savankov and V. M. Doroshenko at the Ivano-Frankovskii Institute of Oil and Gas were doing theoretical and experimental work on the use of hydrochloric acid and magnesium as a means of elevating reservoir temperatures near the wellbore.

This in situ thermochemical treatment of the productive zone combines the effect of hydraulic fracturing, with sand and granulated magnesium, and the exothermal reaction of HCL and magnesium. This is basically the same process which in the U.S.A. is referred to as a "hot acid" treatment and has been used for several years as a means of well rate stimulation, but not specifically as a thermal process for heavy oil reservoirs.

In addition to thermal and mining methods, the oil industry in the U.S.S.R. actively utilizes a variety of other methods for improving oil recovery. Conventional waterflooding has been used for many years. Chemical flooding, carbon dioxide injection, and hydrocarbon miscible displacement are all adding to the daily oil production of the U.S.S.R. Except for some remnants of conventional waterflooding, none of these techniques are apparently being applied commercially in heavy oil reservoirs.

Polymer floods have been successfully field tested in the Kuibyshev region, as well as in Bashkiria and in the deposits of the Tatar region. Typical reservoir characteristics for chemical flooding projects are represented by the Romashkino deposit with porosity and permeability ranging from 24.5 to 28.8 percent and 830 to 6000 md, respectively. The initial oil saturation was from 76 to 81 percent of pore space, and the 31°API oil has a viscosity of 26 cp.

Chemical flooding was initiated in 1967 with a field test in the Nicolo-Bere area of the Arlan field. A surfactant solution was injected in an internal line-drive of 12 wells, with a reported increase in recovery of three to six percent over a comparable control area. Other chemical floods using polymers and alkaline solutions have also been field tested. To date, there appears to be no field-scale application of micellar flooding, but extensive research is being conducted by the All-Union Research Institute in Moscow.

Notable is the absence of miscible phase flooding for heavy oil recovery, and only one area, Abdrakhmanovskaja, appears to be active in carbon dioxide injection. Here, it

Table 5-IV. Summary of U.S.S.R. Improved Recovery Activity

Field	Vis. (cp)	Porosity (%)	Perm. (md)	SG/°API	Project * Type
Tanatar	100	29	8350	0.896/26.4	SI
Karsak	960	23	2500	0.920/22.3	SI
Koschagyl	182	27	3912	0.900/25.7	SI
Teren'uzyuk	87	47	15	0.946/18.1	SI
Teren'uzyuk	78	21	1260	0.943/18.6	SI
Kenkiyak	180	31	502	0.913/23.5	SI
Kopa	520	28	1968	0.900/25.7	SI
Kumsay	1163	36	5000	0.939/19.2	SI
Iskine	121	25		0.923/21.8	SI
Baychunas	76	28	926	0.880/29.3	SI
Karaarna	144	29	1391	0.965/15.1	SI
Shubarkuduk		28		0.910/24.0	SI
Akzhar		29			SI
Karatyube	164	29	2500	0.904/25.0	SI
Yarega (Depth 200 m)	17500	23	1600	0.950/17.5	TM
Romashkino 1	26	28	830-6000	0.872/30.8	PF
Romashkino 5	18	21	592	0.868/31.5	PF
Karsak	252	30	4884	0.91/24.0	FC
Koschagyl	56	28	1045	0.844/36.2	FC
Teren'uzyuk	23	33	1540	0.938/19.4	FC
Martyshi	24	29	1542	0.950/17.4	FC
Karaton	25	37	432	0.900/25.7	FC
Karaarna	70	29	279	0.963/15.4	FC
Sugushlin		23	40		FC

* SI = Steam Injection
TM = Thermal Mining
PF = Polymer Flooding
FC = Forward Combustion

appears that preliminary study and planning are in progress for the recovery of an oil with a gravity in the 30°API range.

Laboratory research has also been conducted on the use of hydraulically-induced vibrations and magnetized water for improving oil recovery; however, these processes have not been proposed for application to heavy crude oil reservoirs. Some microbiological work has also been done.

In general, the oil and gas industry of the U.S.S.R. has suffered from a lack of State emphasis, a shortage of experienced scientists, rigidly imposed technical constraints, a heavily bureaucratic structure, inadequate and insufficient equipment, and an often-hostile natural environment. As a result, the Soviets have had to rely heavily on outside-developed technology and equipment and on recovery processes for heavy oil which are well-known or which can be installed and operated principally by the application of large labor forces. With the exception of the unique thermal mining operations, little or no innovative technology has been applied successfully. By their own admission, otherwise routine projects have been delayed or severely hampered by inadequate and insufficient equipment, lack of necessary parts and maintenance, and various types of malfunctions and bureaucratic red tape.

In the opinion of knowledgeable observers, the Soviet oil and gas producing industry generally lags the western world by 10 to 20 years in terms of process applications, design, equipment, and computational techniques and facilities, in spite of considerable research activities and an awareness of western technical literature. However, rapid strides are being made, and more emphasis on oil production as an international political weapon will undoubtedly hasten this progress.

5.16 VENEZUELA

Presently known deposits of heavy oil in Venezuela trend east-west across the entire northern half of the country. Oil-in-place estimates for the deposits range between one and two trillion barrels, the majority being in the largely undeveloped Orinoco Petroliferous Belt which by some estimates alone contains nearly two trillion barrels. Fields located in the major depositional basins represent deposits of about 10 billion barrels in the Maracaibo basin

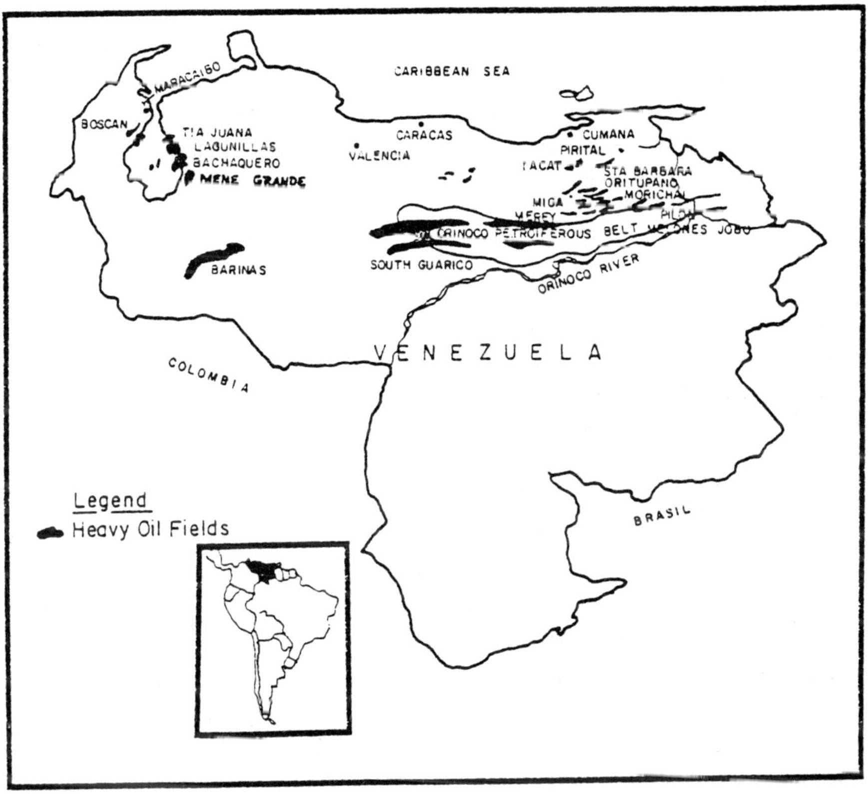

Figure 5-14. Heavy Oil Occurrences in Venezuela

of northwestern Venezuela, 100 million barrels in the Barinas sub-basin of western Venezuela, and 40 billion barrels in the Maturin sub-basin to the north and east of the Orinoco Heavy Oil Belt.

The Venezuelan heavy oil reservoirs contain a wide range of reservoir characteristics and crude oil properties. In the accompanying table are reservoir and fluid parameters representative of fields in two major producing areas, the Maracaibo basin and the Eastern basin at the northeast tip of the Orinoco region, or Maturin sub-basin area. Crudes with gravities as low as 8.7 to 9.0°API are found in the Eastern basin and are suspected to represent most of the crude deposits in the Orinoco Heavy Oil Belt as well. Crudes with gravities ranging up to 20°API are present in some reservoirs, however. Gravities in the Maracaibo basin cover a similar range of 10.3 to 18.0°API, with most being in the 12 to 15°API range.

Most heavy oil reservoirs in Venezuela are somewhat producible by primary means. The significant factor making primary production of the low gravity oils possible is the relatively high reservoir temperatures, especially when compared to most Canadian reservoirs containing similarly low gravity crudes at reservoir temperatures of only 50 to 70°F.

The lowest temperature shown for a Venezuelan reservoir is 100°F., occurring in shallow reservoirs located along the Bolivar Coast in the Maracaibo basin. The lower viscosities associated with the warmer temperatures allow the oil to be mobile at reservoir conditions and thus producible. The highest reported viscosities, ranging to 15,000 cp, are substantially lower than similar crudes in Canadian reservoirs where viscosities at reservoir conditions range from 50,000 to over 1,000,000 cp, making them essentially immobile in the reservoir.

Table 5-V. Reservoir Conditions and Fluid Characteristics Venezuelan Heavy Crudes

FIELD	RESERVOIR DEPTH, FT	RESERVOIR TEMP., °F	°API @ 60°F	VISCOSITY @ RES. COND. cp
Boscan	8000	175-185	10.3-12.0	165-275
Tia Juana Lagunillas Bachaquero	2000-4300	120-138	12.0-18.0	15-650
Tia Juana Lagunillas Bachaquero	800-2500	100-123	11.0-15.2	100-15000
Morichal Jobo Pilon Quiriquire	2900-4000	133-148	8.7-15.0	40-2150
Miga Melones Oveja	2000-6000	130-160	9.0-20.0	20-600

Two principal reservoir production drive mechanisms are present in the Venezuelan heavy oil reservoirs being produced by primary means. One, solution gas drive, is in most cases responsible for 1.5 to seven percent primary recovery. The second, compaction of the formation sand as the reservoir ure is depleted, has accounted for primary recoveries of from nine to 24 percent in several Bolivar Coast fields. Even in the Orinoco Heavy Oil Belt, which was originally called a "tar belt" because of the low API gravities, recent investigations have indicated that much of the oil is mobile at reservoir conditions and therefore producible by primary means. These drive mechanisms are by no means limited to primary operations but are also significant throughout thermal recovery operations.

Production of heavy oil in Venezuela began in 1914 with the discovery and subsequent development of the Mene Grande field. It was not until 1957 that a field test was made at Mene Grande to investigate the effects of applying steam to heat the reservoir in hopes of improving recovery. This early field test gave rise to the "steam soak" or cyclic steam injection process so widely used today. A steam displacement or "steam drive" test followed the Mene Grande testing in 1961. This test, labeled the C3/C4 project, was carried out in the northern part of the Tia Juana field, was conducted for six years, and provided well documented results which showed it to be less favorable economically than the more energy-efficient steam soak process. However, plans for a large scale steam drive project were drawn up in 1972 at the insistence of the Ministry of Mines and Hydrocarbons. Those plans culminated in 1978 with the start-up of full scale injection operations into 19 injectors with an additional 132 producing wells forming seven-spot hexagonal patterns. This project, known as the Tia Juana M-6 Steam Drive Project, is currently in its third year of operation by Maraven, S. A. Meanwhile, during the sixties and seventies, further development and use of the steam soak process was employed in the Tia Juana and other Venezuelan heavy oil fields.

In situ combustion testing was started in 1959, two years after steam stimulation testing began, with the commencement of the East Tia Juana Block K-7 test. Movement of the combustion front was closely monitored throughout the test, and a core hole was drilled behind the front to analyze combustion zone rock and residual fluid properties. A narrow sweep pattern and the less than significant production results prompted the project to be shut down in February 1962.

Another test was made during the mid-to-late sixties in Block J-6 of the East Tia Juana field which involved alternate air and water injection schemes, forming a partially quenched combustion test. Results of the test were hampered by corrosion in the injection wells and related mechanical problems. The project was shut down in 1968 with inconclusive production results.

Other in situ combustion tests were carried out at about the same time in eastern Venezuela and showed more positive results. One, located in the Miga field, began in 1964 and is currently operating. It involves the dry forward combustion process in a reservoir over 4000 feet deep with 14°API oil having a viscosity of 280 to 430 cp. A similar combustion project is currectly being operated by Corpoven in the Morichal field area. A wet combustion project operated by Meneven in the Melones field was temporarily suspended in 1979 due to mechanical operating problems and is being redesigned.

Current efforts involving secondary heavy oil recovery methods in Venezuela are summarized in the accompanying table. Forty projects are in operation, 34 of which are cyclic steam injection or, "steam soak," projects. The remaining six active projects include the previously mentioned in situ combustion operations in the Miga and Morichal fields, two steam drive projects, one hot water flood, and one gas injection operation. Intense interest is currently being focused on heavy oil recovery in Venezuela particularly in the vast deposits of the Orinoco Heavy Oil Belt. Non-thermal processes are not being ruled out for future heavy oil recovery test projects; however, processes such as caustic flooding and emulsion and CO_2 injection will require extensive laboratory investigation before being seriously considered for pilot testing.

Steam soak is expected to continue as the main process for improved heavy oil recovery in all regions of Venezuela. Reasons for this outlook are the proven success and the

Table 5-VI. Summary of Active Projects Recovery of Heavy Oil Venezuela

FIELD	OPERATOR	HEAVY OIL RECOVERY PROJECT TYPE	NUMBER OF PROJECTS	OIL GRAVITY °API @ 60°F	RESERVOIR DEPTH, FT	OIL-IN-PLACE 10^9 STB
Bachaquero	Lagoven	Cyclic Steam Inj.	2	12-18	2000-4300	20.5
	Maraven	Cyclic Steam Inj.	3	11-15	800-2500	
Jobo	Corpoven	Cyclic Steam Inj.	2	9	3400-3850	8.6
	Lagoven	Cyclic Steam Inj.	1	9		
Lagunillas	Lagoven	Cyclic Steam Inj.	2	11-16	1800-3000	23.5
	Maraven	Cyclic Steam Inj.	6			
	Maraven	Steam Drive	1			
Melones	Meneven	Wet Combustion	Temp. Susp.	18	2000-6000	2.5
Miga	Meneven	Combustion	1	13	4000-4350	23.2
Morichal	Corpoven	Cyclic Steam Inj.	2	12	3500-4000	5.0
	Corpoven	Hot Water Flood	1			
	Corpoven	Combustion	1			
Pirital	Corpoven	Steam Drive	1	17	450-1100	0.1
Tia Juana	Lagoven	Cyclic Steam Inj.	2	12	1600-2900	19.0
	Maraven	Cyclic Steam Inj.	1			
	Meneven	Cyclic Steam Inj.	13			
Oveja	Meneven	Gas Injection	1	15-20	3800-4200	0.8

favorable economics of this method. Also, much experience has been gained in field and mechanical operating practices over the past 20 years of steam injection operations.

5.17 MISCELLANEOUS COUNTRIES

5.17.1 Europe

Of this group, the most important heavy oil deposits and recovery activities occur in Hungary. The two best known heavy oil deposits are Bogacs in the east and Nagylengyel in the west. It is within this latter field that alkaline ammonia is being injected to improve recovery of heavy and viscous oil. The Nagylengyel field, now in the final phase of primary production, is the most significant in Transdanubia. It is cut with many major faults to yield a horst and garben structure which is comprised of about 15 independent reservoirs spatially.

Figure 5-15. Heavy Oil Occurrences in Hungary

The main productive interval, an Upper Cretaceous Rudistic limestone, has seven individual pay zones; and the Gryphean (UK) and Helvetion (Mio) each contain one pay zone. All formations, excepting that of Miocene age, are carbonates. Secondary porosity plays a vital role in hydrocarbon storage, since primary porosity is only two to three percent. Permeability is reported to be low in all of these formations, which are also oil-wet and are assumed to have been almost 100 percent oil saturated.

The highly oil-wet reservoir and the physical and chemical fluid characteristics apparently afford the proper combination for enhancing recovery by reaction with ammonium hydroxide. The 10 to 22°API oil contains high molecular weight napthenes, fatty acids, and esters which react with the alkaline medium to initiate the process of emulsification and entrainment.

Field testing began in 1973 with favorable results. This led to a full scale alkaline ammonia flood in 1975. No further field results are reported, but in the laboratory a three percent pore volume slug (25 wt. percent NH_4OH) yielded an eight percent increase in oil recovery when connate water was absent. The same size slug yielded an 11 percent increase in recovery in the presence of connate water.

Recoverable heavy oil reserves in Hungary have been placed at 189 million barrels.

Heavy oil deposits are produced in several other European countries. Spain obtains 17°API oil from the San Carlos Islands I and II. Production between 1975 and 1978 amounted to about 30 million barrels, and remaining recoverable (primary) reserves are estimated to be about five million barrels. Dorado, also located on the east coast of Spain, produces 21.3°API oil. This field was discovered in 1977 and has an estimated seven million barrels of recoverable reserves.

Other European countries known to have deposits of heavy oil and their estimated recoverable reserves are: Austria (four million barrels), Albania (42 million barrels), Bulgaria (six million barrles), and Yugoslavia (one million barrels). No significant recovery efforts have been reported for these countries.

There are no known heavy oil occurrences in Poland. The oil fields of the Carpathian Region and more recent shoreline discoveries are reported to be high quality crudes.

Cultured bacteria have been injected in Poland in field tests involving 35°API crude oil in order to increase recovery. These tests involved a total of twenty wells which were apparently producing at very low rates in the order of two to three barrels per month. The wells were observed to show an increase in procution rate of 20 to 200 percent. These efforts have now ceased.

There are no reported heavy oil deposits in East Germany of any consequence.

In the North Sea, east of the Shetland Islands and 15 miles northwest of the Frigg field, Siebens has discovered a deposit of 11.7°API oil. This is the only known heavy oil reserve in the North Sea, since discoveries by BP and Esso west of the Shetland Islands range from 22 to 25°API and qualify only as low gravity crudes. None of these reserves are of economic potential under today's conditions.

5.17.2 Middle East

Heavy oil has been known to exist in several Middle East countries for centuries, the largest deposits being in Iran, Iraq, and Syria. Very little production occurs from these heavy oil reservoirs, and no known heavy oil recovery processes are being applied to improve or stimulate recovery.

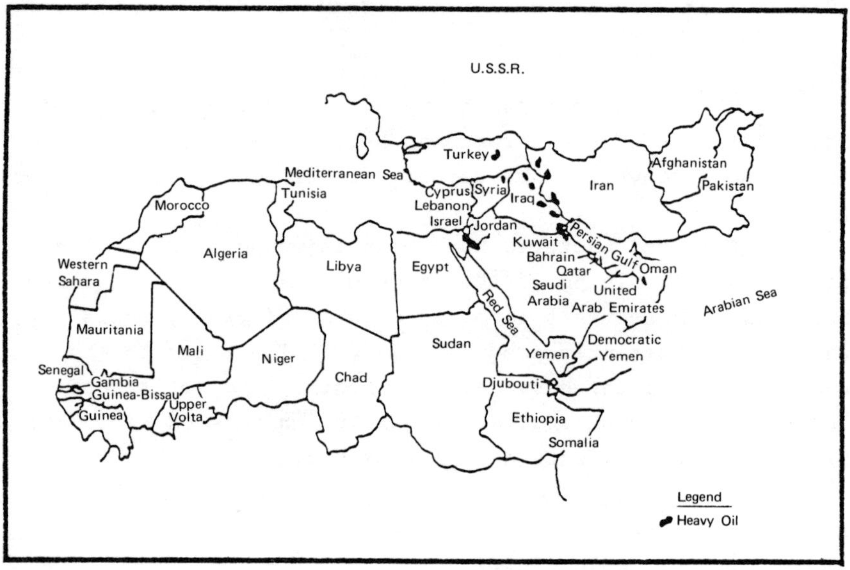

Figure 5-16. Heavy Oil Occurrences in the Middle East

5.17.3 Others

Some quantities of heavy oil have been discovered in Angola and Cameroon in west central Africa. In 1977, the Petrangol Well QF-1 flowed 300 bbl/day of a 16.5°API oil from a depth of 1750 meters. This Albian reservoir is located on the northern coast of Angola, near the Zaire border. In Cameroon, oil of 16 to 20°API has been found in

Table 5-VII. Summary of Middle East Heavy Oil Occurrences

Country	Heavy Oil-In-Place (Barrels)	Recovery Efforts Beyond Primary	Area	Remarks
Bahrain	None	None	North Central Section of Persian Gulf	
Iran	31 Billion	None		
Iraq	± 50 Billion	None	Nation-Wide	Crude oil ranging from 5-20°API gravity has been drilled throughout Iraq.
Israel	Unknown	None	Dead Sea	The Gurim #3 Well, Southwest of Dead Sea produced 2500 BO(20°API) in 1970 (Jurassic) No further development.
Jordan	Unknown	None	Dead Sea	Seepage of heavy oil in and around Dead Sea
Kuwait	Several Billion	None	Bahrah, Burgon, and Miragish	
Kuwait-Saudi Divided Zone	Unknown	None	Wafra	The field at Wafra is under primary recovery.
Lebanon	None	None		
Oman	Unknown	None	Habur, Qarn Alam, Amal, Marmul, and Rulit	Shell oil is currently developing oil fields in Oman. No known heavy oil recovery processes are being implemented.
Qatar	None	None		
Saudi Arabia	Unknown	None		Only one field Abu Sa'fah has known heavy oil. Well logs indicate probable heavy oil in other fields.
Syria	Unknown	None		Nearly all known oil resources in Syria are heavy or near heavy oil. Major reservoirs of heavy oil are at Jubaissa and Karatchok.
U. A. Emirates	None	None		
Yemen & So. Yemen	None	None		

Miocene reservoirs at 900 to 4200 meters by Elf-Serepca and Pecten. These reservoirs are all located in the Rio del Rey area near the northwest coast. No heavy oil recovery processes have been reported for these areas.

In Argentina, according to representatives of the national oil company, a thermal stimulation project, cyclic steam injection, is planned for 1981 in the Llancanelo field in the State of Mendoza. This reservoir lies at a depth of

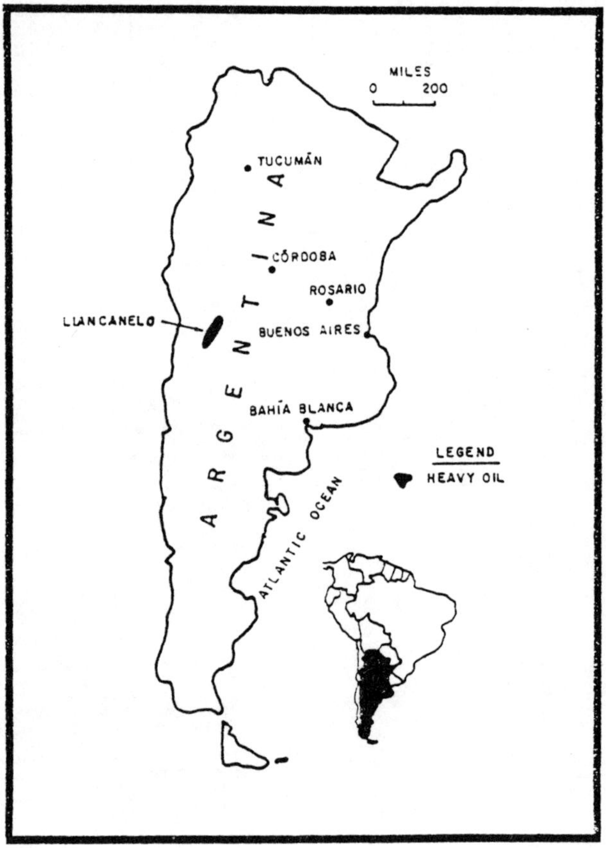

Figure 5-17. Heavy Oil Occurrences in Argentina

less than 600 meters and has a porosity in the upper 20 percent range. Otherwise, there are several known heavy oil deposits in Argentina, but no definitive efforts have been made to determine the quantity of reserves. One estimate, though, has placed these reserves of heavy oil at 171 million barrels.

Heavy oil deposits in Colombia are known to exist in the Meta basin east of Bogota and in the Napo-Putumayo sub-basin which extends south into Ecuador, where more deposits are known in the Marañon sub-basin. These oils are of a 12 to 14°API gravity and are found at depths of from 7500 to 14,000 feet. In addition, a deposit of 10.9°API oil is located at a depth of 6100 feet in Oriente Province of northeastern Ecuador. No significant heavy oil recovery operations are active in these countries.

In southern Barbados, General Crude discovered, in 1978, an onshore deposit of 18°API oil in the Charnocks area south of the Woodbourne field. No known heavy oil recovery processes are in operation here.

Surinam has one reported heavy oil deposit. Available information indicates this oil is at a depth of about 550 feet and has a gravity of 16 to 18°API. No heavy oil recovery processes are reported.

Appendix A
Literature Survey Summaries

TITLE

Correlations for The Performance Prediction of Heavy Oil Pools

AUTHOR(S)

Abougoush, M. S.

REFERENCE

Unitar Report 40, First Intl. Conference on The Future of Heavy Oil and Tar Sands, 1979, Edmonton, Alberta, Canada.

COUNTRY/REGION

Canada

TYPE OF PROJECT

None

SUMMARY OF PAPER

Reservoir simulation studies were made using a two-dimensional, three-phase coning model to simulate water coning behavior of heavy oil reservoirs and a two-dimensional, three-phase black oil simulator to simulate the water flood performance of a typical Lloydminster type heavy oil reservoir. General correlations were then made from the model results showing sensitivity to such parameters as: 1.) viscosity of oil, 2.) oil zone/water zone thickness, 3.) total fluid production rate, 4.) completion interval.

NEW OR DIFFERENT IDEAS/CONCEPTS/PRACTICES

L I T E R A T U R E S U R V E Y F O R M

TITLE

The Application of Improved Oil Recovery Methods

AUTHOR(S)

Anon.

REFERENCE

Unknown

COUNTRY/REGION

Russia

TYPE OF PROJECT

All

SUMMARY OF PAPER

U.S.S.R. is testing, developing, or has operational all of the following combinations of recovery methods: water w/chemical reagents, water-soluble surfactants, thickening agents, alkali, acids w/gases, hydrocarbon gases, carbon dioxide w/heat, w/hot water, steam in situ combustion, solvents, micellar solutions, and alcohol.

This paper was written as an introduction message to visiting professionals and is elementary in scope.

NEW OR DIFFERENT IDEAS/CONCEPTS/PRACTICES

L I T E R A T U R E   S U R V E Y   F O R M

TITLE

Estimating the Possible Oil Output in The Exploitation of the Russian Field by Thermal Methods

AUTHOR(S)

Arzhanov, F. G., Kuvshinov, A. S., Fain, Yu. B., Pravednikov, N. K., Vashurkin, A. I., Tsybul'ko, A. M., Radovskii, N. L., Sadchikov, P. B.

REFERENCE

Neft. Khoz., No. 2, 1977, p. 24-26.

COUNTRY/REGION

Russia/Tyumensk

TYPE OF PROJECT

Thermal

SUMMARY OF PAPER

Results of calculations were presented to illustrate that wet in situ combustion or steam injection is the only viable method to recover the oil from Cretaceous sands in the Tyumensk area. Apparently, laboratory work was the justification for making extensive calculations. Experiments with cold and hot water, steam, and wet insitu combustion resulted in residual oil saturations of 49% (cold water), 27% (hot water-220°C), 23% (steam-245°C), and the residual fuel content after wet in situ combustion was 20 kg/m^3.

NEW OR DIFFERENT IDEAS/CONCEPTS/PRACTICES

LITERATURE SURVEY FORM

TITLE

History of the EMBA Oil Industry

AUTHOR(S)

Ayzenshtadt, G. Ye. A.

REFERENCE

Geologiya Nefti i Gaza, Vol. 8, No. 9, 1964.

COUNTRY/REGION

Russia/Emba

TYPE OF PROJECT

Field Development

SUMMARY OF PAPER

A historical review of the Emba Field as it was developed from 1892 to present.

NEW OR DIFFERENT IDEAS/CONCEPTS/PRACTICES

L I T E R A T U R E   S U R V E Y   F O R M

TITLE

Azerbaijan Underground Oil Mine Work To Start

AUTHOR(S)

Anon.

REFERENCE

Oil and Gas Journal, May 7, 1979.

COUNTRY/REGION

Russia/Baku

TYPE OF PROJECT

Mining

SUMMARY OF PAPER

Very general description of underground mining. Article states that this project is to commence possibly in 1981. Rig to drill vertical shafts can drill hole up to 9 meters in diameter.

NEW OR DIFFERENT IDEAS/CONCEPTS/PRACTICES

LITERATURE SURVEY FORM

TITLE

Economics Effectiveness of New Methods Boosting Stratal Oil Recovery

AUTHOR(S)

Balashova, I. V. et al.

REFERENCE

Ekonomika Neftyanoy Promyshlenosti, Moscow, July, 1979, No. 7, p. 26-30.

COUNTRY/REGION

Russia/General

TYPE OF PROJECT

Thermal, Miscible, Chemical

SUMMARY OF PAPER

An economic comparison of various new recovery methods. Article discusses how to make the comparison and give cost results of various recovery methods.

NEW OR DIFFERENT IDEAS/CONCEPTS/PRACTICES

L I T E R A T U R E S U R V E Y F O R M

TITLE

Thermal Recovery in The Venezuelan Heavy Oil Belt

AUTHOR(S)

Ballard, J. R., Lanfranchi, E. E., Vanags, P. A.

REFERENCE

The Journal of Canadian Petroleum, April-June, 1977, Vol. 16 #2.

COUNTRY/REGION

Venezuela/Orinoco Belt - Jobo Field

TYPE OF PROJECT

Steam Soak

SUMMARY OF PAPER

This paper describes the method and results of a cyclic steam soak process in two reservoirs containing 9° API and 13.5° API crudes. Portable 30 MMBTU-per-hour steam generators are used to inject 80% quality steam at 2,500 Psi. Soak periods vary from one week to one month per treatment. Two to five-fold production rate increases were obtained when wells were returned to production after the soak period.

NEW OR DIFFERENT IDEAS/CONCEPTS/PRACTICES

Pre-stretched casing, two-stage cementing with thermally resistant cement mixtures were used in the downhole completion process. Also, the tubing-casing annulus is packed with a fatty-acid compound (Ken Pak) during the injection cycle for casing insulation.

LITERATURE SURVEY FORM

TITLE

German Producers Pin Hopes on Steam

AUTHOR(S)

Bleakley, W. B.

REFERENCE

Petroleum Engineer International, p. 40-54, June, 1979.

COUNTRY/REGION

Germany

TYPE OF PROJECT

Steamflood and Steam Soak

SUMMARY OF PAPER

General summary of initial efforts of steam soak in the Georgsdorf and Ruhlermoor Fields which were not conclusive. Steamflood applications are being field tested. Thermal reserves in the Emsland area are estimated to be 193 mm barrels, about 12% of oil in place, 60% will go unrecovered, 27° API crude.

NEW OR DIFFERENT IDEAS/CONCEPTS/PRACTICES

None

L I T E R A T U R E   S U R V E Y   F O R M

TITLE

Production Characteristics and Oil Recovery in The Orinoco Oil Belt

AUTHOR(S)

Borregales, Carolos J.

REFERENCE

Unitar Report 30, First Intl. Conference on The Future of Heavy Oil and Tar Sands, 1979, Edmonton, Alberta, Canada.

COUNTRY/REGION

Venezuela

TYPE OF PROJECT

Several conventional: steam soak and drive, in situ combustion. Non-thermal: caustic, emulsion, and CO_2.

SUMMARY OF PAPER

Discussion of all methods being tried or considered for additional heavy oil recovery. Methods include steam soak, steam drive, in situ combustion (including wet combustion). Non-thermal methods being considered only include caustic flooding, emulsion and CO_2 injection. An in situ (wet) combustion project in the Melõnes Field was shut down in 1979 for mechanical problems and is being re-designed.

NEW OR DIFFERENT IDEAS/CONCEPTS/PRACTICES

Steam soak followed by steam drive is being instigated with residue from the heavy-oil upgrading process being used to provide fuel for the high fuel requirements of steam drive.

L I T E R A T U R E   S U R V E Y   F O R M

TITLE

Role of Geomicrobiology in Enhanced Recovery of Oil: Status Quo

AUTHOR(S)

Bubela, B.

REFERENCE

The APEA Journal, 1978, p. 161-166.

COUNTRY/REGION

Australia with references to Poland, Hungary and Russia

TYPE OF PROJECT

Microbiological

SUMMARY OF PAPER

Discussed degradation of crudes by organisms and resulting affect on oil recovery, lower volume, more gas, and etc. Also, it cites work being done in Russia, Poland, Hungary and gives partial field results.

NEW OR DIFFERENT IDEAS/CONCEPTS/PRACTICES

None - This has been known for years, although no known projects exist in the U. S.

L I T E R A T U R E S U R V E Y F O R M

TITLE

Use of a Thermal Gas Generator to Start the Combustion Process in a Bituminous Seam of the Sugushlin Area

AUTHOR(S)

Bulgakov, R. T., Talantov, A. V., Kleev, A. M., Smerkovich, E. S., and Margulis, V. Ya.

REFERENCE

Nefteprom. Delo., 1978, No. 12, p. 5-8.

COUNTRY/REGION

Russia/Sugushlin

TYPE OF PROJECT

Thermal

SUMMARY OF PAPER

Measures related to the testing of a thermal gas generator were implemented in the test area from 1971 through July, 1976. These measures consisted of injecting steam, solvents and sulfuric acid, hydroabrasive and thermal gas generator processing. Air was injected for two hours before igniting generator with A-76 gasoline, then switched to TS-1 fuel. Output temperature of generator can vary from 600-2300°K. Besides CO_2 being a product of combustion, CO_2 was generated by thermal decompositions of the carbonate rock as evidenced by the presence of CaO at the producing well.

NEW OR DIFFERENT IDEAS/CONCEPTS/PRACTICES

Use of down-hole thermal gas generator to ignite combustion and permit variations of O_2 and water content of injection stream.

L I T E R A T U R E   S U R V E Y   F O R M

TITLE

The Influence of Dissolved Gas in Crude Oils on the Low-Tension Flood Process

AUTHOR(S)

Burkowsky, M., Marx C.

REFERENCE

Erdoel-Erdgas-Zeitschrift, International Edition, No. 5 (1), p. 36-44, 1979.

COUNTRY/REGION

Germany

TYPE OF PROJECT

Surfactant

SUMMARY OF PAPER

Laboratory studies using commercial surfactant, degassed crude, methane saturated crude, and natural gas mixture saturated crude. Results indicate little effect of gas in solution but considerable effect of reservoir temperature. Interfacial tension reaches a minimum over a small temperature range.

NEW OR DIFFERENT IDEAS/CONCEPTS/PRACTICES

None

L I T E R A T U R E S U R V E Y F O R M

TITLE

Silverdale Combination Thermal Drive Project

AUTHOR(S)

Cady, G. V., Hoffman, S. J., and Scarborough, R. M.

REFERENCE

Unitar Report 10, First Intl. Conference on the Future of Heavy Oil and Tar Sands, 1979, Edmonton, Alberta, Canada.

COUNTRY/REGION

Canada

TYPE OF PROJECT

In Situ Combustion

SUMMARY OF PAPER

As per title, this paper discusses the planning, design, and three years of performance of an in situ combustion with simultaneous injection of water project in the Sparky pool in the Lloydminster Field. Eight 30-acre inverted seven spot patterns were in operation during 1979, with positive indications of economic success through apparent high pattern productivity. This method is expected to increase recoveries from 5% to 40% of original oil in place. Through the three year project history an overall water-air injection ratio of 210 Bbl/MMscf resulted. The air injected to oil produced ratio was 3.5 Mscf/Bbl.

NEW OR DIFFERENT IDEAS/CONCEPTS/PRACTICES

L I T E R A T U R E   S U R V E Y   F O R M

TITLE

Summary Report on The First International Conference on The Future of Heavy Crude and Tar Sands.

AUTHOR(S)

Prepared by Mr. Tom Campbell

REFERENCE

Heavy Oils World Wide Resources

COUNTRY/REGION

TYPE OF PROJECT

Various - Conference concerned itself w/most types EOR projects.

SUMMARY OF PAPER

The significance of heavy crudes and tar sands throughout the world was highlighted. Companies, governments, etc. involved with research and production participated in conference and many presented papers. Brief description of some projects was given. The apparent most significant locations of deposits in free world was given.

NEW OR DIFFERENT IDEAS/CONCEPTS/PRACTICES

1. Electrical heat generation in situ between two wells
2. R. F. energy
3. Down-hole steam generation for "huff-puff" or continuous process.

LITERATURE SURVEY FORM

TITLE

Experimental Steamflood Cold Lake Oil Sands

AUTHOR(S)

Capeling, R. R., Peggs, J. K.

REFERENCE

Unitar Report 94, First Intl. Conference on The Future of Heavy Oil and Tar Sands, 1979, Edmonton, Alberta, Canada.

COUNTRY/REGION

Canada

TYPE OF PROJECT

Steamflood

SUMMARY OF PAPER

Discusses the performance results of two pilot steamfloods operated during the years 1965-1970. An inverted five-spot pattern was used in phase I and injection and production performance closely monitored. An 11% oil recovery with a 7.6 Bbl/Bbl steam oil ratio was obtained in phase I. Useful qualitative information and operating experience was also gained.

Phase II, carried out on an adjacent lease, utilized a similar 5-spot pattern but with the addition of 4 outside producing wells, and three additional injectors. Operating and completion techniques learned in Phase I were used. However, results obtained in Phase II indicate a less efficient pilot with 8.5% recovery and an SOR of 16 Bbl/Bbl.

L I T E R A T U R E   S U R V E Y   F O R M

TITLE

Heavy Crude Oil Potential of Saskatchewan

AUTHOR(S)

Christopher, J. E., Knudsen, R. H.

REFERENCE

Unitar Report 36, First Intl. Conference on The Future of Heavy Oil and Tar Sands, 1979, Edmonton, Alberta, Canada.

COUNTRY/REGION

Canada

TYPE OF PROJECT

None

SUMMARY OF PAPER

Provides a description of the geologic units and locations where they exist which contain commercial quantities of heavy oil. Reservoir and crude oil properties for the different areas are given, and a discussion of potential recoveries under primary, waterflood, steam and in situ combustion operations is included.

NEW OR DIFFERENT IDEAS/CONCEPTS/PRACTICES

LITERATURE SURVEY FORM

TITLE

Five New Oil Sands Projects Needed in 80's

AUTHOR(S)

Collyer, C. R. (Speech Summary)

REFERENCE

The Northern Miner, Oct. 4, 1979.

COUNTRY/REGION

Canada

TYPE OF PROJECT

None

SUMMARY OF PAPER

Discusses the urgency of developing and implementing economical methods of recovering Canada's heavy crudes.

NEW OR DIFFERENT IDEAS/CONCEPTS/PRACTICES

L I T E R A T U R E S U R V E Y F O R M

TITLE

Petroleum Tomorrow: Resources and Recovery

AUTHOR(S)

Despaires, P., Simandoux, P., Bernard, F., Leduc, J., Putz, A.

REFERENCE

Petrole et Techniques, No. 248, p. 7-46, Nov., 1977.

COUNTRY/REGION

France and others

TYPE OF PROJECT

Several

SUMMARY OF PAPER

General discussion of EOR methods by several specialists, global resume of reserves, present recovery estimates, and EOR prospects. All gravity crudes included. Some general economic considerations of relative costs of new reserves versus EOR.

NEW OR DIFFERENT IDEAS/CONCEPTS/PRACTICES

None

L I T E R A T U R E   S U R V E Y   F O R M

TITLE

Conditions for Miscibility of Petroleum with Carbon Dioxide

AUTHOR(S)

Dunyushkin, I. I., and Hamiot, A. Yu.

REFERENCE

Neft Khoz., 1978, No. 3, p. 59-61.

COUNTRY/REGION

Russia/General

TYPE OF PROJECT

Carbon Dioxide Miscible

SUMMARY OF PAPER

Paper presents results of laboratory experiments to determine the solubility of CO_2 in certain Soviet crude oils.

NEW OR DIFFERENT IDEAS/CONCEPTS/PRACTICES

L I T E R A T U R E   S U R V E Y   F O R M

TITLE

Oil Overflows in a Stratified Formation Upon Cessation of Steam Pumping

AUTHOR(S)

Dzhamalov, I. M. and Matveenko, L. M.

REFERENCE

Neftyanoe Khoz, April, 1977, No. 4, p. 42-44.

COUNTRY/REGION

Russia/General

TYPE OF PROJECT

Steam Injection

SUMMARY OF PAPER

A method is proposed for calculating the amount of oil transfer from the low permeability to the high permeability sections after steam injection has ceased. Calculations were compared with field recovery thereby supposedly confirmed the calculations procedure.

NEW OR DIFFERENT IDEAS/CONCEPTS/PRACTICES

LITERATURE SURVEY FORM

TITLE

Lloydminster Heavy Oil Thermal Recovery Cost Considerations

AUTHOR(S)

Edwards, M. W.

REFERENCE

Unitar Report 90, First Intl. Conference on The Future of Heavy Oil and Tar Sands, 1979, Edmonton, Alberta, Canada.

COUNTRY/REGION

Canada

TYPE OF PROJECT

None

SUMMARY OF PAPER

A detailed analysis of the cost factors involved in producing heavy oil. Also analyzed are the various demand and oil price schedules, taxes, and royalty considerations.

NEW OR DIFFERENT IDEAS/CONCEPTS/PRACTICES

LITERATURE SURVEY FORM

TITLE

Wet Combustion-A Tertiary Recovery Process for The Pembina Cardium Reservoir

AUTHOR(S)

Ejiogu, G. J., Bennion, D. W., Moore, R. G., Donnelly, J. K.

REFERENCE

The Journal of Canadian Petroleum Technology, July-Sept., 1979.

COUNTRY/REGION

Canada

TYPE OF PROJECT

In Situ Combustion (Wet)

SUMMARY OF PAPER

Presents a discussion of the design and results of a laboratory experiment studying the insitu combustion process. The questions proposed and investigated dealt with the economics of the process as a tertiary recovery method in high gravity oil reservoirs with medium to low residual oil saturations. Of particular concern was the ability to sustain combustion at economical air-oil ratios and what effect on recovery varying amounts of water injected would have.

The experimental fire tube was packed with crushed core from the Pembina Cardium Reservoir and saturated with Pembina crude having an average gravity of 37° API. The oil saturated core was then flooded with water to establish oil saturations of 28 to 33 percent.

L I T E R A T U R E   S U R V E Y   F O R M

TITLE

EOR Pilot Advancing in Saskatchewan

AUTHOR(S)

News Article

REFERENCE

Oil & Gas Journal, June 30, 1980.

COUNTRY/REGION

Canada

TYPE OF PROJECT

(Light Oil Reservoirs) Water Injection followed by Slugs of Rich Gas, then Lean Gas.

SUMMARY OF PAPER

Article announcing that Gulf Canada Resources, Inc.'s light oil EOR project in the Willmar Oil Field is entering its third and final phase. The final phase follows water injection and consists of injecting rich gas and water alternately for a 15-month period to be followed with lean gas injection for 15 months to displace the rich gas injected.

NEW OR DIFFERENT IDEAS/CONCEPTS/PRACTICES

L I T E R A T U R E   S U R V E Y   F O R M

TITLE

Drilling Evaluation in The Orinoco Oil Belt, 1973-1977

AUTHOR(S)

Essis. B., A., and Silva, L. F.

REFERENCE

Unitar Report 78, First Intl. Conference on The Future of Heavy Oil and Tar Sands, 1979, Edmonton, Alberta, Canada.

COUNTRY/REGION

Venezuela

TYPE OF PROJECT

None

SUMMARY OF PAPER

Discusses drilling characteristics and problems with special attention to data gathering through fluid sampling, coring and core analysis, log evaluation, DST's, and RFT's. Aim is for better planning and a more useful evaluation of wild-cat drilling programs.

NEW OR DIFFERENT IDEAS/CONCEPTS/PRACTICES

L I T E R A T U R E S U R V E Y F O R M

TITLE

Process Selection Considerations in The Upgrading of Cold Lake Bitumen

AUTHOR(S)

Esso Resources Canada, Limited

REFERENCE

Unitar Report 47, First Intl. Conference on The Future of Heavy Oil and Tar Sands, 1979, Edmonton, Alberta, Canada.

COUNTRY/REGION

Canada

TYPE OF PROJECT

None

SUMMARY OF PAPER

Discusses primarily the factors involved in upgrading heavy crudes to pipeline quality. Those factors discussed are: yield and quality of the upgraded crude, proven technology available, energy efficiency, environmental regulations and targets, and economics.

NEW OR DIFFERENT IDEAS/CONCEPTS/PRACTICES

LITERATURE SURVEY FORM

TITLE

Recovery of Lloydminster and Morichal Crudes by Caustic, Acid and Emulsion Floods

AUTHOR(S)

Farouq Ali, S. M., Figueroa, J. M., Azuaje, E. A., and Farquharson, R. G.

REFERENCE

The Journal of Canadian Petroleum, Jan.-Mar., 1979.

COUNTRY/REGION

Canada and Venezuela

TYPE OF PROJECT

Caustic, acid, and emulsion floods

SUMMARY OF PAPER

As per title, two heavy crudes were lab tested with several acid and caustic emulsifying solutions and compared as to droplet size and stability. Core displacement tests were run with slugs of water-in-oil and oil-in-water emulsions. Water-in-oil emulsion slugs were found most effective in recovering oil. Studies were made using caustic and acid slugs with less efficient oil displacement resulting. No field studies were reported on.

NEW OR DIFFERENT IDEAS/CONCEPTS/PRACTICES

L I T E R A T U R E   S U R V E Y   F O R M

TITLE

Role of Injection-Production Strategy in The Solvent-Steamflooding of the Athabasca Oil Sands

AUTHOR(S)

Farouq Ali, S. M., Look Yee, K. C., Cordero, F. J., Figueroa, J. M.

REFERENCE

The Journal of Canadian Petroleum, Oct.-Dec., 1979.

COUNTRY/REGION

Canada

TYPE OF PROJECT

Solvents in Conjunction with Steam Injection

SUMMARY OF PAPER

Reports on the procedure and results of 37 laboratory tests of solvent-steam injection. A physical unscaled model was built and used for the tests. Athabasca oil sand was used to pack the model. Five operational strategies were investigated whereby the schemes of solvent injection with and without steam, into injection wells versus into production wells, and alternatively with slugs of steam provided the variable conditions. The most effective schemes at displacing the bitumen were two operations that involved injection of all or part of the solvent into the producing wells. The theory being a better diffusion of heat in the area around the producers.

NEW OR DIFFERENT IDEAS/CONCEPTS/PRACTICES

L I T E R A T U R E   S U R V E Y   F O R M

TITLE

A three-Phase, Two-Dimensional Compositional Thermal Simulator for Steam Injection Processes

AUTHOR(S)

Ferrer, J., Farouq Ali, S. M.

REFERENCE

The Journal of Canadian Petroleum Technology, Jan.-Mar., 1977.

COUNTRY/REGION

Venezuela

TYPE OF PROJECT

Steam Injection

SUMMARY OF PAPER

Describes the mathematics and computational procedures utilized in a numerical model for simulating three-phase, three dimensional, multicomponent flow, with interphase mass and heat transfer. Examples are presented which illustrate application of the model to simulate cyclic steam stimulation and steamfloods.

NEW OR DIFFERENT IDEAS/CONCEPTS/PRACTICES

L I T E R A T U R E   S U R V E Y   F O R M

TITLE

Electrical Induction Heating - A New Concept In In Situ Recovery

AUTHOR(S)

Fisher, Sidney T.

REFERENCE

Energy Processing/Canada, September-October, 1977, p.42-48.

COUNTRY/REGION

TYPE OF PROJECT

Electrical Induction Heating

SUMMARY OF PAPER

This new method requires drilling shafts and tunnels plus smaller producing holes and heating deposits inductively. Paper presents results of calculation for recovery cost etc. This is only an idea and as yet has not been field tested, although some lab research is underway.

NEW OR DIFFERENT IDEAS/CONCEPTS/PRACTICES

Inducing heat in the heavy oil deposit with subsequent results.

L I T E R A T U R E S U R V E Y F O R M

TITLE

Vibrational Effects on the Critical Zone of Wells in Multistratal Deposits

AUTHOR(S)

Gadiev, S. M., Lur'e, M. V., Polyanskaya, L. V., Akhundov, R. A., Botvinnik, R. A.

REFERENCE

Izv. Vyssh Uchebn Zaved Neft Gaz, No. 6, p. 31-35, 1977.

COUNTRY/REGION

Russia

TYPE OF PROJECT

Vibrational disruption of viscosity

SUMMARY OF PAPER

Results of laboratory research and computation, are presented on the design of a downhole hydraulic vibrator. High frequency pulses reportedly disrupt the fluid viscosity and allow greater recovery of oil. Conclusion from this effort are: 1.) greatest vibrational effects on depleted stratum are realized when the vibrator is constructed from the most elastic material possible, 2.) greatest vibrational effects on saturated stratum are realized with a maximal increase in inflexibility of the flexible pipe vibrator.

NEW OR DIFFERENT IDEAS/CONCEPTS/PRACTICES

LITERATURE SURVEY FORM

TITLE

Model of Dynamic Effects in The Stratum

AUTHOR(S)

Gadiev, S. M. et al.

REFERENCE

Igv. Vyssh. Uchebu. Zaved. Neft Gay., 1977, No. 9, p. 25-29.

COUNTRY/REGION

Russia - Lab research at I. M. Gubkin Moscow Petrochemical and Gas Industry Institute

TYPE OF PROJECT

Hydraulic Compression Wave - The Liquid Piston

SUMMARY OF PAPER

In lab test confined core was subjected to hydraulic vibrations of 10-100 Hz. A membrane separates hydraulic system from core. Same as sonic except lower frequency and different method of generating vibrations. Concept seems to increase capillary replacement of oil by water. Author reports 10-12% increase in oil recovery.

NEW OR DIFFERENT IDEAS/CONCEPTS/PRACTICES

Method of generating vibrations is by hydraulic action and transfer to porous medium through membrane, no mention of membrane material.

LITERATURE SURVEY FORM

TITLE

The Design of the Peace River In Situ Oil Sands Pilot Project

AUTHOR(S)

Gorril, R. G., Kitzan, P., Komery, D. P.

REFERENCE

Unitar Report 14, First Intl. Conference on The Future of Heavy Oil and Tar Sands, 1979, Edmonton, Alberta, Canada.

COUNTRY/REGION

Canada

TYPE OF PROJECT

Steam Drive

SUMMARY OF PAPER

Provides a review of the planning and design of a pilot steam drive project. As in all pilot operations, the goal was to investigate a sufficient number of parameters to enable future large scale project design for the economical recovery of heavy oil. Some of the parameters studied were: well spacing, type of well pattern, number of patterns, and well completion intervals, etc. Deviated wells and a variety of well completion techniques were used for cost effectiveness comparison.

NEW OR DIFFERENT IDEAS/CONCEPTS/PRACTICES

L I T E R A T U R E   S U R V E Y   F O R M

TITLE

Environmental Considerations in The Cold Lake Project

AUTHOR(S)

Gossen, R. G. and Mann, G. J.

REFERENCE

Unitar Report 23, First Intl. Conference on The Future of Heavy Oil and Tar Sands, 1979, Edmonton, Alberta, Canada.

COUNTRY/REGION

Canada

TYPE OF PROJECT

Steam Stimulation

SUMMARY OF PAPER

Discusses all the possible environmental effects of the operations associated with a steam stimulation project.

NEW OR DIFFERENT IDEAS/CONCEPTS/PRACTICES

L I T E R A T U R E   S U R V E Y   F O R M

TITLE

How Husky Makes a Profit in Heavy Oil Production

AUTHOR(S)

Gurel, D. O.

REFERENCE

World Oil, Sept., 1979.

COUNTRY/REGION

Canada

TYPE OF PROJECT

None

SUMMARY OF PAPER

Discusses primarily operational problems encountered in producing viscous oils and ways employed by Husky to efficiently solve them. Resulting from the effort to minimize costs are strict well completion procedures and specialized surface and sub-surface pumping equipment of high quality.

A portion of these type leases operated by Husky are located in California as well as Canada.

NEW OR DIFFERENT IDEAS/CONCEPTS/PRACTICES

L I T E R A T U R E S U R V E Y F O R M

TITLE

Hot-Mine Method Applied to Yarega Fields

AUTHOR(S)

Gurov, Ye.

REFERENCE

Neftyanik, Moscow, No. 8, Aug., 1979, p. 17-19.

COUNTRY/REGION

Russia/Yarega-Komi (A.S.S.R.)

TYPE OF PROJECT

Hot-Mine or Thermal Shaft

SUMMARY OF PAPER

A scant history of oil mining and more general description of the Hot-Mine method. This paper states that the central chamber is positioned below the production formation and drifts are drilled slanting upward for steam injection and subsequent oil drainage.

NEW OR DIFFERENT IDEAS/CONCEPTS/PRACTICES

Horizontal or up-side-down production wells

LITERATURE SURVEY FORM

TITLE

Occurrence of Heavy Crudes and Tar Sands in Latin America

AUTHOR(S)

Gutierrez, F. J.

REFERENCE

Report 77, First International Conference on the Future of Heavy Crude and Tar Sands, UNITAR, 1979, Edmonton, Alberta Canada.

COUNTRY/REGION

Latin America with Emphasis on Venezuela

TYPE OF PROJECT

None

SUMMARY OF PAPER

Summarizes the occurrences of heavy crudes - gives reserve figures and some production figures.

NEW OR DIFFERENT IDEAS/CONCEPTS/PRACTICES

L I T E R A T U R E   S U R V E Y   F O R M

TITLE

Heavy Crudes of Venezuela: An Alternative for the Future

AUTHOR(S)

Gutierrez, Francisco J.

REFERENCE

Chapter 43, Ministerio de Minase Hidro Carburos.

COUNTRY/REGION

Venezuela

TYPE OF PROJECT

None

SUMMARY OF PAPER

Presents primarily a discussion on the physical properties of heavy crudes, the upgrading technology required at the refinery, and the plan for utilization of the product.

NEW OR DIFFERENT IDEAS/CONCEPTS/PRACTICES

L I T E R A T U R E   S U R V E Y   F O R M

TITLE

Hard Lessons Learned by Pioneers

AUTHOR(S)

Editorial

REFERENCE

Canadian Chemical Processing, June 6, 1979.

COUNTRY/REGION

Canada

TYPE OF PROJECT

Oil Sands Mining

SUMMARY OF PAPER

Discusses in general some of the construction problems and costs faced in getting the Great Canadian Oil Sands Ltd. mining project under way. The project utilizes giant bucketwheel excavators to provide the oil-soaked sand to a 45,000 B/D extraction plant.

Note: This paper contains a summary table of Canadian operators involved in in situ testing programs.

NEW OR DIFFERENT IDEAS/CONCEPTS/PRACTICES

L I T E R A T U R E   S U R V E Y   F O R M

TITLE

Tunnels in Oil Sand

AUTHOR(S)

Harris, M. C., Poppen, S., Morgenstern, N. R.

REFERENCE

The Journal of Canadian Petroleum, Oct.-Dec., 1979.

COUNTRY/REGION

Canada

TYPE OF PROJECT

Large Diameter Tunnels Constructed in the Oil Sand Layer.

SUMMARY OF PAPER

Discusses engineering and soil mechanics design and problems encountered in drilling the Saline Creek tunnel. Principal considerations in the process are: The deformation (swelling) that occurs in the oil sand as it is exposed; presence of weak clay seams, gas pockets, water bearing sands and fractured rocks resulting from sinkholes and faults; and ventilation.

NEW OR DIFFERENT IDEAS/CONCEPTS/PRACTICES

L I T E R A T U R E   S U R V E Y   F O R M

TITLE

Reservoir Description of The Clearwater Oil Sand
Cold Lake, Alberta, Canada

AUTHOR(S)

Harrison, D. B., Glaister, R. P., and Nelson, H. W.

REFERENCE

Unitar Report 21, First Intl. Conference on The Future of Heavy oil and Tar Sands, 1979, Edmonton, Alberta, Canada.

COUNTRY/REGION

Canada

TYPE OF PROJECT

None

SUMMARY OF PAPER

Presents the results of a depositional environment study of the Mannville sand group aimed at obtaining a better reservoir description to answer the where's and why's of bitumen deposits. Results are presented which show an apparent relation between bitumen saturation and a number of parameters including depositional environments, mean grain size, amount of fines and the presence of diagenetic clays.

NEW OR DIFFERENT IDEAS/CONCEPTS/PRACTICES

L I T E R A T U R E S U R V E Y F O R M

TITLE

Selective Electric Reservoir Heating

AUTHOR(S)

Harvey, A. H.; Arnold, M. D.; El-Feky, S. A.

REFERENCE

The Journal of Canadian Petroleum Technology, July-Sept., 1979.

COUNTRY/REGION

TYPE OF PROJECT

Resistive Heating

SUMMARY OF PAPER

Paper is presentation of mathematical models and laboratory research on heating oil in reservoirs by passing electric current. Laboratory work and design permits selective interval heating within the productive zone. Two hypothetical field cases are presented and include electrical requirements, increase in temperature and oil recovery.

NEW OR DIFFERENT IDEAS/CONCEPTS/PRACTICES

The idea is to generate heat within productive interval by use of electric current from one electrode (well) to another.

L I T E R A T U R E S U R V E Y F O R M

TITLE

Some Innovative Approaches which May Facilitate Production of Heavy Crudes

AUTHOR(S)

Havlena, Z. G.

REFERENCE

Unitar Report 39, First Intl. Conference on The Future of Heavy Oil and Tar Sands, 1979, Edmonton, Alberta, Canada

COUNTRY/REGION

Non-National

TYPE OF PROJECT

General

SUMMARY OF PAPER

A general review of the possibilities and problems associated with EOR. Special attention was drawn to such processes as wet in situ combustion, special solvents, bacteria and enzymes, and use of 85% + pure O_2 in in situ combustion. Some work with bacteria has been done in Poland, Hungary, and California, however, no people or field names were given.

NEW OR DIFFERENT IDEAS/CONCEPTS/PRACTICES

No new ideas but the paper urges the use of concepts and ideas previously tried on a limited scale. Suggested laboratory and field research to improve on technology and application at some of the more remote means of heavy oil EOR.

L I T E R A T U R E   S U R V E Y   F O R M

TITLE

Heavy Oil Clogs Soviet Output Growth

AUTHOR(S)

Anon.

REFERENCE

Oil and Gas Journal, Jan. 4, 1980, p. 38-40.

COUNTRY/REGION

Russia/General

TYPE OF PROJECT

SUMMARY OF PAPER

Article reports that during 1971-75, 98.5% of the oil found in Tataria was heavy oil. In northern sector of Western Siberia the 500 million recoverable giant (18°) discovered in 1968 still is not on commercial production. On northeast side of Caspian Sea, Mangyshlak Peninsula, the fields Uzen, 1961, and Zhetybai have heavy oil reserves of 3.5-5.3 billion bbl. and 1.1 billion bbls. respectively. In Uzen untreated Caspian Sea water was injected for pressure maintenance, but bacteria ate the oil and left a residue of H_2S. Article relates vast waste and inefficiency of materials and manpower as reported by Pravda.

L I T E R A T U R E   S U R V E Y   F O R M

TITLE

Steam-Drive Recovery Being Used on Lake Maracaibo Coastal Field

AUTHOR(S)

Herrera, Armando J.

REFERENCE

The Oil & Gas Journal, July 17, 1978.

COUNTRY/REGION

Venezuela, M-6 Project - Tia Juana Field

TYPE OF PROJECT

Steam-Drive

SUMMARY OF PAPER

Describes the engineering and geological considerations of developing a steam drive project as a follow-up to steam soak. Pilot test and lab studies are cited for expected additional recovery. Design of initial inverted seven-spot injection pattern allows flexibility to go to a line drive pattern if proven necessary.

NEW OR DIFFERENT IDEAS/CONCEPTS/PRACTICES

LITERATURE SURVEY FORM

TITLE

Physical Characterization and Microbiological Settling-Rate Modification of Aqueous Suspensions from Hot-Water Process Oil-Sands Extraction

AUTHOR(S)

Hocking, Martin B.

REFERENCE

Fuel, Vol. 56, July, 1977.

COUNTRY/REGION

Canada

TYPE OF PROJECT

Surface Extraction of Oil from Mined Tar Sands

SUMMARY OF PAPER

Discusses primarily the nature of the water effluent from the hot-water-process-tar-sands-extraction method. Several experimental studies with microbiological agents to accelerate sedimentation of suspended hydrocarbon and mineral fines are reported on for the treatment and reuse of the effluent water stream.

NEW OR DIFFERENT IDEAS/CONCEPTS/PRACTICES

LITERATURE SURVEY FORM

TITLE

Secondary and Tertiary Oil Recovery in the Federal Republic of Germany

AUTHOR(S)

Hoefling, B. von

REFERENCE

Erdoel-Erdgas-Zeitschrift, v. 94, No. 4, p. 116-118, Apr., 1978.

COUNTRY/REGION

Germany

TYPE OF PROJECT

Several

SUMMARY OF PAPER

Summary of waterflood efforts in Germany. Also notice of 5 field projects with carbon dioxide, in situ combustion, polymer, steam and hot water (no details). Statistics show 129 fields under waterflood, approximately 75 percent of total production, average water cut of 86%. Twenty five percent of production is by natural waterdrive.

NEW OR DIFFERENT IDEAS/CONCEPTS/PRACTICES

None

LITERATURE SURVEY FORM

TITLE

Developing Automated Oil Field in Balakhany

AUTHOR(S)

Il ina, S.

REFERENCE

Baku Vyshka, Mar., 1979, p. 2.
U. S. Joint Publications Service, May 4, 1979, p. 80.

COUNTRY/REGION

Russia/Baku

TYPE OF PROJECT

Mining-no mention of Thermal Stimulation

SUMMARY OF PAPER

More P. R. about underground mining in Russia.

There seems to be an endless supply of general papers on this method, all of which are designed to elevate the enthusiasm of the Russian oil field worker.

NEW OR DIFFERENT IDEAS/CONCEPTS/PRACTICES

LITERATURE SURVEY FORM

TITLE

Twenty Years' Operation of an In Situ Combustion Project

AUTHOR(S)

Jenkins, George R., Kirkpatrick, John W.

REFERENCE

The Journal of Canadian Petroleum Technology, Jan.-March, 1979.

COUNTRY/REGION

Canada

TYPE OF PROJECT

"COFCAW" (Combination of Forward Combustion and Waterflood Process)

SUMMARY OF PAPER

Discusses the problems faced as an operator of an in situ combustion project and many of the mechanical equipment and techniques developed as the need became apparent. Special well completions are required for three different types of wells, production wells, injection, and observation. Data collection methods were computerized as computer technology became available to aid in compiling more and better data from the wells.

NEW OR DIFFERENT IDEAS/CONCEPTS/PRACTICES

LITERATURE SURVEY FORM

TITLE

Field Pilot Project Optimization Through Single Well, Pre-Pilot Tests

AUTHOR(S)

Jesperson, P. J.

REFERENCE

Unitar Report 44, First Intl. Conference on the Future of Heavy Oil and Tar Sands, 1979, Edmonton, Alberta, Canada.

COUNTRY/REGION

Canada

TYPE OF PROJECT

Steam Stimulation

SUMMARY OF PAPER

This paper presents a proposed plan for carrying out a single well test to evaluate major factors affecting pilot project design of a thermal recovery method such as steam stimulation. The single well test should involve close monitoring of steam injection rates, quality, pressure, and temperatures during the injection cycle and production response, composition of produced fluids, pressures, and temperatures during the production phase.

NEW OR DIFFERENT IDEAS/CONCEPTS/PRACTICES

L I T E R A T U R E   S U R V E Y   F O R M

TITLE

Application of Microbiological Methods for Secondary Oil Recovery from the Carpathian Crude Oil Reservoirs

AUTHOR(S)

Karaskiewicz, Jerzy

REFERENCE

Pr. Inst. Naft., Cracow, 1974, 3-67

COUNTRY/REGION

Poland

TYPE OF PROJECT

Microbiological Field Test

SUMMARY OF PAPER

Twenty wells were involved in field tests of bacteria culture injection at depths of 500-1500 meters, wells were treated with 200 kg of molasses and closed in for 3-6 months. Producing rates increased by 20-200%, however, total rates were not given. Viscosity reductions between 16.95 and 10.54 cps. at 20°C were reported.

NEW OR DIFFERENT IDEAS/CONCEPTS/PRACTICES

L I T E R A T U R E   S U R V E Y   F O R M

TITLE

Importance of Reservoir Description in Evaluating In Situ Recovery Methods for Cold Lake Heavy Oil, Part I - Reservoir Description, Part II - In Situ Application

AUTHOR(S)

Kendall, G. H. and Towson, D. E.

REFERENCE

Bulletin of Canadian Petroleum Geology, Vol. 25, No. 2, May, 1977, p. 314-340.

COUNTRY/REGION

Canada

TYPE OF PROJECT

None

SUMMARY OF PAPER

Part I discusses the need and importance of good geological control and description in the planning and implementation of heavy oil recovery processes.

Part II discusses these major factors associated with heavy oil recovery that must be overcome by an in situ process. They are: 1.) high oil viscosity, 2.) oil saturated sands lack reservoir communication, 3.) the reservoir sands have little or no internal reservoir energy. Four recovery processes are mentioned in general: a) steam soak, b) steam drive, c) in situ combustion, and d) solvent injection.

L I T E R A T U R E   S U R V E Y   F O R M

TITLE

The Potential of An Oil Bed

AUTHOR(S)

Khalimov, E., Deputy Oil Minister

REFERENCE

Oil Extraction Methods Discussed by Deputy Minister U. S. Joint Publications Research Service, March 12, 1979, p. 63-66.

COUNTRY/REGION

Russia

TYPE OF PROJECT

SUMMARY OF PAPER

A very general and elementary discussion on oil production with mention of thermal recovery. The main thrust, however, was that the equipment needed is not being built on schedule and that which is delivered is not complete or tested. Steam injection is operational in the Baku region and in the Karazhanbazskoye deposit in Krasnodarskiy Kray where primary recovery yielded 7% and recovery by steam is predicted to be 40%.

NEW OR DIFFERENT IDEAS/CONCEPTS/PRACTICES

L I T E R A T U R E   S U R V E Y   F O R M

TITLE

History of Development of the Oil and Gas Industry in Ukhta

AUTHOR(S)

Krems, A. Y.

REFERENCE

Geologiya Nefti i Gaza, Vol. 8, No. 9, 1964.

COUNTRY/REGION

Russia

TYPE OF PROJECT

SUMMARY OF PAPER

This paper is purely historical relative to the title.

NEW OR DIFFERENT IDEAS/CONCEPTS/PRACTICES

LITERATURE SURVEY FORM

TITLE

Sediment-Geological Study of A Steam-Drive Project in The Deltaic River Sands

AUTHOR(S)

Kruit, Dr. C.

REFERENCE

Unitar Report 33, First Intl. Conference on The Future of Heavy Oil and Tar Sands, 1979, Edmonton, Alberta, Canada.

COUNTRY/REGION

Venezuela

TYPE OF PROJECT

Steam Drive

SUMMARY OF PAPER

Discusses methodology and types of data used to define sand-shale distribution within the project area. Sediments in the area consisted of deltaic/fluviatile sands overlain by a transgressive sequence of sand and shale lenses.

NEW OR DIFFERENT IDEAS/CONCEPTS/PRACTICES

The author prepared injection-hexagon correlation sections to show the detailed sand/shale distributions useful in comparisons with production information to better understand and monitor fluid movements.

L I T E R A T U R E S U R V E Y F O R M

TITLE

Caspian Seawater Treated for Soviet Steam Project

AUTHOR(S)

Leitner, G. F.

REFERENCE

Petroleum Engineer, Nov., 1979.

COUNTRY/REGION

Russia/Caspian Sea

TYPE OF PROJECT

Steam

SUMMARY OF PAPER

Nine reverse osmosis (RO) units were constructed by Water Services of America, Inc., to treat Caspian sea water which is used in the steam flood method in the heavy oil reservoirs near the Caspian Sea. Each RO unit has capacity of 9500 B/D yielding an operating capacity of about 80,000 B/D for the entire plant.

NEW OR DIFFERENT IDEAS/CONCEPTS/PRACTICES

L I T E R A T U R E   S U R V E Y   F O R M

TITLE

Report on On-Site Visits

AUTHOR(S)

Lewin & Associates

REFERENCE

Company Report

COUNTRY/REGION

France, Germany, Venezuela, and Canada

TYPE OF PROJECT

Steam, Miscible, and Polymer Floods in general.

SUMMARY OF PAPER

Reports finding on visits to IFP, Germany (Inst. of Pet. Eng'r.), Intevep, Venezuela, and AOSTRA, Canada. Paper includes current efforts and future research projects planned by above entities. Most activities relate to steam via additives, completion practices, etc. However, IFP is involved with polymer and surfactant use.

NEW OR DIFFERENT IDEAS/CONCEPTS/PRACTICES

None

L I T E R A T U R E   S U R V E Y   F O R M

TITLE

The Orinoco Belt-Venezuela's Oil Bonanza

AUTHOR(S)

Lowenstein, Roger

REFERENCE

The Energy Daily, Monday, July 30, 1979.

COUNTRY/REGION

Venezuela

TYPE OF PROJECT

Mention of steam soak

SUMMARY OF PAPER

Summarizes an interview with Carlos Borregales, heavy oil co-ordinator for Petroleos de Venezuela. Discussion centers on Venezuela's heavy oil reserves with emphasis on refining techniques to upgrade the oil. Discussion also mentions the cyclic steam soak process as major enhancement to increase recoveries.

NEW OR DIFFERENT IDEAS/CONCEPTS/PRACTICES

LITERATURE SURVEY FORM

TITLE

OPEC Prices Make Heavy Oil Look Profitable

AUTHOR(S)

Marshall, Eliot

REFERENCE

Science, Vol. 204, June 22, 1979.

COUNTRY/REGION

Canada

TYPE OF PROJECT

None

SUMMARY OF PAPER

Discusses the worldwide economic picture as it relates to the higher cost of extracting heavy crudes, both in Canada and in other areas of the world.

NEW OR DIFFERENT IDEAS/CONCEPTS/PRACTICES

LITERATURE SURVEY FORM

TITLE

Estimate of Oil Resources, Lloydminster Area, Alberta

AUTHOR(S)

McCrossan, R. G., Procter, R. M., and Ward, W. J.

REFERENCE

Unitar Report 87, First Intl. Conference on The Future of Heavy Oil and Tar Sands, 1979, Edmonton, Alberta, Canada.

COUNTRY/REGION

Canada

TYPE OF PROJECT

None

SUMMARY OF PAPER

As per title a discussion of the geology, occurrence of oil, and statistical evaluation of pool size distribution is presented in this paper. Thickness of pay zone in different pools was used to categorize along with presence of bottom water the different possible applications of EOR possibilities.

NEW OR DIFFERENT IDEAS/CONCEPTS/PRACTICES

L I T E R A T U R E   S U R V E Y   F O R M

TITLE

Enhanced Recovery of Athabasca Oil Sands

AUTHOR(S)

McNally, Rich

REFERENCE

Petroleum Engineer, Nov., 1979.

COUNTRY/REGION

Canada - Alberta

TYPE OF PROJECT

In Situ Combustion

SUMMARY OF PAPER

Discusses Amoco-Canada's pilot in situ process experiment on a five acre test plot at Gregoire Lake. A three step process was employed: 1.) formation preheating via dry forward combustion, 2.) pressure blow-down by stopping air injection and allowing temperature to equalize in the formation, and 3.) restarting burning and air injection with water included to aid in heat utilization. Results were good enough for consideration of a large scale economical operation.

NEW OR DIFFERENT IDEAS/CONCEPTS/PRACTICES

Air injection at fracturing pressures is a prerequisite to successful in situ combustion operations.

L I T E R A T U R E S U R V E Y F O R M

TITLE

World Geography of Heavy Crude Oils

AUTHOR(S)

Meyer, R. F., Dietzman, W. D.

REFERENCE

Unitar Report 68, First Intl. Conference on the Future of Heavy Oil and Tar Sands, 1979, Edmonton, Alberta, Canada.

COUNTRY/REGION

Worldwide

TYPE OF PROJECT

None

SUMMARY OF PAPER

Paper makes an initial attempt to give location and quantity of heavy oil reserves throughout the world; production figures when available were also given.

NEW OR DIFFERENT IDEAS/CONCEPTS/PRACTICES

LITERATURE SURVEY FORM

TITLE

Change in Oil Yield of A Stratum Model With Displacement of Oil With Magnetized Water, 1977.

AUTHOR(S)

Mikhelman, A. I., Fedoriv, L. V., and Radish, V. N.

REFERENCE

Izv. Vyssh. Uchebn. Zaved. Neft. Gaz, No. 9, 1977, p. 89-90.

COUNTRY/REGION

Russia - Lab Research, Ivanov-Frankov Institute of Oil & Gas

TYPE OF PROJECT

Injection of Magnetized Water

SUMMARY OF PAPER

Injecting electromagnetized water resulted in 4-5% increase in oil recovery from lab test on cores of SS.

NEW OR DIFFERENT IDEAS/CONCEPTS/PRACTICES

Use of magnetic water as a displacement medium

L I T E R A T U R E S U R V E Y F O R M

TITLE

Mining of Viscous Crude Claimed Viable

AUTHOR(S)

Anon.

REFERENCE

Oil and Gas Journal, Jan. 5, 1976.

COUNTRY/REGION

Russia

TYPE OF PROJECT

Thermal (Steam) Mining

SUMMARY OF PAPER

Paper describes Russian method of sinking shaft to above pay, drilling radial holes from there so as to inject steam. Paper gives cost and some other details of process. This is very similar to the HOP Process used By Barber Oil in Kern River Field in California. Some differences are obvious, however, looks like Barber is going to bottom of pay w/chamber.

NEW OR DIFFERENT IDEAS/CONCEPTS/PRACTICES

Drilling horizontal shafts for steam injection and oil production. A working chamber is created in and below pay and from there the horizontal shafts are drilled. This appears to be the "Thermal-Shaft" in Yarega Method patent description.

LITERATURE SURVEY FORM

TITLE

Current Policies of The Government of Saskatchewan Regarding Heavy Crude Oil Development

AUTHOR(S)

Moncur, R. H., Sully, R. R.

REFERENCE

Unitar Report 53, First Intl. Conference on The Future of Heavy Oil and Tar Sands, 1979, Edmonton, Alberta, Canada.

COUNTRY/REGION

Canada

TYPE OF PROJECT

None

SUMMARY OF PAPER

Discusses the future market demand for heavy crude domestically (in Canada) and internationally and what considerations should be made from a governmental standpoint with regard to differential pricing allowed and export regulations. Within this framework general mention is made of the costs and problems associated with enhanced recovery methods and crude upgrading required in the heavy oil production process.

NEW OR DIFFERENT IDEAS/CONCEPTS/PRACTICES

LITERATURE SURVEY FORM

TITLE

Priorities For Non-Conventional Hydrocarbon Liquid Production

AUTHOR(S)

Mowll, R., Hambling, J. K.

REFERENCE

Unitar Report 93, First Intl. Conference on The Future of Heavy Oil and Tar Sands, 1979, Edmonton, Alberta, Canada.

COUNTRY/REGION

General

TYPE OF PROJECT

SUMMARY OF PAPER

Concerns with economical recovery - not a very helpful paper - Discusses resources and talks in general terms on recovery methods.

NEW OR DIFFERENT IDEAS/CONCEPTS/PRACTICES

LITERATURE SURVEY FORM

TITLE

Numerical Simulation of Steam Drive in The Tia Juana M-6 Project

AUTHOR(S)

Munoz, J. Diaz

REFERENCE

Unitar Report 34, First Intl. Conference on The Future of Heavy Oil and Tar Sands, 1979, Edmonton, Alberta, Canada.

COUNTRY/REGION

Venezuela

TYPE OF PROJECT

Steam Drive

SUMMARY OF PAPER

Describes parameters studied for sensitivity affecting recovery from a steam drive project. Ultimately, from model results, down-dip wells completed in the lower portions of the productive interval and in the highest permeable areas of the field recovered the highest percentage of incremental oil. Steam injectors were placed in low quality reservoir areas.

NEW OR DIFFERENT IDEAS/CONCEPTS/PRACTICES

L I T E R A T U R E   S U R V E Y   F O R M

TITLE

The Ways of Development of Heavy Oils and Bitumens in the U.S.S.R.

AUTHOR(S)

Muslimov, R. Kh.

REFERENCE

Unitar Report 95, First Intl. Conference on The Future of Heavy Oil and Tar Sands, 1979, Edmonton, Alberta, Canada.

COUNTRY/REGION

Russia

TYPE OF PROJECT

Chemical, Thermal (Steam, Insitu and Electrical-Resistive) Underground mining and a combination of the above

SUMMARY OF PAPER

Briefly states methods of EOR being tried in Russia. The above processes/methods are being tried extensively in Siberia, Tatar and Bashkir Republics, Kuybyshev region, and other areas. H_2S acid and solvents, high frequency electrical current heating, cyclic and continuous steam injection and in situ combustion have been carried out in the Tatar Republic.

NEW OR DIFFERENT IDEAS/CONCEPTS/PRACTICES

Possible new ideas are injection of H_2S acid w/solvent and also resistive type electrical heat. The underground mining as referred to in Yarega is mentioned.

L I T E R A T U R E   S U R V E Y   F O R M

TITLE

Make the Road Wider to Tertiary Production Methods

AUTHOR(S)

Nechipurenko, O., May, 1979.

REFERENCE

Baku Vyshka, May, 1979, p. 103-106.
U. S. Joint Publications Research Service, July 19, 1979, p. 103-106.

COUNTRY/REGION

Russia

TYPE OF PROJECT

In Situ and Steam

SUMMARY OF PAPER

A bureaucratic explanation of the fantastic possibilities of EOR and why success has not been achieved. It was stated that "how to insulate the steam pipelines, has not been solved". Foam-cement was mentioned as part of the process to "provide the normal combustion". Problems in specific districts were mentioned i.e., either worker or equipment.

Specialists of the Laboratory of Aznipineft instituted an in situ combustion pilot project, under the sea (probably the Caspian).

NEW OR DIFFERENT IDEAS/CONCEPTS/PRACTICES

L I T E R A T U R E S U R V E Y F O R M

TITLE

Increase the Yield of Oil Strata "Shaft in Balakhany" (Experimental Underground Oil Field)

AUTHOR(S)

Nechipurenko, O.

REFERENCE

Vyshka, Baku, Aug., 1979, No. 28, p. 2.

COUNTRY/REGION

Russia

TYPE OF PROJECT

SUMMARY OF PAPER

Points out that there are differences in various aspects from one field to another, i.e., well length, spacing also talks about delays of schedule relative to construction of underground mines at Balakhany. Two main problems appear to be construction of substation and no electricity to site. No mention about using thermal stimulation in conjunction with mine.

NEW OR DIFFERENT IDEAS/CONCEPTS/PRACTICES

Underground Thermal-Shaft

L I T E R A T U R E   S U R V E Y   F O R M

TITLE

The Occurrence of Heavy Oil in The Middle East

AUTHOR(S)

Nehring, Richard

REFERENCE

Unitar Report 42, First Intl. Conference on The Future of Heavy Oil and Tar Sands, 1979, Edmonton, Alberta, Canada.

COUNTRY/REGION

Middle East - So. Yemen to Turkey

TYPE OF PROJECT

None

SUMMARY OF PAPER

Gives locations of heavy and near heavy petroleum occurrences and the efforts (none beyond primary) put forth to recover same.

NEW OR DIFFERENT IDEAS/CONCEPTS/PRACTICES

None

L I T E R A T U R E   S U R V E Y   F O R M

TITLE

New Oil Extraction Technique

AUTHOR(S)

News Letter Item

REFERENCE

U. S. Joint Publications Research Service, March 5, 1979, p. 90.

COUNTRY/REGION

Russia/Bashkria

TYPE OF PROJECT

In Situ Combustion

SUMMARY OF PAPER

Paper states that successful tests have been run in the Bashkria Oil Fields using in situ combustion. The paper states that a new manometer and thermometer have been designed to monitor the "in-stratum" combustion process. The manometer and thermometer are probably high temperature devices.

NEW OR DIFFERENT IDEAS/CONCEPTS/PRACTICES

Possibly the manometer and thermometer for monitoring high temperature bottom-hole conditions.

L I T E R A T U R E   S U R V E Y   F O R M

TITLE

Heavy Oil Sand In Situ Pilot Plants in Alberta (Past and Present)

AUTHOR(S)

Nicholls, J. H., and Luhning, R. W.

REFERENCE

The Journal of Canadian Petroleum Technology, July-Sept., 1977.

COUNTRY/REGION

Canada

TYPE OF PROJECT

Wet and Dry In Situ Combustion, Steam Soak, Steam Drive, and Electrical Preheating

SUMMARY OF PAPER

Provides a description of the reservoir properties and problems faced in producing the heavy crudes in Alberta. Also included is a project by project description of all previous, current and planned (as of 1977) thermal recovery pilots, and the company(s) doing the work.

Variations in the steam drive methods described are such as: 1.) establishing interwell communication through horizontal fractures, 2.) using surfactants to enhance permeability through the bitumen for better steam distribution, and 3.) changing steam injection pressures to above and below fracturing pressure during alternate phases of a project.

Electrical preheating followed by steam drive is being tested by Petro-Canada Exploration Ltd. in Athabasca.

LITERATURE SURVEY FORM

TITLE

Heavy Crudes in The Maranon Basin Oriental Peru

AUTHOR(S)

Pardo, Alfredo

REFERENCE

Unitar Report 52, First Intl. Conference on The Future of Heavy Oil and Tar Sands, 1979, Edmonton, Alberta, Canada.

COUNTRY/REGION

Peru/Oriental

TYPE OF PROJECT

None

SUMMARY OF PAPER

About 1.5 billion bbl. in Eastern Basin Margin. Exploration just now getting started. Heavy oils from 8-20° API.

NEW OR DIFFERENT IDEAS/CONCEPTS/PRACTICES

L I T E R A T U R E   S U R V E Y   F O R M

TITLE

Transport of Heavy Crudes

AUTHOR(S)

Parra, Douglas

REFERENCE

Unitar Report 81, First Intl. Conference on The Future of Heavy Oil and Tar Sands, 1979, Edmonton, Alberta, Canada.

COUNTRY/REGION

Venezuela

TYPE OF PROJECT

None

SUMMARY OF PAPER

Discusses field locations of heavy crudes (with maps) and distribution problems and facilities required.

NEW OR DIFFERENT IDEAS/CONCEPTS/PRACTICES

L I T E R A T U R E S U R V E Y F O R M

TITLE

The Geology of The Orinoco Heavy Oil Belt an Integrated Interpretation

AUTHOR(S)

Petroleos de Venezuela, S. A., Coordination of Exploration

REFERENCE

Unitar Report 92, First Intl. Conference on The Future of Heavy Oil and Tar Sands, 1979, Edmonton, Alberta, Canada.

COUNTRY/REGION

Venezuela

TYPE OF PROJECT

None

SUMMARY OF PAPER

Discusses the sedimentary environments and faulting responsible for the regional stratigraphy and structure of the Orinoco belt.

NEW OR DIFFERENT IDEAS/CONCEPTS/PRACTICES

L I T E R A T U R E   S U R V E Y   F O R M

TITLE

The Petroleum Reservoir's Potential

AUTHOR(S)

Editorial

REFERENCE

Greater Use of Thermal Methods in Azerbaijan Oilfields Urged. U. S. Joint Publications Research Serv., April 19, 1979, p. 54-56.

COUNTRY/REGION

Russia

TYPE OF PROJECT

Steam - In Situ Combustion

SUMMARY OF PAPER

This editorial states the possibilities of thermal and other recovery methods. Specifically it points out that thermal methods have been tried in four districts, but only in one district, Leninneft, have good results been obtained. The paper states that "packing the bottom-hole zone of the bed with a carbonate-cement formulation, restricting water flow with the use of a foam-cement solution" is a good method of shutting off water.

No explanation of the foam-cement is given. Workers in Leninskiy Rayon are setting up equipment to start a new in situ combustion project.

NEW OR DIFFERENT IDEAS/CONCEPTS/PRACTICES

Possible foam-cement

LITERATURE SURVEY FORM

TITLE

Projection and The Results of Application of The Selective System of Waterflooding in The Oil Fields (With the Ashtashskaya Area)

AUTHOR(S)

Anon.

REFERENCE

Unknown

COUNTRY/REGION

Russia

TYPE OF PROJECT

Waterflood

SUMMARY OF PAPER

This paper is poorly translated (or written). No new concepts are here. The paper proposes to select injection wells, which in relation to producing wells, will give greatest reservoir conformance. Injection well selected on the basis of:

1. greatest reservoir thickness
2. highest productivity index
3. best connection coefficient

The effect is to increase the volume of reservoir contacted by water.

NEW OR DIFFERENT IDEAS/CONCEPTS/PRACTICES

LITERATURE SURVEY FORM

TITLE

Economic Evaluation of Steam-Heat Stimulation of Usinskoye Field Heavy Crude

AUTHOR(S)

Pukhova, L. A.

REFERENCE

Ekonomika Neftyanoy Promyshlennosti, Moscow, July, 1979, No. 7, p. 34-37

COUNTRY/REGION

Russia/Komi (ASSR)

TYPE OF PROJECT

Economics

SUMMARY OF PAPER

Purely an economic evaluation with no calculations or tables. Article indicates variants cost similar to the methods used in any economic analysis.

Article states that the Yaregskoye Field is being developed by the heat-shaft method.

NEW OR DIFFERENT IDEAS/CONCEPTS/PRACTICES

LITERATURE SURVEY FORM

TITLE

In Situ Combustion with Oxygen Combined with Water Injection (ISCOWI)

AUTHOR(S)

Pusch, Guenter

REFERENCE

Erdol und Kohle - Erdgas - Petrochemie-Chemie, v. 30, No. 1, p. 13-25, Jan., 1977.

COUNTRY/REGION

Germany

TYPE OF PROJECT

In Situ Combustion

SUMMARY OF PAPER

In depth description of the oxygen-enriched wet combustion process. Primarily a technical discussion of probable field application results and problems. Probable application of the process to coal gasification in the future is suggested.

NEW OR DIFFERENT IDEAS/CONCEPTS/PRACTICES

L I T E R A T U R E   S U R V E Y   F O R M

TITLE

Combustion Initiation at The Bottom Water-Oil Sands Interface

AUTHOR(S)

Raisbeck, J. M.

REFERENCE

Petroleum Society of CIM, Paper 79-30-15, presented at 30th Annual Tech. Mtg., May 8-11, 1979.

COUNTRY/REGION

Canada

TYPE OF PROJECT

In Situ Combustion

SUMMARY OF PAPER

This paper describes the procedure and results of a two-well in situ combustion test in the Wabasca oil sands.

Prior to the combustion test the two test wells had been used in a steam stimulation pilot test. Also, in preparation for the combustion test, the production well received an 11-day steam injection slug and the air injection well was perforated in a two foot interval immediately above the oil-water contact. The primary goal of the test was obtained in that a successful burning front was created near the bottom water zone of the reservoir and propogated toward the producing well.

L I T E R A T U R E S U R V E Y F O R M

TITLE

Problems at Embaneft' Association Detailed

AUTHOR(S)

* Rakhimgaliyev, K., Gora, N.
* Supervisor of Heavy Industry, Central Committee, Communist Party of Kazakhstana

REFERENCE

Alma-Ata Narodnoye Khozyaystvo Kazakhstana, No. 12, Dec. 1978, p. 28-33.

COUNTRY/REGION

Russia/Emba

TYPE OF PROJECT

SUMMARY OF PAPER

Paper deals mainly with inefficiencies existing within certain ministries which permit problems that ultimately lower the oil production rate. This paper blames the field operating unit (engr., mgrs. workers, and etc.) in the Emba area oil fields. Paper cites many examples of shoddy work which causes delays and other problems. The paper states that the Dossorneft collective is directing it's initiative toward "thermal treatment of the bottom-hole zone, electrical heating of formation, and others."

Steam injection suffers at the Kenkiyak (Emba) Field due to use of low capacity generators. In situ combustion is used in Azerbaydzhan and Mangyshlak with positive results.

NEW OR DIFFERENT IDEAS/CONCEPTS/PRACTICES

The geological division of Emba plans to seal off the waterflooded formations and interstratifications using gipan, a new method.

L I T E R A T U R E   S U R V E Y   F O R M

TITLE

Summary of Foreign EOR Projects (Letter Form)

AUTHOR(S)

Dr. Richards

REFERENCE

Letter

COUNTRY/REGION

Non-U.S.

TYPE OF PROJECT

All Conventional, Steam, CO_2, Miscible, Caustic, etc.

SUMMARY OF PAPER

Points out emphasis that Canada, Venezuela and U.S.S.R. are placing on heavy oil recovery processes. U.S.S.R. has technique of underground drilling, production chambers, etc. which it is trying to sell to Canada.

NEW OR DIFFERENT IDEAS/CONCEPTS/PRACTICES

L I T E R A T U R E S U R V E Y F O R M

TITLE

Heavy Oil In Western Europe

AUTHOR(S)

Rosing, D. E.

REFERENCE

Unitar Report 79, First Intl. Conference on the Future of Heavy Oil and Tar Sands, 1979, Edmonton, Alberta, Canada.

COUNTRY/REGION

Western Europe - W. Germany, France, Netherlands, Italy

TYPE OF PROJECT

SUMMARY OF PAPER

Per title, author gives reserves and presents map showing heavy oil belt of W. Germany plus offshore deposits. Author discusses tertiary recovery methods in Schoonebeck (Netherlands).

NEW OR DIFFERENT IDEAS/CONCEPTS/PRACTICES

L I T E R A T U R E S U R V E Y F O R M

TITLE

Russia Aims to Hike Heavy-Oil Recovery

AUTHOR(S)

Anon.

REFERENCE

The Oil and Gas Journal, April 12, 1976.

COUNTRY/REGION

Russia/Baku

TYPE OF PROJECT

SUMMARY OF PAPER

Old Baku-area fields in Azerbaijan are to become the testing ground for strip and underground mining to recover heavy oil. Quantities of crushed rock are put into containers where either kerosine, alkali water, or caustic soda are used to strip the oil from the rock. Underground mining will result in production chambers where gravity will force crude to flow downward into these production chambers and then pumped to the surface. Baku's underground mines are to be similar to those already in operation, the Yarega heavy-oil field near Ukhta in the Komi Autonomous Republic. Steam injection and in situ combustion projects are in operation in the Yarega Field, however, a shortage of instrumentation for constant control and monitoring has hindered particularly the thermal recovery process.

LITERATURE SURVEY FORM

TITLE

Possibility of Heating a Fractured Formation of Highly Viscous Oil Underlaid by Water

AUTHOR(S)

Ruzin, L. M., Migunov, V. I., Chuprov, I. F.

REFERENCE

Neft Khoz., No. 9, 1977, p. 36-38.

COUNTRY/REGION

Russia/General

TYPE OF PROJECT

Thermal

SUMMARY OF PAPER

Authors suggested the possibility of heating oil bearing stratum by injecting steam into the stratum below the oil-water contact. Thermal conductivity through fractures is suggested as the means of heat transfer. Calculations and graphs were presented; no laboratory or field test efforts were mentioned or suggested.

NEW OR DIFFERENT IDEAS/CONCEPTS/PRACTICES

Heating oil bearing stratum by injecting steam into the water zone of a fractured formation.

L I T E R A T U R E S U R V E Y F O R M

TITLE

Emba Oil Fields Chosen for Thermal Stimulation Identified

AUTHOR(S)

Sagingaliyev, B. S. Dodonova, I. A., Rakovskiy, N. L., and Shelar, O. G.

REFERENCE

VNII OENG, 1979.

COUNTRY/REGION

Russia

TYPE OF PROJECT

Steam and In Situ Combustion

SUMMARY OF PAPER

This paper presents the criteria for selecting reservoirs which are suitable for steam injection and in situ combustion.

Steam criteria are:

$\phi_{sd} > 18\%$, $\phi_{ls} > 12\%$
$k > 100$ md, $\mu > 50$ cp
$h \geq 10$ m, depth ≤ 700 m

In Situ Combustion Criteria same as Steam except:

$h \geq 3.0$ m, depth ≤ 2000 m
$\mu \geq 10$ cp

Paper gives fields/reservoirs which are planned for Thermal development in Emba region.

LITERATURE SURVEY FORM

TITLE

Method of Projecting Technological Parameters of the Process of In-Situ Thermochemical Treatment

AUTHOR(S)

Savenkov, G. D., Doroshenko, V. M.

REFERENCE

Izv. Vyssh. Uchebn. Zaved., Neft. Gaz., No. 4, 1978, p. 35-37.

COUNTRY/REGION

Russia

TYPE OF PROJECT

In Situ Thermochemical

SUMMARY OF PAPER

The thermochemical process simply relates to the artificial fracturing of the productive interval and propping the fracture open with sand and magnesium granules. This action is followed by acid (15% HCl) which reacts with the magnesium. The reaction is exothermic and consequently negates the usual effects of cooling and also melts any paraffin or other semisolids which may have a tendency to plug the permeability.

NEW OR DIFFERENT IDEAS/CONCEPTS/PRACTICES

Use of an exothermic chemical reaction to heat up the area of the wellbore thereby reducing viscosity of semisolids allowing a better wellbore clean up.

L I T E R A T U R E   S U R V E Y   F O R M

TITLE

Performance History Improved Recovery by Gas Injection in Heavy Oil Reservoirs Oveja Field-Eastern Venezuela

AUTHOR(S)

Servello, J. P., Meneven, S. A.

REFERENCE

Unitar Report 31, First Intl. Confererence on The Future of Heavy Oil and Tar Sands, 1979, Edmonton, Alberta, Canada.

COUNTRY/REGION

Venezuela

TYPE OF PROJECT

Gas Injection-Pressure Maintenance

SUMMARY OF PAPER

Discusses field histories of four heavy-oil reservoirs in which fluid, rock, and production mechanism were favorable to gas injection for pressure maintenance to improve oil recovery. Some basic conditions are: 30+ percent porosities; initial water saturations of about 10 percent, and permeabilities of over 2.5 darcies. During primary operations good gas-oil segregation in the gas cap and saturated oil zone was evident.

NEW OR DIFFERENT IDEAS/CONCEPTS/PRACTICES

L I T E R A T U R E   S U R V E Y   F O R M

TITLE

Predicting Bitumen Recovery from Steam Stimulation

AUTHOR(S)

Shepherd, D. W.

REFERENCE

World Oil, Sept., 1979.

COUNTRY/REGION

Canada

TYPE OF PROJECT

Steam

SUMMARY OF PAPER

Discusses empirically derived techniques for predicting heavy oil recovery in a steam-soak process. The empirical data was obtained from an initial 23 well pilot in the Clearwater reservoir unit at Cold Lake, and followed up with the Leming pilot incorporating 80 wells.

NEW OR DIFFERENT IDEAS/CONCEPTS/PRACTICES

L I T E R A T U R E   S U R V E Y   F O R M

TITLE

Enhanced Oil Recovery: The French Contribution

AUTHOR(S)

Simandoux, Pierre

REFERENCE

L'Industrie du Petrole, v. 47, No. 508, p. 75-91, 1979.

COUNTRY/REGION

France

TYPE OF PROJECT

Several

SUMMARY OF PAPER

Presents a general summary of enhanced recovery methods, limitations of application, and general economic comparisons. Notes French efforts: high pressure miscible gas, Hassi-Messaoud Field, Algeria; enriched gas slug, Rainbow Field, Alberta; polymer injection, Chateaurenard Field, France; microemulsion (6 wells), same field; steam injection, Lacq Superior Field, France; in situ combustion, Suplacu de Barcau Field, Rumania (Cooperative agreement, French Petroleum Institute.

NEW OR DIFFERENT IDEAS/CONCEPTS/PRACTICES

None

L I T E R A T U R E S U R V E Y F O R M

TITLE

Physical Model Study of Inverted Seven Spot Steamfloods in A Pool Containing Conventional Heavy Oil

AUTHOR(S)

Singhal, A. K.

REFERENCE

Petroleum Society of CIM, Paper 79-30-13, Presented at 30th Annual Tech. Mtg., May, 1979.

COUNTRY/REGION

Canada

TYPE OF PROJECT

Steam Injection

SUMMARY OF PAPER

Discusses the many types of partially scaled physical models that have been used to study various parameters affecting the outcome of a steam injection project. The author then proceeds to discuss his model study and design in which he used a triangular model constructed to represent 1/12 of an inverted 7-spot pattern. Variables evaluated in the author's model include: 1.) the effects of injection rates, steam quality and enthalpy input, 2.) relationship of steam volume and in situ emulsification to oil recovery, 3.) effects of completion intervals, 4.) effects of different production strategies, including cold water injection following steam slugs, steam flood following a water flood, and pressure cycling and steam soak.

L I T E R A T U R E   S U R V E Y   F O R M

TITLE

Studies on The Use of Tunnels and Horizontal Wells for The Recovery of Heavy Crudes

AUTHOR(S)

Smith, G. W., Butler, R. M.

REFERENCE

Unitar Report 62, First Intl. Conference on The Future of Heavy Oil and Tar Sands, 1979, Edmonton, Alberta, Canada

COUNTRY/REGION

Canada

TYPE OF PROJECT

Mining

SUMMARY OF PAPER

As per title, this paper discusses the possible application of steam injection into horizontally drilled wells for additional heavy oil recovery in Canada. The reference process and procedure described is that being used in the Yarega reservoir in Siberia.

The advantage of such processes is obtaining a higher density of injection and production wells at a lower cost.

NEW OR DIFFERENT IDEAS/CONCEPTS/PRACTICES

LITERATURE SURVEY FORM

TITLE

Chemical Properties and Reservoir Characteristics of Bitumen and Heavy Oil from from Canada and Venezuela

AUTHOR(S)

Starr, John, Prats, Jose M.

REFERENCE

Unitar Report 65, First Intl. Conference on The Future of Heavy Oil and Tar Sands, 1979, Edmonton, Alberta, Canada.

COUNTRY/REGION

Venezuela and Canada

TYPE OF PROJECT

None

SUMMARY OF PAPER

Presents location maps of heavy oil fields, physical property comparisons, and compositional comparisons of heavy hydrocarbons in Canada and Venezuela.

NEW OR DIFFERENT IDEAS/CONCEPTS/PRACTICES

L I T E R A T U R E   S U R V E Y   F O R M

TITLE

The State of the Study and Application of the Method for Displacement of Oil by Steam Combined with Waterflooding

AUTHOR(S)

Anon.

REFERENCE

Unknown

COUNTRY/REGION

Russia

TYPE OF PROJECT

Steam and Thermal Mining

SUMMARY OF PAPER

Historical summary of oil displacement by heat carrier. In 1947-50 steam and flue gas was injected at the Shodnitsa Oil Field in the Ukrain. During the latter part of the sixties VNII was actively engaged in the study of steam injection into reservoirs. Cyclic steam injection was applied in the heavy oil sands of Krasnodar and Sakhalin area in 1965-66. Continuous steam drive was initiated in 1958 in the oil fields Okha (Sakhalin area) and Yarega (Komi A.S.S.R.) VNII also started work on the thermal slug method during this time. The thermal slug method has also been successfully implemented in parts of the Okha and Yarega Oil Fields in addition to the oil fields, Katangli, Borislav, Kenkiyak, Sibsa, etc. Additional oil recovery from Okha due to thermal stimulation is reported to be 37%.

VNII, oil combine "Komineft", and Institute Pechor-Nipineft offered, tested and applied on a commercial

SUMMARY OF PAPER (Continued)

scale in Yarega, the thermal mining method. The area of field where the thermal mine process was developed has the following characteristics, depth 200 m, perm 0.5 to 5 darcies, initial press 14 to 15 atm, BHT 6 to 8° C, viscosity 15-20 X 10^3 cp, specific gravity 0.94 to 0.95. Primary recovery was about 2%, thermal mining has yielded up to 50-60% recovery.

LITERATURE SURVEY FORM

TITLE

Steam, Fireflood Delays Thwart Soviet Oil Flow

AUTHOR(S)

Anon.

REFERENCE

The Oil and Gas Journal, Mar. 26, 1979, p. 62-64.

COUNTRY/REGION

Russia

TYPE OF PROJECT

Thermal

SUMMARY OF PAPER

Paper is a result of briefing by E. Khalimov (Deputy Minister of Oil) who states the technology of thermal recovery is available but that equipment such as steam generators is grossly lacking. He cites some specific problems with design testing, and building of equipment for oil field needs.

The U.S.S.R. is badly lacking in being able to apply 1960's U.S.A. technology due to equipment needs. Based on other papers there is probably much truth to what Khalimov says; however, the Russians are famous for blaming someone else for their shortcomings. This fact is born out in other papers.

LITERATURE SURVEY FORM

TITLE

The Present Status, Principles and Efficiency of Waterflooding in the Oilfields of the Soviet Union

AUTHOR(S)

Stosur, G.

REFERENCE

Unpublished Report

COUNTRY/REGION

Russia

TYPE OF PROJECT

Waterflooding

SUMMARY OF PAPER

This article presents an elementary review of waterflooding in Russia.

NEW OR DIFFERENT IDEAS/CONCEPTS/PRACTICES

LITERATURE SURVEY FORM

TITLE

The Electric Pre-Heat Recovery Process

AUTHOR(S)

Towson, Dr. D. E.

REFERENCE

Unitar Report 12, First Intl. Conference on The Future of Heavy Oil and Tar Sands, 1979, Edmonton, Alberta, Canada.

COUNTRY/REGION

Canada

TYPE OF PROJECT

Electric Resistance Heating

SUMMARY OF PAPER

Discusses the physics involved in heating an oil sand electrically, to reduce bitument viscosity. The advantage to electric heating is that fluid contact with the bitumen is not required as in steam, solvent, or diluent floods. A major disadvantage is the inefficiency of converting fossil fuel to electricity and the high cost of electricity. Currently, in addition to an active laboratory research program, a preliminary pilot is under construction near Fort McMurray.

NEW OR DIFFERENT IDEAS/CONCEPTS/PRACTICES

L I T E R A T U R E   S U R V E Y   F O R M

TITLE

The Oil Sands of Alberta: Source of Canada's Future Oil Supply?

AUTHOR(S)

Towson, D. E.

REFERENCE

The Journal of Canadian Petroleum, Oct.-Dec., 1979.

COUNTRY/REGION

Canada - Alberta

TYPE OF PROJECT

Mining, In Situ, and Steam

SUMMARY OF PAPER

Summarizes mining projects by Suncor, Syncrude, and the Alsands project being planned by Shell, et al. These projects all involve similar processes of mining, collecting, separation, and upgrading of the oil sands.

Several in situ projects are mentioned as being planned or pilot projects currently going, by a number of international companies.

NEW OR DIFFERENT IDEAS/CONCEPTS/PRACTICES

L I T E R A T U R E   S U R V E Y   F O R M

TITLE

Trapping of Oil at Intra-Mannville (Lower Cretaceous) Disconformity in Lloydminster Area, Alberta and Saskatchewan

AUTHOR(S)

Vigrass, Laurence W.

REFERENCE

AAPG Bulletin, Vol. 61, No. 7, July, 1977.

COUNTRY/REGION

Canada

TYPE OF PROJECT

None

SUMMARY OF PAPER

A geologic description of structure and stratigraphy of the heavy-oil bearing formations in the subject area (per title).

NEW OR DIFFERENT IDEAS/CONCEPTS/PRACTICES

L I T E R A T U R E   S U R V E Y   F O R M

TITLE

Planning and Operation of Steam Boiler Units in Thermal Recovery Projects

AUTHOR(S)

Von Schlemm, F.

REFERENCE

Erdoel-Erdgas-Zeitschrift, v. 95, No. 7, p. 238-247, July, 1979.

COUNTRY/REGION

Germany

TYPE OF PROJECT

Steamflood

SUMMARY OF PAPER

Summary of considerations in planning to insure most trouble-free operation of steam boilers (both single and two stage). Installations in Georgesdorf and Ruhlermoor Fields (N. Germany) are shown as examples.

NEW OR DIFFERENT IDEAS/CONCEPTS/PRACTICES

None

LITERATURE SURVEY FORM

TITLE

Water Conservation in A Steam Stimulation Project

AUTHOR(S)

Whalley, M. J. and Wilson, T. M.

REFERENCE

Unitar Report 61, First Intl. Conference on The Future of Heavy Oil and Tar Sands, 1979, Edmonton, Alberta, Canada.

COUNTRY/REGION

Canada/Cold Lake

TYPE OF PROJECT

Steam Drive

SUMMARY OF PAPER

Discusses the nature of the produced water resulting from a steam drive project and how it may be treated and processed for reuse. By recycling the produced water, the projected gross water requirements for Esso's planned Cold Lake Heavy Oil Recovery Project could be reduced by as much as 60 percent.

NEW OR DIFFERENT IDEAS/CONCEPTS/PRACTICES

L I T E R A T U R E S U R V E Y F O R M

TITLE

Heavy Crude Oil and Tar Sand Resources and Reserves of the United States, Emphasis on The State of Texas

AUTHOR(S)

Whiting, Robert L.

REFERENCE

Supplemental to Unitar Report 68, First Intl. Conference on The Future of Heavy Oil and Tar Sands, 1979, Edmonton, Alberta, Canada.

COUNTRY/REGION

U. S. A.

TYPE OF PROJECT

None

SUMMARY OF PAPER

Gives location and amount of known heavy oil reserves in the United States, by state (110 X 109 STB) and considers everything up to 25° API as heavy.

NEW OR DIFFERENT IDEAS/CONCEPTS/PRACTICES

L I T E R A T U R E   S U R V E Y   F O R M

TITLE

40 Years of Heavy Oil Exploration and Production

AUTHOR(S)

Williams, James A.

REFERENCE

Unitar Report 17, First Intl. Conference on The Future of Heavy Oil and Tar Sands, 1979, Edmonton, Alberta, Canada.

COUNTRY/REGION

Western U. S. and Canada

TYPE OF PROJECT

Polymers, Fireflood and Steam

SUMMARY OF PAPER

Summarizes the exploration and production history of Husky Oil in the Western U. S. and Canada, in their search for heavy oil. Until recent economic changes in the price of oil, producing heavy oil was very marginal and required strict operating costs control. Husky is currently operating two firefloods successfully in the Lloydminster area and is continuing to evaluate the effects and economics of steam displacement. The major problem in producing Lloydminster wells is in controlling sand production at economical production rates.

NEW OR DIFFERENT IDEAS/CONCEPTS/PRACTICES

LITERATURE SURVEY FORM

TITLE

Work Still Lags on Soviet Heavy Oil Underground Mine

AUTHOR(S)

Anon.

REFERENCE

Oil and Gas Journal, Oct. 15, 1979, p. 93.

COUNTRY/REGION

Russia/Baku

TYPE OF PROJECT

Underground Mine

SUMMARY OF PAPER

Russia lags behind schedule at Balakhany Field in the Baku Region. Say more gas is present and longer drifts are needed than in the Yarega Field near Ukhta.

NEW OR DIFFERENT IDEAS/CONCEPTS/PRACTICES

L I T E R A T U R E   S U R V E Y   F O R M

TITLE

Effectiveness of Experimental Flooding Using Polymer Solutions

AUTHOR(S)

Zaidullin, G. N., Saidfullin, Z. G., Fatkullin, A. Kh., Lerman, B. A., and Muslimov, R. Kh.

REFERENCE

Neft Khoz., 1977, No. 2, p. 31-34

COUNTRY/REGION

Russia/Kuibyshev, Bashkiria, Tataria

TYPE OF PROJECT

Polymer Flooding

SUMMARY OF PAPER

Field experiments were carried out in the above regions using polyacrylamide reagent (AMF). In Nov., 1973, 182 tons of 7% AMF was injected into the fifth deposit of the Bobriki Horizons of the Romashkino Deposit. Results show 15.9% improved recovery over conventional waterflood at comparable injected volumes. Study is now underway to determine the possibility of injecting industrial waste water at the same location. Laboratory experiments were conducted using powdered Polyacrylamide (PAARK 123-sh), hydroxyethylcellulose (HEC), methyl cellulose and K-9, AMF and Pusher 500 reagents. Experiments were carried out in Transparent 50 X 50 X .85 cm model. Concentrations were used to yield a viscosity of 5-6 cp. Results show the PAARK and Pusher-500 to be equal in quality but superior to the other polymers used. Recoveries over conventional waterflooding were 10.6% and 6.4-8.2% for the PAARK and Pusher-500, and the other polymers, respectively.

Appendix B
List of Personal Communications

1. Ace, Allen; UK/North Sea; British Petroleum.
2. Anz, James; Indonesia; Core Laboratories, Inc.
3. Arkenberg, Pablo; Argentina; Quimica Hoechst.
4. Austin, Ellis; Venezuela; Desha Consulting.
5. Bader, B.; Downhole Steam Generators; Sandia Laboratories, DOE.
6. Ballard, Jim; Thermal Insulation, Venezuela; AMOCO.
7. Bonet, M.; Brazil; Petroleos Brasilieros.
8. Brimhall, Ronald M.; Thermal Mining, U.S.S.R.; Texas A & M University.
9. Carrillo, Augustiń; Mexico; Petroleos Mexicanos.
10. Ciapuscio, Hector; Argentina; SETEC.
11. Clark, J. B.; Biological; University of Oklahoma.
12. Decan, Noel; Venezuela; Core Laboratories, Inc.
13. Dogru, Sevgi; Turkey; Consultant.
14. Earlougher, R. C.; Thermal Mining, U.S.S.R.; Resource Sciences Corporation.
15. Fisher, Sidney T.; Induction Heating; F. T. Fisher's Sons.
16. Fox, Ron; Downhole Steam Generator; Sandia Laboratories, DOE.
17. Frick, John R., Jr.; Biological; Natomas.
18. de la Fuente, Gialo; Mexico; Instituto Mexicano de Petroleo.
19. Funck, Jacqueline; France; Institut Francais du Petrole.
20. Garms, Kenneth G.; Italy; Core Laboratories, Inc.
21. Geffen, Theodore M.; General; AMOCO Research.

22. Hays, J. M.; Otis Power Lift; Otis Engineering.

23. Heiberg, Siguard; Norway/North Sea; Norwegian Petroleum Directorate.

24. Howard, John; General/Steam; Greenwich Oil Corp.

25. Hujsak, Karol; Venezuela; Hyco Corporation.

26. Jackson, Richard; Germany; Mobil.

27. Joiner, David; General; ARCO.

28. Kirkpatrick, J. W.; Canada In-Situ; AMOCO Research.

29. Kokcu, M.; Bati Raman, Turkey; Mobil.

30. Marberry, James; Germany; Mobil.

31. Marshall, J. P.; UK/North Sea; Core Laboratories, Inc.

32. Matheny, Shannon; General; The Oil and Gas Journal.

33. Mazza, Jorge; Argentina; SETEC.

34. Menzie, Donald E.; Biological; University of Oklahoma.

35. Meyerhoff, A. A.; Soviet Sphere Heavy Crude Activities; Consultant.

36. Muchacho, Luis; Venezuela; Lagoven.

37. Mungan, Nick; Canada; Mungan Petroleum Consultants.

38. Nehring, Richard; Mid-East; The Rand Corporation.

39. Parrish, David; General; AMOCO Research.

40. Pato, Sergio; Brazil; Petroleos Brasilieros.

41. Perez, Jaime; Steam Injection, Mexico; Petroleos Mexicanos.

42. Ray, Jack; USSR: Tenneco.

43. da Silva, Haroldo; Brazil/Iraq; Braspetro.

44. Simandoux, Pierre; Europe/France; Institut Francais du Petrole.

45. Stacey, Don; General; AMOCO Research.

46. Teyssier, Joaquin; Steam Injection, Mexico; Instituto Mexicano de Petroleo.

References and Bibliography

2. BASIC RECOVERY PROCESSES FOR HEAVY OILS

1. Bentsen, R. G., and Donohue, D.A.T., "A Dynamic Programming Model of the Cyclic Steam Injection Process," Journal of Petroleum Technology, Dec. 1969.

2. Boberg, T. C., and Lantz, R. B., "Calcuation of the Production Rate of a Thermally Stimulated Well," Journal of Petroleum Technology, Dec. 1966.

3. Gogarty, W. B., "Status of Surfactant or Micellar Methods," Journal of Petroleum Technology, Jan. 1976.

4. Holm, L. W., "Status of CO_2 and Hydrocarbon Miscible Oil Recovery Methods," Journal of Petroleum Technology, Jan. 1976.

5. Johnson, C. E., Jr., "Status of Caustic and Emulsion Methods, Journal of Petroleum Technology, Jan. 1976.

6. Martin, W. L., Alexander, J. D., and Dew, J. N., "Process Variables of In Situ Combustion," Journal of Petroleum Technology, Feb. 1958.

7. Marx, J. W., and Langenheim, R. H., "Reservoir Heating by Hot Fluid Injection," Trans of SPE-AIME, Vol. 216, 1959.

8. Parrish, D. R., Rausch, R. W., et al., "Underground Combustion in the Shannon Pool, Wyoming," Journal of Petroleum Technology, Feb. 1962.

9. Warren, J. E., Reed, R. L., and Price, H. S., "Theoretical Consideration of Reverse Combustion," Paper Presented at the 34th Annual Technical Conference of the Society of Petroleum Engineers, Dallas, Texas, Oct. 1959.

3. MODIFIED AND INNOVATIVE PROCESSES FOR HEAVY OILS

1. Burger, J. G., "Enhanced Recovery in Heavy Crude and Bitumen Deposits," Paper, EEC Symposium on New Technologies for the Exploration and Exploitation of Oil and Gas Resources, Luxemburg, Apr. 18-20, 1979.

2. Havlena, Z. G., Some Innovative Approaches Which May Facilitate Production of Heavy Crudes, UNITAR Conference on the Future of Heavy Crude and Tar Sands, Edmonton, Alberta, Canada, 1979.

3. Mowll, R., and Hambling, J. K., "Priorities for Non-Conventional Hydrocarbon Liquids Production," UNITAR Conference on the Future of Heavy Crude and Tar Sands, Edmonton, Alberta, Canada, 1979.

Section 3.1 Thermal Recovery Methods

1. "Enhanced-Recovery Test Using Electricity Slated In Texas," The Oil and Gas Journal, June 23, 1975.

2. Towson, D. E., "The Electric Pre-Heat Recovery Process," UNITAR Conference on the Future of Heavy Crude and Tar Sands, Edmonton, Alberta, Canada, 1979.

3. Harvey, A. Herbert, Arnold, M. D., and Samy A. El-Feky, "Selective Electric Reservoir Heating," Journal of Canadian Petroleum Technology, July-September, 1979.

4. Fisher, Sidney T., "Electrical Induction Heating - A New Concept in In Situ Recovery," Energy Processing/Canada, September-October, 1977.

5. Fisher, Sidney T., "Processing of Coal, Oil Sand, and Heavy Oil In Situ by Electric and Magnetic Fields," Canadian Electrical Engineering Journal, Vol. 4, No. 4, 1979.

6. Fisher, Sidney T., "The Solid Fossil Fuels: Long-Term Source of All the Energy Commodities," UNITAR Conference on Long-Term Energy Sources, Montreal, Canada, 1979.

7. Fisher, Sidney T., "Processing of Coal, Oil Sand, and Heavy Oil In Situ by Electric and Magnetic Fields," Canadian Electrical Engineering Journal, Vol. 5, No. 3, 1980.

8. Fisher, Sidney T., "Advances in Induction Heating," The Oil and Gas Journal, June 16, 1980.

9. Fisher, Sidney T., "Draft Pilot In Situ Plan for Enhanced Recovery from Viscous Oil,: Personal Correspondence, 1980.

10. "Unique Oil-from-Shale Process Under Development," The Oil and Gas Journal, Apr. 14, 1980.

11. "Balakhany Oil Mine Under Construction," Baku Vyshka, Mar. 13, 1979.

12. Danysh, D. V. and Gurov, Y. I., "Thermal Method of Re-Covering Petroleum from Mines," Neft. Khoz., No. 9, Sept. 1979.

13. "Experimental Underground Oil Field," Baku Vyshka, Aug. 28, 1979.

14. "Hot-Mine Method Applied to Yarega Fields," Moscow Neftyanik, No. 8, Aug. 1979.

15. "Mining of Viscous Crude Claimed Viable," The Oil and Gas Journal, Jan. 5, 1976.

16. "Patent of the Yarega Method - Ukhta," U.S. Joint Publications Research Service, No. 857, March 12, 1979.

17. Uren, L. C., Petroleum Production Engineering - Exploitation, McGraw-Hill Book Company, Inc., New York, 1953, pp. 540-545.

18. Dobson, W. F. and Seelye, D. R., "Mining Technology Assists Oil Recovery from Wyoming Field," Paper SPE 9418 presented at 55th Annual Fall Technical Conference, Dallas, Texas, Sept. 1980.

19. Smith, G. W., and Butler, R. M., "Studies on the Use of Tunnels and Horizontal Wells for the Recovery of Heavy Crudes," UNITAR Conference on the Future of Heavy Crude and Tar Sands, Edmonton, Alberta, Canada, 1979.

20. Stramp, R. L., "The Use of Horizontal Drainholes in the Empire Abo Unit," Paper SPE 9221 presented at the 55th Annual Fall Technical Conference, Dallas, Texas, Sept. 1980.

21. "Field Work Begins on Unique Heavy Oil Process," The Oil and Gas Journal, April 7, 1980.

22. Babijan, G. A., et al., "Results of Theoretical, Experimental and Oil Field Tests of Chemical Agents (SAS, CO_2, PAA) Utilization Efficiency to Improve Oil Recovery," Paper presented at Joint Soviet - U.S. Waterflood Symposium, Tyumen, Siberia, U.S.S.R., May 1978.

23. Cady, G. V., Hoffman, S. J., and Scarborough, R. M., "Silverdale Combination Thermal Drive Project," UNITAR Conference on the Future of Heavy Crude and Tar Sands, Edmonton, Alberta, Canada, 1979.

24. "Emba Oil Fields Chosen for Thermal Stimulation Identified," Moscow, Neft. Geol. Geof., No. 3, 1979.

25. Farouq Ali, S. M., Look Yee, K. D.,Cordero, F. J., and Figueroa, J. M., "Role of Injection-Production Strategy in the Solvent-Steamflooding of the Athabasca Oil Sands," Journal of Canadian Petroleum Technology, Oct.-Dec.,1979.

26. Nechipurenko, O., "Make the Road Wider to Tertiary Production Methods," Baku Vyshka, May 29, 1979.

27. "New Oil Extraction Technique," Kiev Rabochaya Gazeta, Dec. 26, 1978.

28. Patton, J. T., Coats, K. H., and Spence, K., "A Parametric Study of the CO_2 HUF-n-PUF Process," Paper SPE 9228, presented at the 54th Annual Fall Technical Conference, Las Vegas, Nevada, Sept. 1979.

29. Savenkov, G. D., and Doroshenko, V. M., "Method of Projecting Technological Parameters of the Process of In Situ Thermochemical Treatment," Izv. Vyssh. Uchebn. Zaved, Neft. Gaz., No. 4, 1978.

30. "The State of the Study and Application of the Method for Displacement of Oil by Steam Combined with Water Flooding," Unknown.

31. Vakhitov, G. G., Bokserman, A. A., et al., "Method of Increasing the Oil Output of Formations by Combining In Situ Combustion with Flooding," Neft. Khoz., No. 8, Aug. 1977.

Section 3.2 Miscellaneous Other Methods

1. Bubela, B., "Role of Geomicrobiology in Enhanced Recovery of Oil: Status Quo," APEA Journal, 1978.

2. Frick, J. R., Jr., "Bacterial Processes for Improved Oil Recovery," Journal of Petroleum Technology, Dec. 1972.

3. Jaranyi, I., Kiss, L., et al., "Changes in the Properties of Oil Sands Resulting from Microbiological Treatment," Vortrage der III Intern. Wissen. Konferenz for Geochemie, Mikrobiologie und Erdolchemie, II, Budapest, 1963.

4. Karaskiewicz, Jerzy, "Application of Microbiological Methods for Secondary Oil Recovery from the Carpathian Crude Oil Reservoirs," Pr. Inst. Naft., Cracow, 1974.

5. Koznetsov, S. I., Ivanov, M. W., and Lalikova, N. N., "Introduction to Geological Microbiology," Akad. Nauk SSSR (U.S.S.R. Academy of Sciences), Moscow, 1962.

6. Lazar, I., "Use of Bacteria in the Recovery Procedures and Migration of Oil from Deposits," Mine, Petrol si Gaze, 1976, pp. 475-480.

7. Wellington, S. L., "Biopolymer Solution Viscosity Stabilization - Polymer Degradation and Antioxidant Use," Paper SPE 9296 presented at 55th Annual Fall Technical Conference, Dallas, Texas, Sept. 1980.

8. Gadiev, S. M., Gimatudinov, S. K., and Kuzmichev, Y. A., "Model of Dynamic Effects in the Stratum," Izv. Vyssh Uchebn Zaved Neft Gaz, No. 9, 1977.

9. Gadiev, S. M., Polyanskaya, L. V., et al., "Vibration Effects on the Critical Zone of Wells in Multi-Stratal Deposits," Izv. Vyssh Uchebn Zaved Neft Gaz, No. 6, 1977.

10. "A Magnetic 'Believe It or Not'," Science News, Vol. 118, Sept. 6, 1980.

11. Mikhelman, A. I., Fedoriv, L. V., and Radish, V. N., "Change in Oil Yield of a Stratum Model with Displacement of Oil with Magnetized Water," Izv. Vyssh. Uchebn Zaved Neft. Gaz, No. 9, 1977.

4. SURFACE AND DOWNHOLE EQUIPMENT

Section 4.1 Thermal Insulation

1. Farouq Ali, S. M., and Meldau, R. F., "Current Steamflood Technology," Journal of Petroleum Technology, October 1979.

2. Johnson, D. R., and Fox, R. L., "examination of Techniques for Thermally Efficient Delivery of Steam to Deep Reservoirs," Paper SPE 8820 presented at First Joint SPE/DOE Symposium on Enhanced Oil Recovery, Tulsa, Oklahoma, April 1980.

3. Satter, A., "Heat Loses During Flow of Steam Down a Wellbore," Journal of Petroleum Technology, July 1965.

Section 4.2 Otis Heavy-Crude Lift System

1. "Innovation Holds Promise for Heavy-Crude Recovery," Enhanced Recovery Week, Sept. 22, 1980.

2. "Otis Heavy-Crude Lift System," Otis Engineering, Dallas, Texas, 1980.

Section 4.3 Steam Generation Equipment

1. "Solid Fuel Steam Generator Planned," The Oil and Gas Journal, May 5, 1980.

2. Schlemm, V. F., "Planning and Operation of Steam Boiler Units in Thermal Recovery Projects," Erdol-Erdgas-Z., July 1979.

3. "Sandia: Downhole Steam Generator Works Well," The Oil and Gas Journal, Sept. 15, 1980.

4. "Solar Energy for Enhanced Oil Recovery," Acurex Corporation, Mt. View, California, 1980.

5. NATIONAL EFFORTS IN HEAVY OIL RECOVERY

1. Amato, F. L., "Petroleum Developments in South America, Central America, Mexico, and Caribbean Area in 1977," Bull. AAPG, Vol. 62, October 1978.

2. "EOR Methods Help Ultimate Recovery," The Oil and Gas Journal, March 31, 1980.

3. "Friendship Tour Covers Three Countries," AAPG Explorer, March 1980.

4. Gutierrez, F. J., "Occurrence of Heavy Crudes and Tar Sands in Latin America," UNITAR Conference on the Future of Heavy Crude and Tar Sands, Edmonton, Alberta, Canada, 1979.

5. Hatfield, L. E., and Neff, C. H., "Developments in South America and Caribbean Area in 1975," Bull. AAPG, Vol. 60, October 1976.

6. Meyer, R. F., "World Geography of Heavy Crude Oils," UNITAR Conference on the Future of Heavy Crude and Tar Sands, Edmonton, Alberta, Canada, 1979.

7. "Summary Report," UNITAR Conference on the Future of Heavy Crude and Tar Sands, Edmonton, Alberta, Canada, 1979.

8. Walters, E. J., "Review of World's Major Oil Sands Deposits," Oil Sands - Fuel of the Future, Canadian Society of Petroleum Geologists, Calgary, Sept. 1974.

9. "WPC Program Spans Subjects Industry-Wide," The Oil and Gas Journal, Aug. 27, 1979.

10. Abougoush, M. S., "Correlations for the Performance Prediction of Heavy Oil Pools," UNITAR Conference on the Future of Heavy Crude and Tar Sands, Edmonton, Alberta, Canada, 1979.

11. "Big Alberta Carbonate Triangle Resource Seen," The Oil and Gas Journal, June 9, 1980.

12. Cady, G. V., Hoffman, S. J., and Scarborough, R. M., "Silverdale Combination Thermal Drive Project," UNITAR Conference on the Future of Heavy Crude and Tar Sands, Edmonton, Alberta, Canada, 1979.

13. Campbell, T., "Cold Lake Project - If and When?," Canadian Petroleum, May 1979.

14. "Canada Will Spend $8.5 Million to Drill 25 New Wells in an Experiment to Extract Heavy Oil at Eyehill Pool, Sask.," Technical Survey, Vol. 35, No. 33, Aug. 18, 1979.

15. Capeling, R. R., and Peggs, J. K., "Experimental Steam-flood Cold Lake Sands," UNITAR Conference on the Future of Heavy Crude and Tar Sands, Edmonton, Alberta, Canada, 1979.

16. Christopher, J. E., and Knudsen, R. H., "Heavy Crude Oil Potential of Saskatchewan," UNITAR Conference on the Future of Heavy Crude and Tar Sands, Edmonton, Alberta, Canada, 1979.

17. "Companies Slate Heavy Oil Work in North, East Alberta," The Oil and Gas Journal, Aug. 18, 1980.

18. "Cost of Heavy Oil Projects Could Exceed $14 Billion," The Northern Miner, Oct. 4, 1979.

19. "Decision Pending on Cold Lake Project," The Oil and Gas Journal, May 15, 1978.

20. Edwards, M. W., "Lloydminster Heavy Oil Thermal Recovery Cost Considerations," UNITAR Conference on the Future of Heavy Crude and Tar Sands, Edmonton, Alberta, Canada, 1979.

21. Ejiogu, G. J., Bennion, D. W., Moore, R. G., and Donnelly, J. K., "Wet Combustion - A Tertiary Recovery Process for the Pembina Cardium Reservoir," Journal of Canadian Petroleum Technology, July-Sept. 1979.

22. "EOR Pilot Advancing in Saskatchewan," The Oil and Gas Journal, June 30, 1980.

23. Farouq Ali, S. M., Figueroa, J. M., and Azuaje, E. A., "Recovery of Lloydminster and Morichal Crudes by Caustic, Acid, and Emulsion Floods," Journal of Canadian Petroleum Technology, Jan.-March 1979.

24. Farouq Ali, S. M., and Meldau, R. F., "Current Steamflood Technology," Journal of Petroleum Technology, October 1979.

25. Farouq Ali, S. M., Look Yee, K. C., Cordero, F. J., and Fiqueroa, J. M., "Role of Injection-Production Strategy in the Solvent-Steamflooding of the Athabasca Oil Sands," Journal of Canadian Petroleum Technology, Oct.-Dec. 1979.

26. "Feasiblity of Heavy Oil Plans Will be Set in 1980," Oil Daily, April 19, 1978.

27. Ferrer, J., and Farouq Ali, S. M., "A Three-Phase, Two-Dimensional Compositional Thermal Simulator for Steam Injection Processes," Journal of Canadian Petroleum Technology, Jan.-March 1977.

28. "Five New Oil Sands Projects Needed in 1980's," The Northern Miner, Oct. 4, 1979.

29. Gorrill, R. G., Kitzan, P., and Komery, D. P., "The Design of the Peace River In-Situ Oil Sands Pilot Project," UNITAR Conference on the Future of Heavy Crude and Tar Sands, Edmonton, Alberta, Canada, 1979.

30. Gurel, D. O., "Operating Practices to Improve Profitability of Conventional Heavy Crude Production," UNITAR Conference on the Future of Heavy Crude and Tar Sands, Edmonton, Alberta, Canada, 1979.

31. Gurel, D. O., "How Husky Makes a Profit in Heavy Oil Production," World Oil, Sept. 1979.

32. Harrison, D. B., Glaister, R. P., and Nelson, H. W., "Reservoir Description of the Clearwater Oil Sand, Cold Lake, Alberta, Canada," UNITAR Conference on the Future of Heavy Crude and Tar Sands, Edmonton, Alberta, Canada, 1979.

33. "Japanese Advance Canadian Oil Sands Acquisition," The Oil and Gas Journal, September 22, 1980.

34. Jardine, D., "Cretaceous Oil Sands of Western Canada," Oil Sands - Fuel of the Future, Canadian Society of Petroleum Geologists, Calgary, Sept. 1974.

35. Jenkins, G. R., and Kirkpatrick, J. W., "Twenty Years' Operation of an In-Situ Combustion Project," Journal of Canadian Petroleum Technology, Jan.-March 1979.

36. Jespersen, P. J., "Field Pilot Optimization Through Single Well, Pre-Pilot Tests," UNITAR Conference on the Future of Heavy Crude and Tar Sands, Edmonton, Alberta, Canada, 1979.

37. "Joint Heavy-Oil, Tar-Sand Research Planned," The Oil and Gas Journal, June 18, 1979.

38. Kendall, G. H., "Importance of Reservoir Description in Evaluating In Situ Recovery Methods for Cold Lake Heavy Oil," Bull., Canadian Petroleum Geology, Vol. 25, No. 2, May 1977.

39. McCrossan, R. G., Procter, R. M., and Ward, W. J., "Estimate of Oil Resources, Lloydminster Area, Alberta," UNITAR Conference on the Future of Heavy Crude and Tar Sands, Edmonton, Alberta, Canada, 1979.

40. McNally, R., "Enhanced Recovery of Athabasca Oil Sands," Petroleum Engineer International, Nov. 1979.

41. Moncur, R. H., and Sulley, R. R., "Current Policies of the Government of Saskatchewan Regarding Heavy Crude Oil Development," UNITAR Conference on the Future of Heavy Crude and Tar Sands, Edmonton, Alberta, Canada, 1979.

42. Nicholls, J. H., and Luhning, R. W., "Heavy Oil Sand In-Situ Pilot Plants in Alberta (Past and Present)," Journal of Canadian Petroleum Technology, July-Sept. 1977.

43. "Oil Sands and Mining Report," Oil Week, March 24, 1980.

44. "OPEC Prices Make Heavy Oil Look Profitable," Science, Vol. 204, June 22, 1979.

45. Raisbeck, J. M., "Combustion Initiation at the Bottom Water-Oil Sands Interface," Paper CIM 79-30-15 presented at 30th Annual Technical meeting of the Petroleum Society of CIM, Banff, May 1979.

46. Sheperd, D. W., "Predicting Bitumen Recovery from Steam Stimulation," World Oil, Sept., 1979.

47. Shepherd, D. W., "Steam Stimulation Recovery of Cold Lake Bitumen," UNITAR Conference on the Future of Heavy Crude and Tar Sands, Edmonton, Alberta, Canada, 1979.

48. Singhal, A. K., "Physical Model Study of Inverted Seven-Spot Steamfloods in a Pool Containing Conventional Heavy Oil," paper CIM 79-10-13 presented at 30th annual technical meeting of The Petroleum Society of CIM, Baniff, May, 1979.

49. Smith, G. W. and Butler, R. M., "Studies on the Use of Tunnels and Horizontal Wells for the Recovery of Heavy Crudes," UNITAR Conference on the Future of Heavy Crude and Tar Sands, Edmonton, Alberta, Canada, 1979.

50. Starr, J. and Prats, J. M., "Chemical Properties and Reservoir Characteristics of Bitumen and Heavy Oil from Canada and Venezuela," UNITAR Conference on the Future of Heavy Crude and Tar Sands, Edmonton, Alberta, Canada, 1979.

51. Towson, D. E., "The Oil Sands of Alberta: Source of Canada's Future Oil Supply?" Journal of Canadian Petroleum Technology, Oct.-Dec., 1979.

52. Vigrass, L. W., "Trapping of Heavy Oil at Lloydminster," Journal of Canadian Petroleum Technology, Jan.-Mar., 1978.

53. Vigrass, L. W., "Trapping of Oil at Intra-Mannville (Lower Cretaceous) Disconformity in Lloydminster Area, Alberta and Saskatchewan," Bull. AAPG, Vol. 61, No. 7, July, 1977.

54. Whalley, M. J. and Wilson, T. M., "Water Conservation in a Steam Stimulator Project," UNITAR Conference on the Future of Heavy Crude and Tar Sands, Edmonton, Alberta, Canada, 1979.

55. White, W. I., "Heavy Oil Occurrences of the Kindersley Area, Saskatchewan," Oil Sands - Fuel of the Future, Canadian Society of Petroleum Geologists, Calgary, Sept., 1974.

56. Williams, J. A., "40 Years of Heavy Oil Exploration and Production," UNITAR Conference on the Future of Heavy Crude and Tar Sands, Edmonton, Alberta, Canada, 1979.

57. "China Oil Production Prospects," U. S. Central Intelligence Agency, ER77-10030U, June, 1977.

58. Zengxiong, Tang. "A Field Pilot Test to Demonstrate the Production Characteristics of Later Stage of Waterflooding in the Daqing Oilfield," paper, SPE Annual Fall Technical Conference, Dallas, Texas, Sept. 1980.

59. Desprairies, P., Simandoux, P., et.al. "Petroleum Tomorrow: Resources and Recovery," Petrole et Techniques, No. 248, Nov., 1977.

60. Sahuquet, B. C., and Ferrier, J. J., "Steam Drive in a Fractured Carbonated Reservoir Lacq Superieur Field," paper SPE 9453 presented at 55th Annual Technical Conference of Society of Petroleum Engineers, Dallas, Texas, Sept., 1980.

61. Simandoux, Pierre, "Enhanced Oil Recovery: The French Contribution," L' Industrie du Petrol, No. 508, 1979.

62. Bleakley, W. B., "German Producers Pin Hopes on Steam," Petroleum Engineer International, June, 1979.

63. Burkowsky, M. and Marx, C., "The Influence of Dissolved Gas in Crude Oils on the Low-Tension Flood-Process," Erdoel-Erdgas. Z., No. 5, May, 1979.

64. von Hoefling, B., "Potential of Enhanced Oil Recovery Methods in the Federal Republic of Germany," Erdoel-Erdgas-Zeitschrift, 95. Jg., December, 1979.

65. von Hoefling, B., "Secondary and Tertiary Oil Recovery in the Federal Republic of Germany," Erdol-Erdgas.Z, No. 4, Apr., 1978.

66. Hofmann, B., "Development of Methods for Production of Tar-Oil in the Federal Republic of Germany," paper, EEC Symposium on New Technologies for the Exploration and Exploitation of Oil and Gas Resources, Luxemburg, Apr. 18-20, 1979.

67. Pusch, Guenter, "Tertiary Oil Recovery Methods," Erdol und Kohle - Erdgas, January, 1977.

68. Schlemm, V. F., "Planning and Operation of Steam Boiler Units in Thermal Recovery Project," Erdol-Erdgas-Z., July, 1979.

69. "Steam Soak Increases Recovery in Indonesia," The Oil and Gas Journal, Aug. 1, 1977.

70. Chierici, G. L., Dalla Casa, G., and Terzi, L., "Enhanced Oil Recovery by Gas Injection in a Heavy-Oil, Fractured Reservoir - Gela Field, Italy," paper, EEC Symposium on New Technologies for the Exploration of Oil and Gas Resources, Luxemburg, Apr. 18-20, 1979.

71. Lapre, J. F. and Pulga, M., "Development of a Recovery Method for Adriatic Sea Heavy Oil Reservoirs," paper EEC Symposium on New Technologies for the Exploration and Exploitation of Oil and Gas Resources, Luxemburg, Apr. 18-20, 1979.

72. "Japanese Advance Canadian Oil Sands Aquisition," The Oil and Gas Journal, Sept. 22, 1980.

73. Uren, L. C., "Petroleum Production Engineering - Exploration," McGraw-Hill Book Company, Inc., New York, 1953, PP 540-545.

74. Andrianasolo, H., Andrianasolo, M. D., and Raveloson, E., "The Bitumen and Heavy Oil Deposits in Madagascar - Feasibility and Prospects of the Bemolanga Deposit," UNITAR Conference on the Future of Heavy Crude and Tar Sands, Edmonton, Alberta, Canada, 1979.

75. Rivera, J., and Rodriquez, R., "CO_2 - The Most Suitable Method to Increase Oil Recovery in Mexico," Mexican Petroleum Institute Review, July, 1977.

76. Pardo, P., "Heavy Crudes in the Maranon Basin, Oriental, Peru," UNITAR Conference on the Future of Heavy Crude and Tar Sands, Edmonton, Alberta, Canada, 1979.

77. "Peru Aims Funds at Exploration, Development," The Oil and Gas Journal, June 16, 1980.

78. Antonica, I., Stoica, A., and Pacoste, G., "The Romanian Petroleum Industry," Republican Committee for Geological Reserves, Romania.

79. Carcoana, A., "Achievements in Increasing the Final Recovery Rate," Petrol si Gaze, Vol. 29, No. 9, Sept., 1978.

80. Carcoana, A., "Enhanced Oil Recovery in the Romanian Oil Fields," ICPPG - Compina, Romania, Sept., 1979.

81. Lazar, I., "Use of Bacteria in the Recovery Procedures and Mignation of Oil from Deposits," Mine, Petrol si Gaze, 1976, pp 475-480.

82. Sandru, L., Carpeniseanu, D., and Ionescu, I., "Improvement of Crude Oil Recovery by Mining Methods," Research and Design Institute for Crude Oil and Gas, Cimpina, Romania.

83. "WPC Host Country Romania Swaps Tools and Technology for Crude," The Oil and Gas Journal, Aug. 27, 1979.

84. Genca, M., Gurses, B., Kantar, K., Sener, I., and Tek, M. R., "Prospects for Enhanced Recovery from Dati Raman Field, Turkey," UNITAR Conference on the Future of Heavy Crude and Tar Sands, Edmonton, Alberta, Canada, 1979.

85. Abramovich, M. V., "History of the Oil Industry of Azerbaijan," Geologiya Nefti Gaza, Vol, 8, No. 9, 1964.

86. "The Application of Improved Oil Recovery Methods" (in USSR), Unknown.

87. Arzhanov, F. G., Kuvshinov, A. S., et. al. "Estimating the Possible Oil Output in the Exploitation of Russian Fields by Thermal Methods," Neft. Khoz, No. 2, 1977.

88. Ayzenshtadt, G. Y., "History of the Emba Oil Industry," Geologiya Nefti Gaza, Vol. 8, No. 9, 1964.

89. "Azerbaijan Underground Oil Mine Work to Start," The Oil and Gas Journal, May 7, 1979.

90. Babaijan, G. A., et. al. "Results of Theoretical, Experimental and Oil Field Tests of Chemical Agents (SAS, CO_2, PAA) Utilization Efficiency to Improve Oil Recovery," paper presented at Joint Soviet-US Waterflood Symposium, Tyumen, Siberia, USSR, May, 1978.

91. "Balakhany Oil Mine Under Construction," Baku Vyshka, Mar. 13, 1979.

92. Bulgakov, R. T., Talantov, A. V., et. al. "Use of a Thermal Gas Generator to Start the Combustion Process in a Bituminous Seam of the Sequshlin Area," Nefteprom. Delo., No. 12, 1978.

93. Danysh, D. V. and Gurov, Y. I. "Thermal Method of Recovering Petroleum from Mines," Neft. Kohz., No. 9, Sept. 1979.

94. "Developing Automated Oil Field in Balakhany," Baku Vyshka, Mar. 1, 1979.

95. Devlikamov, V. V. "Study of Oil Formation Development Carried Out at the Ufa Petroleum Institute," Izv. Vyssh. Uchebn Zaved Neft Gaz, No. 11, 1977.

96. Dunyushkin, I. I., and Hamiot, A. Y. "Conditions for Miscibility of Petroleum with Carbon Dioxide," Neft Khoz., No. 3, 1978.

97. Dzhamalov, I. M. and Matveenko, I. M. "Oil Overflows in a Stratified Formation upon Cessation of Steam Pumping," Neft. Khoz., No. 4, Apr. 1977.

98. "Economic Effectiveness of New Methods Boosting Stratal Oil Recovery," Moscow Ekon. Neft. Promysh., No. 7, July, 1979.

99. "Emba Oilfields Chosen for Thermal Stimulation Identified," Moscow, Neft. Geol. Geof., No. 3,]979.

100. "Experimental Underground Oil Field," Baku Vyshka, Aug. 28 1979.

101. Gadiev, S. M., Gimatudinov, S. K., and Kuzmichev, Y. A. "Model of Dynamic Effects in the Stratum," Izv. Vyssh Uchebn Zaved Neft Gaz, No. 9, 1977.

102. Gadiev, S. M., Polyanskaya, L. V., et. al. "Vibration Effects on the Critical Zone of Wells in Multi-Stratal Deposits," Izv.Vyssh Uchebn Zaved Neft Gaz, No. 6, 1977.

]03. "Greater Use of Termal Methods in Azerbaijan Oilfields Urged," Baku Vyshka, Mar. 14, 1979.

104. "Heavy Oil Clogs Soviet Output Growth," The Oil and Gas Journal, Jan. 14, 1980.

105. "Hot-Mine Method Applied to Yarega Fields," Moscow Neftyanik, No. 8, Aug. 1979.

106. Ivanov, V. A., Chislov, A. D., et. al. "Results of Pumping a Polymer Solution in the Karazhanbas Field," Neft Khoz, No. 9, Sept., 1978.

107. Kisilenko, B. F., Kennevi, F. A., and Ryzhik, V. M. "Injection of Hot Water to Improve the Oil Porous Formations Containing Highly Viscous Oil," Neft Khoz, No. 12, Dec., 1977.

108. Kisilenko, B. E., Kennevi, F. A., et. al. "Possibility of Improving the Extraction of High-Viscosity Petroleum by Modification of the Mining System," Neft. Khoz, No. 6, June, 1978.

109. Koznetsov, S. I., Ivanov, M. W. and Lalikova, N. N. "Introduction to Geological Microbiology," Akad. Nauk SSSR (USSR Academy of Sciences), Moscow, 1962.

110. Krems, A. Y. "History of Development of the Oil and Gas Industry in Ukhta," Geologiya Nefti Gaza, Vol. 8, No. 9, 1964.

111. Leitner, G. F. "Caspian Seawater Treated for Soviet Steam Flood," Petroleum Engineer International, Nov. 1979.

112. "Mining of Viscous Crude Claimed Viable," The Oil and Gas Journal, Jan. 5, 1976.

113. Muslimov, R. H., and Khammadeev, and Soltanov, S. A. "Practice of Romashinko Oil Field Development in Tatar Republic," paper presented at Joint Soviet - U.S. Waterflood Symposium, Tyumen, Siberia, U.S.S.R., May 1978.

114. Muslimov, R. K., "The Ways of Development of Heavy Oils and Bitumens in the U.S.S.R.," UNITAR Conference on the Future of Heavy Crude and Tar Sands, Edmonton, Alberta, Canada, 1979.

115. Nagornyi, L. A., "Possibility of Intrastratal Combustion in the Tatariya Fields," Neft Khez, No. 10, Oct. 1978.

116. "New Oil Extraction Technique," Kiev Rabochaya Gazeta, December 26, 1978.

117. "Oil Extraction Methods Discussed by Deputy Minister," Moscow Sotsialist. Industriya, Jan. 31, 1979.

118. "Patent of the Yarega Method - Ukhta," U.S. Joint Publication Research Service, No. 857, March 12, 1979.

119. "The Present Status, Principles and Efficiency of Waterflooding in the Oil Fields of the Soviet Union," Unknown, 1979.

120. "Problems at Embaneft' Association," Alma Ata Narodnoye Khoz. Kazak., No. 12, Dec. 1978.

121. "Projection and Results of Application of the Selective System of Water Flooding in the Oil Fields of the Akhtashskaya Area," Unknown.

122. Pukhova, L. A., "Economic Evaluation of Steam-Heat Stimulation of Usinskoye Field Heavy Crude," Ekonomika Neft. Prom., No. 7, July 1979.

123. "Russia Aims to Hike Heavy Oil Recovery," The Oil and Gas Journal, April 12, 1976.

124. Ruzin, L. M., Miqunov, V. I., and Chuprov, I. F., "Possibility of Heating a Fractured Formation of Highly Viscous Oil Underlaid by Water," Neft. Khoz., No. 9, 1977.

125. Savenkov, G. D., and Doroshenko, V. M., "Method of Projecting Technological Parameters of the Process of In-Situ Thermochemical Treatment," Izv. Vyssh. Uchebn. Zaved, Neft. Gaz., No. 4, 1978.

126. Starr, J., and Prats, J. M., "Chemical Properties and Reservoir Characteristics of Bitumen and Heavy Oil from Canada and Venezuela," UNITAR Conference on the Future of Heavy Crude and Tar Sands, Edmonton, Alberta, Canada, 1979.

127. "The State of the Study and Application of the Method for Displacement of Oil by Steam Combined with Water Flooding," Unkown.

128. "Steam, Fire Flood Delays Thwart Soviet Oil Flow," The Oil and Gas Journal, March 26, 1979.

129. Vakhitov, G. G., Bokserman, A. A., et al., "Method of Increasing the Oil Output of Formations by Combining In Situ Combustion with Flooding," Neft. Khoz., No. 8, August 1977.

130. "Work Still Lags on Soviet Heavy Oil Underground Mine," The Oil and Gas Journal, Oct. 15, 1979.

131. Zaidullin, G. N., Saifullin, Z. G., et al., "Effectiveness of Experimental Flooding Using Polymer Solutions," Neft. Khoz., No. 2, 1977.

132. Ballard, J. R., Lanfranchi, E. E., and Vanags, P. A., "Thermal Recovery in the Venezuelan Heavy Oil Belt," Journal of Canadian Petroleum Technology, April-June 1977.

133. Borregales, C. J., Kadi, N., et al., "Espaciamento de Pozos," Paper, Simposio de Crudos Extra-Pesados, Maracay, Venezuela, October 1976.

134. Borregales, C. J., "Inyeccion Alternada de Vapor en la Costa Bolivar," paper, Simposio de Crudos Extra-Pesados, Maracay, Venezuela, October 1976.

135. Borregales, C. J., "Production Characteristics and Oil Recovery in the Oninoco Oil Belt," UNITAR Conference on the Future of Heavy Crude and Tar Sands, Edmonton, Alberta, Canada, 1979.

136. Escojido, D., "Subsidence in the Bolivar Coast," UNITAR Conference on the Future of Heavy Crude and Tar Sands, Edmonton, Alberta, Canada, 1979.

137. Escojido, D., "Summary of In-Situ Combustion Field Project," Maraven, S.A., Caracas, Venezuela, Sept. 1976.

138. Essis, A., and Silva, L. F., "Drilling Evaluation in the Orinoco Oil Belt, 1973-1977," UNITAR Conference on the Future of Heavy Crude and Tar Sands, Edmonton, Alberta, Canada, 1979.

139. Gamsin, S., "Prices Change Orinoco from Neglected Field into Venezuela Bonanza," Oil Daily, Jan. 17, 1980.

140. Herrera, A. J., "Steam Drive Recovery Being Used on Lake Maracaibo Coastal Field," The Oil and Gas Journal, July 17, 1978.

141. Kriut, C., "Sediment-Geological Study of a Steam-Drive Project in the Deltaic River Sands," UNITAR Conference on the Future of Heavy Crude and Tar Sands, Edmonton, Alberta, Canada, 1979.

142. Munoz, J. D., "Numerical Simulation of Steam Drive in the Tia Juana M-6 Project," UNITAR Conference on the Future of Heavy Crude and Tar Sands, Edmonton, Alberta, Canada, 1979.

143. "The Orinoco Belt - Venezuela's Oil Bonanza," Energy Daily, July 30, 1979.

144. Parra, D., "Transport of Heavy Crudes," UNITAR Conference on the Future of Heavy Crude and Tar Sands, Edmonton, Alberta, Canada, 1979.

145. Petroleos de Venezuela, "The Geology of the Orinoco Heavy Oil Belt - An Integrated Interpretation," UNITAR Conference on the Future of Heavy Crude and Tar Sands, Edmonton, Alberta, Canada, 1979.

146. Petroleos de Venzuela, "Practicas de Perforacion, Completacion, Reparacion, e Inyeccion de Vapor," Paper, Simposio de Crudos Extra-Pesados, Maracay, Venezuela, October 1976.

147. Petroleos de Venezuela, "Proyectos Pilotos y Mayores," Paper, Simposio de Crudos Extra-Pesados, Maracay, Venezuela, October 1976.

148. Servello, J. P., "Performance History Improved Recovery by Gas Injection in Heavy Oil Reservoirs, Oveja Field - Eastern Venezuela," UNITAR Conference on the Future of Heavy Crude and Tar Sands, Edmonton, Alberta, Canada, 1979.

149. Terwilliger, P. L., Clay, R. R., et al., "Fireflood of the P2-3 Sand Reservoir in the Miga Field of Eastern Venezuela, Journal of Petroleum Technology, Jan. 1975.

150. Van der Knapp, W., "M-6 Steam Drive Project. Prelimanary Results of a Large Scale Field Test," paper, SPE 9452 presented at 55th Annual Technical Conference of the Socity of Petroleum Engineers, Sept. 21-24, 1980.

151. "Venezuela Turns to Enhanced Recovery to Protect Dwindling Oil Reserves," Energy Daily, Oct. 5, 1979.

152. Ban, A., Balint, V., and Poch, F., "Ammonia Enhances Production from Hungarian Field," The Oil and Gas Journal, July 11, 1977.

153. "BP Finds More Oil Shows West of Shetlands," The Oil and Gas Journal, July 26, 1979.

154. Harmsen, G. J., "Steamflooding in a Water Drive Reservoir in the Schoonbeek Field in the Netherlands," paper presented at the 10th World Petroleum Congress, Bucharest, Romania, Sept. 1979.

155. Jaranyi, I., Kiss, L., et al., "Changes in the Properties of Oil Sands Resulting from Microbiological Treatment," Vortrage der III Intern. Wissen. Konferenz for Geochemie, Mikrobiologie und Erdolchemie, II, Budapest, 1963.

156. Karaskiewicz, Jerzy, "Application of Microbiological Methods for Secondary Oil Recovery from the Carpathian Crude Oil Reservoirs," Pr. Inst. Naft., Cracow, 1974.

157. King, R. F., "Petroleum Exploration and Development in Europe in 1975," Bull. AAPG, Vol. 60, October 1976.

158. King, R. E., "Petroleum Exploration and Production in Europe in 1977," Bull. AAPG, Vol. 62, October 1978.

159. King, R. E., "Petroleum Exploration and Production in Europe in 1978," Bull. AAPG, Vol. 63, October 1979.

160. Nehring, R., "The Occurrence of Heavy Oil in the Middle East," UNITAR Conference on the Future of Heavy Crude and Tar Sands, Edmonton, Alberta, Canada, 1979.

161. Petrovic, K., and Petrovic, J., "Low Quality Heavy Crude as Beneficial Refinery Feedstock Blending Component," UNITAR Conference on the Future of Heavy Crude and Tar Sands, Edmonton, Alberta, Canada, 1979.

162. Rosing, D. E., "Heavy Oil in Western Europe: A Survey," UNITAR Conference on the Future of Heavy Crude and Tar Sands, Edmonton, Alberta, Canada, 1979.

163. Scorcelletti, P. G., "Petroleum Developments in Central and Southern Africa in 1977," Bull. AAPG, Vol. 62, October 1978.

164. Scorcelletti, P. G., and Abbott, B. M., "Petroleum Developments in Central and Southern Africa in 1978," Bull. AAPG, Vol. 63, October 1979.

Part II

Tar Sands

The information in Part II is from *An Assessment of International Tar Sands Recovery and Upgrading Processes* (UCRL Report 15323), prepared by Booz-Allen & Hamilton Inc., for International Energy Technology Assessment Program, Energy and Resource Planning Group, under Glenn Werth, Lawrence Livermore National Laboratory, Livermore, California, under the auspices of the U.S. Department of Energy, December 1980.

I. Introduction

Currently 46 percent of the world's primary energy demands are supplied from petroleum sources with world consumption totaling over 50 million barrels per day. At present production rates, proven petroleum reserves will last only 30 to 40 years, although future discoveries and improved recovery methods are expected to extend production by a number of years. However, most future petroleum production will have to come from increasingly marginal deposits in difficult locations, which will result in increasingly higher production costs.

At present, a severe and costly imbalance exists between petroleum supply and demand because the industrialized countries with high petroleum demand either have minimal oil production or are rapidly depleting their oil reserves. For example, Japan, the United States and Western Europe now consume 60 percent of world oil production but produce only 18 percent of the world's supplies and possess only 10 percent of the world's proven oil reserves. The United States alone imports nearly half of the 17 million barrels it consumes daily which results in an annual out-flow of $90 billion.

This production-consumption imbalance has resulted in the formation of a strong producer's cartel - OPEC - that has the power to set both world oil prices and total production rates. Imported oil prices have increased by a factor of 12 during the last 7 years and unstable political situations in the producing areas have increasingly interrupted oil supplies. Thus, the oil consuming countries are now faced with economic instability and possible future threats to their national security. These forces have combined to give increased urgency to the consuming countries efforts to develop alternative sources of liquid fossil fuels that are based on indigenous resources. Most of the easily developed deposits of higher quality oil have already been developed, so it will be necessary in the future to exploit lower grade oil resources, such as the very heavy oil in tar sands.

However, recovery of bitumen from tar sand has proven technically difficult and financially costly. The very heavy oil that is recovered from tar sands must be moderately upgraded to reduce the oil's viscosity so that it can be pipelined to the refinery. Moreover, in the refinery, the crude must be severely upgraded to produce the required product slate of specification grade liquid products—also a costly and technically difficult process.

With the above in mind, Lawrence Livermore Laboratories requested that Booz-Allen & Hamilton highlight and evaluate foreign technologies that might aid in the future development of similar heavy oil deposits in the United States. This information on foreign technology developments would aid in formulating and directing policy decisions on future research programs within the United States. Emphasis of the study was on post-1978 developing technologies.

The scope of this effort included sizing the resource base and reviewing and evaluating past, present, and planned research and field developments on processes for mining, producing, extracting, and upgrading very heavy oils recovered from tar sands. On the production side, very heavy oil was defined as having a gravity less than 10-12° API and greater than 100,000 centipoise viscosity at 50°F, i.e., bitumen recovery from tar sands where primary production was impossible because of the oil's high viscosity. This definition was not meant to include kerogen in oil shales or the equivalent. On the upgrading side, hydrocarbons whose characteristics dictated additional processing prior to conventional refining into salable products (1050°+ material) were included, regardless of origin in order to encompass all pertinent upgrading technologies.

This report is presented in three major sections with supporting appendices and an executive summary. The characteristics and location of the world's principal tar sands resources are presented in Chapter III. Production methods, both above ground and in-situ, factors inherent in choosing between methods and the commercial and pilot application of those methods are discussed in Chapter IV. Chapter V then presents the factors which affect the process selection for upgrading bitumen and very heavy oils so they can be pipelined and/or refined to a desired mix of consumer products. Chapter V also contains a review of conventional upgrading methods and recent developments, both foreign and domestic. A more detailed discussion of selected U.S. upgrading processes is presented in Appendix I as background. Appendix II contains relevant abbreviations and conversion factors; Appendix III presents the project's bibliography.

II. Executive Summary

The severe and costly imbalance between petroleum supply and demand within the industrialized countries has increased the need for them to develop alternative sources of liquid fossil fuels from indigenous resources. For example, the over 1,440 billion barrels of bitumen in-place in the form of tar sands and heavy oil, which constitutes a significant potential alternative source of liquid fossil fuels, merits close examination, especially as oil prices continue to rise.

Canada has the world's largest tar sand resource base with an estimated 980 billion barrels of bitumen in-place, principally north of Edmonton and in the "carbonate triangle" of Alberta. The second largest resource base is in Venezuela, located largely in the Orinoco Heavy Oil Belt in the eastern part of the country. It contains some 700 billion barrels of in-place heavy oil with quantities ranging from 8° to 18° API. Other principal tar sands resources are located in parts of the United States (especially Utah), Malagasy, Albania, Trinidad, Rumania, and the U.S.S.R.

The geologic and chemical properties of the world's tar sands deposits vary considerably. Similarly deposits vary in their overburden thickness -- a key to selecting the production process. As illustrated in Exhibit II-1, and discussed further in Chapter IV, deposits with less than 150 feet of overburden can usually be surfaced-mined and the bitumen recovered by the hot water extraction process. Deep deposits with more than 500 feet overburden must be produced with in-situ recovery methods. In-situ recovery methods usually involve some method for heating the formation (such as steam injection or in-situ combustion) to reduce the bitumen viscosity. Moderately deep deposits (150-500 feet) are frequently not suitable for either surface mining or in-situ recovery. Underground mining for producing bitumen methods from moderately deep deposits are being developed in the U.S.S.R. The status of foreign development of these various production methods is shown in Exhibit II-2.

Numerous but common generic variables must be considered when selecting the optimum recovery method for a country's specific tar sand deposit. The selection criteria fall into three general groups: site-specific factors, process-related factors, and external factors, all of which are discussed in detail in Chapter IV. Site-specific factors include physical characteristics, geographic location, and bitumen quality of the deposit, and the availability of fuel and water. Process-related factors, which must be weighed in relation to the deposit, include process efficiency, reliability and development status, required investment, and estimated operating costs. Assuming the site-specific and process-related variables are favorable, the project must next be evaluated carefully against a number of external factors.

EXHIBIT II-1 Classification of Bitumen and Very Heavy Oil Production Processes

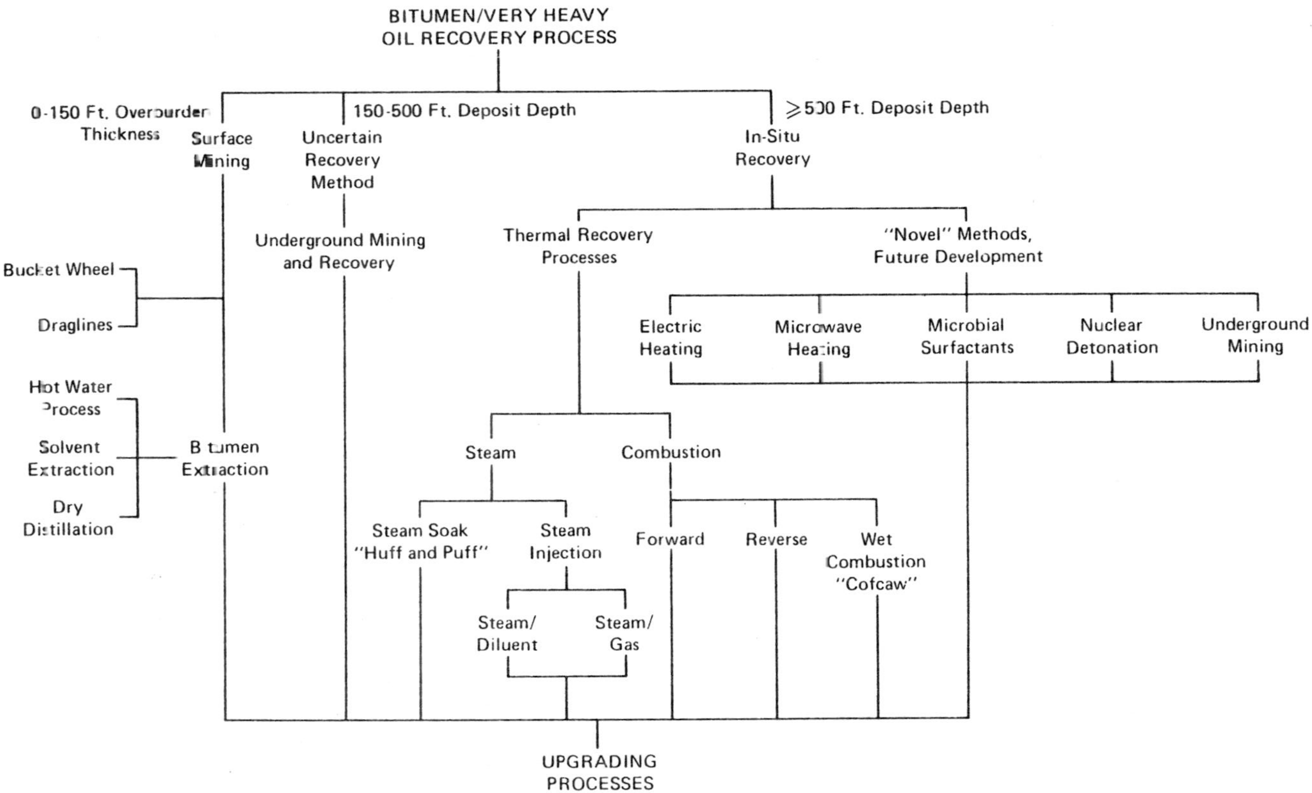

EXHIBIT II-2

Bitumen and Very Heavy Oil Production Methods: Status of Foreign Development

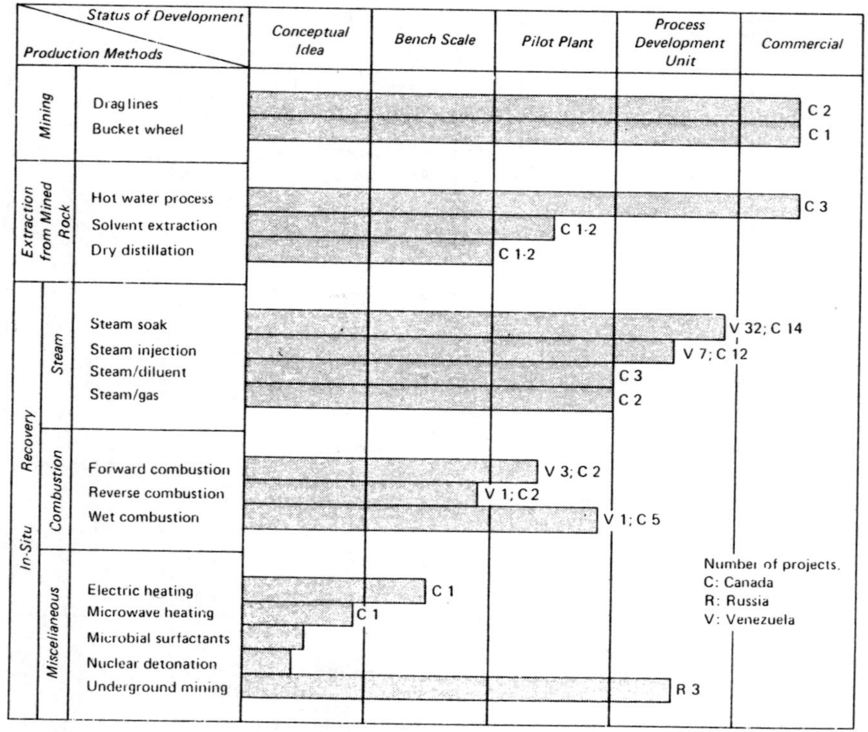

The external factors are determined by the overall international and domestic business and economic climate that would either accelerate or retard tar sands development. These include inflation, taxes and royalties, projected world oil prices, interfuel competition, future environmental constraints, financial risk, and alternative investment opportunities.

Of the nations surveyed, currently only Canada has commercial tar sands plants operating. The two plants, Great Canadian Oil Sands (now called Suncor), 65,000 b/d and Syncrude Canada, 125,000 b/d, are based on surface mining and hot water extraction. Two additional commercial projects are in the construction or planning stage: Alsands 140,000 b/d and Esso Cold Lake 160,000 b/d. Continuation of these projects is presently in doubt because of a disagreement between the federal and provincial governments regarding the price of oil and royalties as well as project economics. For example, the investment required for tar sands projects has escalated drastically from $8,000 to $63,000 per daily barrel capacity from 1968 to 1987 (current dollars), largely as a result of inflation and the more severe environmental regulations.

Only about 5 percent of the world's tar sands deposits can be recovered by surface mining and the remaining 95 percent must be recovered by in-situ recovery methods or by other methods. As a result, it is not surprising that several large scale in-situ pilot projects are underway in Canada, Venezuela, and the U.S. These pilot projects are based principally on thermal recovery methods including steam injection, in-situ combustion, or heating the formation with electric power. In-situ pilot projects require significant investments and considerable time before production results are fully known. For example, the Canadian in-situ pilot projects require investments of $20 to $70 million and have time schedules ranging from 3 to 8 years between the project planning stage and proven production results. U.S. projects are expected to follow similar patterns. The U.S.S.R. has relied extensively on underground mining but is expected to shift to thermal recovery methods.

Heavy oil and tar sands bitumen as defined in the introduction, regardless of country of origin, differ from conventional refinery feedstocks in that they are lower in gravity and contain significantly less hydrogen and more carbon than most crudes. Heavy oils are characterized by a high asphaltene fraction that contains a disproportionately high level of contaminants such as sulfur, nitrogen, and heavy metals. The near crystalline structure of the asphaltenes makes removal of these contaminants particularly difficult. The low gravity and hydrogen-to-carbon ratio contribute to the oils' high viscosity and high pour point. This often necessitates heating, dilution, or other treatments just to allow the oil to be transported.

Conventional refinery processes such as distillation, coking, and asphalt blowing, which typically convert the bottom of the barrel to useful products, are not well suited to heavy oil processing. The high proportion of non-distillables in heavy oils necessitates larger capacities for residue processing than would be needed for a conventional refinery with an equivalent volume of throughput. An increased volume of products from these processes would saturate the already limited asphalt and cement industry markets. In addition, the high contaminant levels in the residue streams result in product coke qualities below acceptable market levels or in residue fuels that require costly flue gas desulfurization. Additional processing or upgrading steps are required before these heavy oils can be converted to valuable liquid products. The generic processing options available are discussed in Chapter V and are presented schematically in Exhibit II-3 with the possible disposition of heavy oil upgrading processes depicted in Exhibit II-4.

Upgrading increases the hydrogen/carbon ratio in bitumen and very heavy oil to the point where the material can be sold or used as finished products or processed further via conventional refining techniques. Most upgrading processes are conventional bottom-of-the-barrel processes that have been designed or modified to cope with the poor characteristics of heavy oil. Upgrading can be performed by either removing carbon (Exhibit II-5) or by adding hydrogen to remove contaminants (Exhibit II-6), or both.

An overview of tar sands and heavy oil upgrading processing is presented in Exhibit II-7, which shows that traditional bottom of the barrel processes are generally either catalytic or noncatalytic. Physical separations such as distillation and solvent deasphalting, thermal cracking processes like visbreaking and coking, and gasification processes are typically noncatalytic. Fluid cat cracking and most hydrogen addition processes are catalytic. Noncatalytic processes in general fail to mitigate the problem of finding a disposition for the large volumes of residual by-product. Catalytic processes have had only limited success in achieving continuous cost effective operation for the range of feedstocks that covers heavy oils and bitumen.

New approaches to upgrading, both foreign and domestic in origin, usually incorporate variations and combinations of conventional processing methods to overcome the traditional barriers imposed by low-quality carbon by-products and highly sensitive catalysts. The Japanese Eureka process, for example, uses steam to prevent coking in its thermal cracking process. The resultant liquid pitch may have more uses than low-quality coke. Direct coking separates the bitumen from tar sand and yields upgraded liquid products in one step. This method eliminates the need for hot water extraction and consumes the coke it produces as process fuel. Alternatively, Flexicoking and Dynacracking (U.S. processes) and a new Japanese process, RAROP Resid Cracking, each combine variations of coking with coke gasification to solve the coke disposition problem and provide liquid fuels.

EXHIBIT II-3

Generic Processing Options for Converting Very Heavy Oils to Finished Products

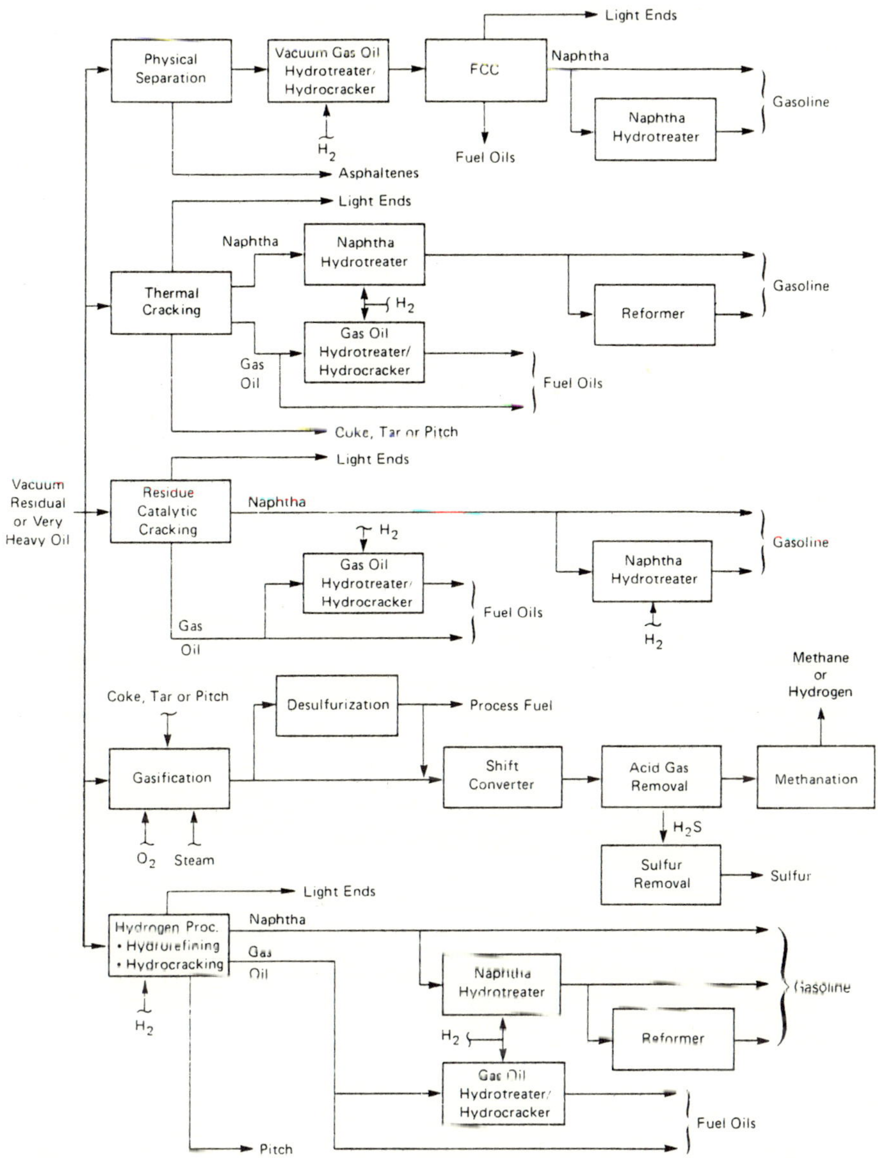

EXHIBIT II-4

Possible Dispositions of Heavy Oil Upgrading Products

	DISPOSITION	
UPGRADING PRODUCT	WITHIN PLANT	MARKET
Light Ends (C_1, C_2, C_3, C_4)	• Process Fuel • Gas Plant • Hydrogen Production	Petrochemical Manufacture, LPG, Natural Gas – –
Light Distillates (C_4, Naphtha, Reformate)	• Gasoline Blending • Petrochemicals Manufacture • Process Fuel	Gasoline Petrochemicals –
Middle Distillates (Kerosene, Diesel, No. 2 Oil)	• Distillate Blending	Fuel[1]
Heavy Distillates (Gas Oils, Resins)	• FCC/Hydrocracker Feed • Fuel Oil Blending • Lube Plant • Process Fuel	Fuels, Mineral Oils & Lighter Products Waxes & Lubricating Oils –
Heavy Residues (Resins & Asphaltenes)	• Fuel Blending • Lube Plant • Asphalt Blending • Gasification • Process Fuel	Fuel Oils Lubricating Oils, Petrolatum Asphalt, Road Oils Methane[2] –
Coke, Tars, Pitch	• Petrochemicals Manufacture • Gasification • Process Fuel	Fuels, Electrode Coke, Carbon Black Methane[2] –
Sulfur	–	Sales
Other (Catalysts, Drag Streams)	• Metals Recovery	Sales Disposal

(1) Fuel includes Jet Fuels, Kerosene, Diesel, No. 2 oil.

(2) Methane represents possible export product, others (i.e., hydrogen and fuel gas) are more suited to process use.

EXHIBIT II-5 Carbon Removal Processes

PROCESS NAME	PROCESS TYPE	DEVELOPER	STAGE OF DEVELOPMENT[1] EXISTING TYPE[2]	EXISTING (NUMBER)	EXISTING START-UP DATE[3]	PLANNED TYPE	PLANNED (NUMBER)	PLANNED START-UP DATE	FEEDSTOCKS	KEY PRODUCTS
PHYSICAL SEPARATION										
Vacuum Distillation	Vacuum Distillation	Many, Non-Proprietary	C	Many					Atmospheric Residue	Vacuum Residue
Solvent Deasphalting	Solvent Deasphalting	Many	C	(41)					Vacuum Residue Whole Bitumen	Bright Stocks Cracking Feedstocks
Demex	Solvent Deasphalting	UOP Process Division	C	(3)		C	(2)		Vacuum Residue	Demetallized Oil
Residuum Oil Supercritical Extraction (Rose	Solvent Deasphalting	Kerr McGee	C	(2)	1979	C	(1)	1981	Atmospheric & Vacuum Residue (shale oil & tar sands w/pilot plt.)	Asphalt, Lube Oils Resins, Cat Cracking Feed
Asphalt Blowing/Bitumen Manufacture	Bitumen Manufacture	Snamprogetti S.p.A., BP Trading Limited	C						Vacuum Residue	Bitumen, Asphalt
THERMAL CRACKING[4]										
Visbreaking	Mild Thermal Cracking	Non-Proprietary	C	(20+)		C	(Several)		Atmospheric & Vacuum Residue	Middle Distillates & Fuel Oil, Improved Pourpt. & Lower Viscosity
Delayed Coking	Coking	Many	C	(50+)	Pre-1960				Vacuum Residue, Tar Sands Liquids	Coke, Gas, LPG, Naphtha & Gas Oils
Fluid Coking	Coking	Exxon Research & Engineering Co.	C	(9)	Pre-1960				Vacuum Residue, Tar Sands Liquids	Coke, Gas, LPG, Naphtha & Gas Oils
Eureka Vacuum Residue Cracking	Thermal Cracking/Steam Stripping	Eureka Chemical and Chiyo Chiyoda Chemical Engineering & Construction Company	C	(1)	1975	C	(1)	1982	Vacuum Residue	Aromatic Pitch & Cracked Oils
ERCO Coker	Direct Fluid Coking	ERCO, Bechtel	BS	(1)	1979	C	(1)	1984	Tar Sand, Water	Synthetic Crude, Putty-like Coke Slurry
LR Gas	Direct Coking	Lurgi	PP	(1)	1979				Tar Sand, Heavy Oil	Med. Btu Gas, Naphtha, Middle Distillates
Taciuk	Direct Coking	UMATAC Industrial Processes Ltd.	PP	(1)		D	(1)	1982	Tar Sand, Heavy Oil	Med. Btu Gas, Naphtha, Middle Distillates
Mobil Ether Solvent	Etherification/Blending/Cracking	Mobil	BS		1976				Bitumen, Vacuum Residue, and Methyl Tertiary Butyl Ether	Butanes, Gasoline
MCC	Fluid Cracking	Mitsubishi Petrochemical and Alberta Research Council	BS, PP	(1), (1)					Bitumen, Vacuum Residue	Olefins (i.e., ethylene, propylene, benzene)
COSMOS	Thermal Cracking with Steam & Molten Salts	Mitsui Petrochemical Ind. & Mitsui Engineering & Shipbuilding Company	BS	(1)	1977	PP	(1)		Low Sulfur Crudes	Olefins

(continued)

EXHIBIT II-5 (Cont.)

PROCESS NAME	PROCESS TYPE	DEVELOPER	STAGE OF DEVELOPMENT[1] EXISTING TYPE[2]	EXISTING (NUMBER)	EXISTING START-UP DATE[3]	PLANNED TYPE	PLANNED (NUMBER)	PLANNED START-UP DATE	FEEDSTOCKS	KEY PRODUCTS
CATALYTIC CRACKING[5]										
Heavy Oil Cracking (HOC)	FCC	Pullman Kellogg and Phillips Petroleum	C	(1)	1961	C	(1)	1980	Atmospheric & Vacuum Residue	Gasoline, Fuel Blending Stock, LPG Carbon Black Feed
Resid Cat Cracking	FCC	Total Petroleum Inc.	C	(1)	1976	C	(2)	1981, 1982	Atmospheric Residue	Gas, Gasoline, Middle Distillates
Reduced Crude Conversion (RCC)	FCC	Ashland Oil Company, UOP	D	(1)	1978	C	(1)	1982	Atmospheric Residue	LPG, Gasoline, Fuel Oils
RAROP Resid Cat Cracking	FCC	Nippon Oil Company	BS	(1)	1979	PP	(1)		Atmospheric Residue	LPG, Gasoline, Fuel Oils
GASIFICATION										
Shell Gasification Process	Partial Oxidation	Shell Development	C	(5+)	1958	C			Vacuum Residue, Asphalt	Hydrogen, Med. Btu Gas
Texaco Partial Oxidation	Partial Oxidation	Texaco	C	(35+)	1956	C			Vacuum Residue, Asphalt	Hydrogen, Med. Btu Gas
M-Gas	Catalytic Gasification	Mitsui Engineering & Ship Bldg. and Mitsui Kozan Coking Ind.	BS	(1)		PP	(1)		Vacuum Residue, Asphalt	Medium Btu Fuel Gas
Flexicoking	Coking & Gasification	Exxon Research & Engineering Co.	C	(1)	1976	C	(1-3)	1982	Vacuum Residue, Asphalt	LPG, Naphtha, Gas Oils, Low BTU Gas
Dynacracking	Coking, Gasification, Cracking	Hydrocarbon Research Inc.	PP	(1)	Pre-1955	D	(1)	1982	Vacuum Residue High Sametals	Distillates Naphthas & High Quality Synfuel Gas
RAROP Resid Cracking and Gasification	Coking, Gasification, Cat Cracking	Nippon Mining Co., Kashima Oil Co., Hitachi Ltd.	BS	(1)	1979	PP	(1)	1981	Vacuum Residue	Middle Distillates & Hydrogen
DIRECT COMBUSTION										
	Steam Generation								Vacuum Residue, Heavy Oil	Steam
	Bunker Fuel Marine Transport								Bunker C	
Exxon Chemically Active	Fluidized Bed Combustion	Exxon Research & Engineering Co.				PP			Vacuum Residue	

(1) Known locations and sizes of existing and planned facilities are included in process descriptions.
(2) Plant types: BS – Bench Scale: Laboratory unit, usually not continuous, capacity <1 b/d.
PP – Pilot Plant: Continuous, capacity varies from 0.5 - 1,000 b/d.
D – Demonstration: Minimum commercial scale, product markets needed, operation focus is in test mode to demonstrate commercial reliability, capacity varies from 1,000 - 10,000 b/d.
C – Commercial: Economic commercial scale operation for profit, capacity $>5{,}000$ b/d.
(3) Date of earliest plant built to accommodate high-carbon feedstocks.
(4) Gasification and direct combustion process are categorized separately.
(5) Many companies are developing catalysts and/or process modifications to handle heavier feeds in existing FCC units, and in new process configurations.

EXHIBIT II-6 Hydrogen Addition Processes

			STAGE OF DEVELOPMENT							
			EXISTING			PLANNED				
PROCESS NAME	PROCESS TYPE	DEVELOPER	TYPE[1]	(NUMBER)	START-UP DATE[2]	TYPE	(NUMBER)	START-UP DATE	FEEDSTOCKS	KEY PRODUCTS
HYDROREFINING										
RDS & VRDS	Desulfurization	Chevron Research Company	C	(1)	1976	C	(2))	1982 1983	Atmospheric & Vacuum Resid & Whole Crude	Light & Heavy Naphthas & Gas Oils & Resid
Resid HDS	Fixed Bed Catalytic	Gulf Research & Development Co.	C	(6)	1970				Atmospheric & Vacuum Residue	Fuel Oil, Cat Cracking Feedstock/ Upgraded Resid
Residfining	Fixed Bed Catalytic	Exxon Research & Engineering Co.	C	(2)	1977				Atmospheric & Vacuum Residue	Distillates, Gas Oil, FCC Feedstock
Unicracking/HDS	Fixed Bed Catalytic	Union Oil of California	C	(2)	1976	C	(2)	1983	Atmospheric Residue	Low S Fuel Oil, Cat Cracker Feedstocks
BOC Unibon	Fixed Bed Catalytic	UOP Process Division	C	(4)	1967				Vacuum Residue, All Crude Fractions	Lt., & Hvy. Gas Oil & Pitch
RCD Unibon	Fixed Bed Catalytic	UOP Process Division	C	(1)	1973				High Metal Vacuum Resid	Hvy. Fuel Oil, Cat Cracking Feedstocks
Shell HDS	Moving Bed Catalytic	Shell Development Company	C	(2)	1973	D[3]	(1)	1981	Atmospheric & Vacuum Residue High Metals & Sulfur	Fuel Oils, Upgraded Resid
Aurabon		UOP Process Division	PP	(1)		D	(1)	1983	Heavy Oils, High Metals & Asphaltene	High Quality Syncrude
HYDROCRACKING										
LC Fining	Expanded Bed Catalytic	CE Lummus & Cities Service	C	(2)	1963				High Metals Vacuum Resid & Bitumen	Fuels & Synthetic Crude
H-Oil	Expanded Bed Catalytic	Hydrocarbon Research Inc.	C	(4)	1963	C	(1)		High Metals Vacuum Resid & Bitumen	Synthetic Crude, Liquid Fuels, Pitch
Kombie Kammer	Non-Catalytic	Veba Oil S.A., Lurgi Corp	C	(1)	Pre-1945				Coal Liquids	Liquid Hydrocarbons
Canmet	Expanded Bed Non-Catalytic	Canadian Energy Mines & Resources Dept./Petro Canada	PP	(1)	1970	D	(1)		Heavy Oil	Fuels & Syncrude
Asphaltene Cracking	Fixed Bed Catalytic	Chiyoda Chemical Eng. & Const. Co.	BS	(1)	1978				Asphaltene Residue	Pour Point Reduction Synthetic Crude
HFC	Moving Bed	Niigata Engineering Co., Ltd.	PP	(1)	Pre-1977				Atmospheric & Vacuum Residue	Syncrude
RAROP Hydro cracker	Fixed Bed	Idemitsu Kosan Company, Ltd.	BS			PP	(1)	19[illegible]	Vacuum Residue	Fuel Gas, Naphtha, Middle Distillates, Residual Oil
OTHER HYDROGEN ADDITION										
Molten Salt Cracking	Catalytic Cracking w/Zinc Chloride Melt	Various	BS						Coal	Gasoline

(1) Plant Types
- BS – Bench Scale: Laboratory unit, usually not continuous, capacity < 1 b/d.
- PP – Pilot Plant: Continuous, capacity varies from 0.5 - 1,000 b/d.
- D – Demonstration: Minimum commercial scale, product markets needed, operation focus is in test made to demonstrate commercial reliability, capacity - 1,000 - 10,000 b/d.
- C – Commercial: Economic commercial scale operation for profit capacity > 5,000 b/d.

(2) Date of earliest plant built to accommodate high-carbon feedstocks.

(3) Demonstration as Hydrodemetallization Process in Venezuela's Carbon Refinery.

EXHIBIT II-7 Tar Sands and Heavy Oil Upgrading Process Comparisons

PROCESS	LIGHT ENDS YIELD (WT %)	LIQUIDS YIELD (WT %)	LIQUIDS QUALITY[1]	RESIDUE YIELD (WT %)	RESIDUE QUALITY	GAS CONVERSION (WT %)	GAS QUALITY	PRIMARY CHARACTERISTICS	RELATIVE ADVANTAGES	RELATIVE DISADVANTAGES
Solvent Deasphalting	None	55-80	L	20-45	L	None		Feed preparation step required	• High liquid yield • Separates metals from DAO • Simple	• Residue disposition required • DAO below pipeline quality, requires additional processing • DAO high in sulfur • High fuel cost • Not proven on bitumen
Visbreaking	<5	10-20	L	70-80	L	None			• Simple	• Low liquid yield • Substantial additional processing needed
Delayed Coking	8-12	60-70	L	22-32	L	None		Semi batch, thermal cracking; coke by-product	• Proven reliability • Full range feedstock flexibility	• Low liquid yield • Coke high in impurities and difficult to use • Liquids may require hydrotreating to saturate olefins • Complex coke removal
Fluid Coking	10-13	56-74	L	9-24[2]	VL	None		Most amenable to large projects; orientation to field construction	• Proven reliability • Full range feedstock flexibility • Energy self sufficient • Less coke than delayed coking	• Coke high in impurities and difficult to use • Flue gas desulfurization needed • Complex operating
Eureka	5	70-76	M	19-26	M	None		Thermal cracking plus steam stripping yields a liquid pitch	• High liquid yield • Proven reliability • Full range feedstock flexibility • Pitch may have advantages over coke • Simple	• Pitch high in impurities • Liquids may require hydrotreating
Lurgi Ruhrgas	5-15	65-80	L	-	-	-	-	Direct coking	• Replaces several separation steps in bitumen recovery	• Equipment required to cool sand • Unproven in bitumen recovery
Taciuk	5-15	65-80	L	-	-	-	-	Direct coking	• Replaces several separation steps in bitumen recovery • Lower Temperatures	• Unproven • Equipment required to cool sand • Unproven
Flexicoking	10-13	56-74	L	1-4[3]	VVL	9-26	Very low Btu	Coking plus gasification, fuel gas produced rather than coke; most amenable to very large projects; orientation to field construction	• No significant coke by-product • Demonstrated commercially • Fuel gas produced may back out valuable fuels • Wide range feedstock flexibility	• Fuel gas is low quality; requires modification of combustion equipment • Liquids may require hydrotreating • Flue gas desulfurization may be required • Unproven on bitumen • Complex operating
Dynacracking		20-75	M-H	Very Little		20-80	Med. Btu	Coking and gasification + hydrocracking, fuel gas and hydrogen produced rather than coke, liquid products more stable	• No significant coke by-product • High naphtha production • Fuel gas is usable with few equipment modifications • Wide range feedstock flexibility • Potential investment savings over flexicoking • Liquid products may not require hydrotreating	• Unproven • Oxygen plant needed • Significant compression required • High light end production, limited middle distillate production • Residue by-product?
RAROP Catalytic Cracking and Gasification	15-25	50-65	M-H	-	-	20-25	-	Fluid cracking plus gasification; produces hydrogen	• High distillate yield • No apparent coke by-product • Hydrogen production contributes to process self-sufficiency	• Unproven • Flue gas desulfurization needed • High light ends production • Significant compression required

[1] Quality Classifications: VL Very low; L Low; M Medium; H High

[2] Net coke make, 25-40% of gross is consumed as fuel.

[3] Net coke make, 95-99% of gross coke is gassified or consumed as fuel

(continued)

EXHIBIT II-7 (Cont.)

PROCESS	LIGHT ENDS YIELD (WT %)	LIQUIDS YIELD (WT %)	LIQUIDS QUALITY	RESIDUE YIELD (WT %)	RESIDUE QUALITY	GAS CONVERSION (WT %)	GAS QUALITY	PRIMARY CHARACTERISTICS	RELATIVE ADVANTAGES	RELATIVE DISADVANTAGES
MCC Thermal Cracking	–	–	H[4]	>10	–	–	–	Fluid thermal cracking; produces olefins	• Produces petrochemical feedstocks • Little or no coke by-product	• Special use products; no fuels • Unproven commercially • Flue gas desulfurization needed • Special product handling needed • Limited to small projects or in conjunction with other upgrading processes
Cosmos	30-40	–	H[4]	>10	–	15-20	–	Produces olefins; coke is burned in internal combustion	• Produces petrochemical feedstocks • Little or no coke by-product	• Special use products; no fuels • Unproven commercially • Flue gas desulfurization needed • Special product handling needed • Limited to small projects or in conjunction with other upgrading processes • Strong feed sulfur limitation
HFC	1-3	60-70	H	25-35	M	None		Hydrocracking with spent desulfurization catalyst	• Good liquid yield • Low hydrogen consumption • Wide range of feedstock flexibility; high metal tolerance • May not require secondary processing • Cheap source of catalyst	• Higher conversions may require additional processing • Hydrogen required • Unproven • H_2S removal needed • High capital and operating costs
CANMET	2-6	70-94	H	4-15	M	None		Non catalytic hydrocracking employs additive	• High liquid yield • Wide range of feedstock flexibility • High conversion of heavy residual material • Pitch drag stream may have application as binder for metallurgical coke	• Drag stream must be disposed of • Hydrogen required • H_2S removal needed
Asphaltene Cracking	–	80-98	H	0-15	M	None		Catalytic hydrocracking; recycles asphaltenes to extinction with the help of solvent deasphalting	• No carbon residue[5] • High liquid yields • Wide range of feedstock flexibility • May be used as sole process in simple upgrading application • Moderate hydrogen consumption	• Hydrogen required • H_2S removal needed • Catalyst replacement and disposal needed • Not demonstrated commercially • High capital and operating cost
RAROP Hydrocracker	–	–	–	–	–	–	–	Fixed bed hydrocracker; experimental	• Assumed high liquid yield	• Hydrogen required • Catalyst required • Unproven • Assumed high capital and operating cost (from hydrogen, catalyst and compression)
Kombie Kainmer	–	–	–	–	–	–	–	Hydrocracker, comparable to Canmet; developed in the 50's	• Assumed high liquid yield • Demonstrated commercially at one time	• Little data
H-Oil/LC-Fining	5-10	70-87	H	5-20	L	None		Expanded bed hydrocracking; lighter products than with fixed bed for same sulfur levels	• High liquid product yield • May not require secondary processing • Reasonable feedstock flexibility • Moderate H_2 consumption • Provision for catalyst replacement • Proven commercially	• High capital and operating costs • Liquid products unstable at high conversions • Hydrogen source requires stability • Pitch by-product with questionable stability • Catalyst susceptibility to metals poisoning • Complete operation • Some feedstock limitations • Catalyst disposal • Not commercially proven on bitumen • H_2S recovery needed

[4]Specialty product: petrochemicals [5]With solvent deasphalting to recycle asphaltenes

(continued)

EXHIBIT II-7 (Cont.)

PROCESS	LIGHT ENDS YIELD (WT %)	LIQUIDS		RESIDUE		GAS		PRIMARY CHARACTERISTICS	RELATIVE ADVANTAGES	RELATIVE DISADVANTAGES
		YIELD (WT %)	QUALITY	YIELD (WT %)	QUALITY	CONVERSION (WT %)	QUALITY			
Fixed Bed HDS	2	40-65	H	29-55	M	None		Hydroprocessing higher desulfurization and less gravity improve-e-ment than expanded bed	• Simple • Good liquid yield • High degree of desulfurization • Residue is suitable as fuel or quality coker feedstock	• Additional processing necessary for high liquid yields • Susceptible to plugging desalting and filtration are necessary pretreatment steps • High capital and operating costs • Not proven on Bitumen • H_2S recovery needed • Catalyst Disposal
Shell HDS/HDM	2	40-70	H	25-55	M	None		Similar to fixed bed system Incorporates "trickle flow" or moving bed to allow on stream catalyst replacement.	• Good liquid yield • Provision for catalyst replacement • High degree of desulfurization • Residue is suitable as fuel or coker feedstock • Commercially proven as residue desulfurization	• Additional processing necessary for high liquid yield • Hydrogen required • High capital and operating costs • Not proven on bitumen or as demetallization process • H_2S removal needed
Aurabon	2-4	40-55	H	30-50	M[6]	None		Primarily feed pretreatment step for secondary processing, feed metals provide catalyst	• Metals in feed act as catalyst • Product suitable for wide range of processing flexibility	• Hydrogen needed • May be economically limited to hgih metal feeds • Unproven • High metal and sulfur drag stream; 2-5 wt % of feed
Texaco Partial Oxidation	None	None		Little		–	Varies	Gas production; most suitable for residue by-products	• Proven technology • Full range feedstock flexibility • No residue by-product • Useful fuel or hydrogen	• No liquid yield • Oxygen plant required • Unproven on bitumen • H_2S removal needed
Shell Gasification Process	None	None		Little		–	Varies	Gas production; most suitable as secondary process for residue by-products	• Proven technology • Full range feedstock flexibility • No residue by-product • Useful fuel or hydrogen	• No liquid yield • Oxygen plant required • Unproven on bitumen • H_2S removal needed
M-Gas	None	None		Little		65-75	H	Gas production; most suitable as secondary process for residue by-products	• Potential source of quality hydrogen	• No liquid yield • Unproven • H_2S removal needed • Oxygen plant required
Heavy Oil Cracking	5-8	65-75	M	None		(Steam)	–	FCC; steam produced rather than coke	• No coke by-product • High naphtha yield • Commercial	• Feed pretreatment necessary • Flue gas desulfurization needed • Unproven on bitumen • Use for steam needed • Catalyst required • Liquids may require hydrotreating
RAROP Resid Cat Cracking	–	–	–	–	–	–	–	FCC; steam produced rather than coke	• May have novel catalyst regeneration method	• Flue gas desulfurization needed • Liquid yield limited • Unproven

[6] High metal high sulfur drag stream

Developments in catalytic processes include more tolerant catalysts, additives to reduce the harmful effects of contaminants on catalysts and improved methods for regenerating catalysts. Upgrading can take place in stages with each stage using a specialized, highly selective catalyst. Metals and solids are removed in the first stage, sulfur in the second, and conversion takes place in subsequent stages.

As seen in Exhibits II-5, II-6, and II-7, U.S. upgrading technology has played a primary role to date. Of the over 15 foreign technologies reviewed, Canada, Japan, and Germany have made the greatest strides in developing very heavy oil processing technology, while Venezuela is relying heavily on U.S. technology. Canada has been motivated in part by its vast reserves of tar sands, and the experience it has gathered in developing specific deposits. Venezuela also possesses large reserves but has limited technical resources and views the implementation of foreign technology as the quickest, most reliable method for developing its heavy oil potential. Japan is very dependent on heavy foreign crudes and is developing the processing flexibility to cope with a greater range of crude qualities. Germany has leveraged historic processing knowledge and markets its technology abroad. Little is known of Eastern Bloc contributions.

Regardless of country, the proper upgrading method for a particular application is the process, per Exhibit II-8, that provides for the maximum recovery of marketable petroleum products, for a specific tar sand deposit, at a minimum cost. Selection must be based on a detailed evaluation of the combined production, upgrading, and product distribution and marketing scheme. In most cases, neither the production method, nor the upgrading process can be selected independently of each other. As in production, the factors that influence the upgrading process selection can be categorized in three areas. Site-specific factors refer to the production site and the production process requirements; market-specific factors define the product market needs and the existing refinery constraints, process factors match the resources and requirements of the site with the needs and requirements of the market.

Exhibit II-9 summarizes and compares the numerous variables that must be analyzed and weighed before proper selection of either a foreign or domestically developed upgrading process can be made for a firm's particular product-market feedstock combination. The ideal process must overcome feedstock, product, resource, technology, environmental and economic barriers within the confines of public policy. The factors that will most influence the selection of one process over another are process reliability, product market price differentials, production site, recovery method employed and overall risk as perceived by the project's investors. While there are many promising foreign technology processing options under development, particularly in Japan, few beyond the Shell HDS/HDM and Canadian CANMET have breached the barriers to commercialization.

EXHIBIT II-8

Heavy Oil Upgrading Process Selection

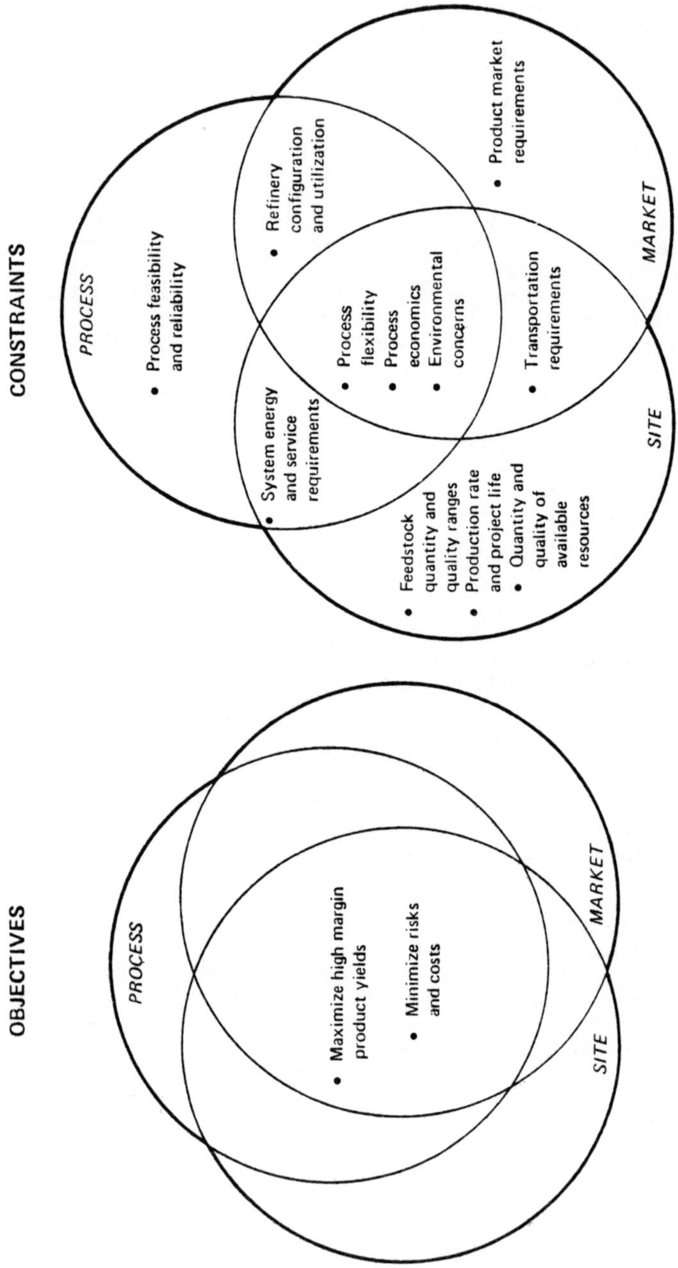

EXHIBIT II-9
Effectiveness of Processes in Overcoming Barriers to Commercialization

BARRIERS TO CONVENTIONAL REFINING / PROCESS	FEEDSTOCK BARRIERS[1]				PRODUCT BARRIERS			ENVIRONMENTAL BARRIERS	
	SULFUR	NITROGEN	METALS	ASPHALTENES	LIQUID YIELD	DISTILLABLE PRODUCT QUALITY	RESIDUAL PRODUCT QUALITY	SO_2	PARTICULATES
Solvent Deasphalting	●	●	●	●	◐	○	○	◐	●
Visbreaking	●	●	●	◐	○	○	○	◐	●
Delayed Coking	●	●	●	●	◐	○	○	◐	●
Fluid Coking	●	●	●	●	◐	○	○	○	◐
Eureka	●	●	●	●	◐	○	◐	◐	●
Lurgi/Ruhrgas	●	●	●	●	◐	○		○	○
Taciuk	●	●	●	●	◐	○		○	○
MCC	◐	◐	◐	●	●	○		○	●
COSMOS	○	◐	◐	○	●	○		◐	●
Flexicoking	●	●	●	●	◐	○	○	○	◐
Dynacracking	●	●	●	●	◐	●		◐	◐
RAROP Catalytic Cracking	?	?	◐	●	●	●		○	?
Heavy Oil Cracking	●	●	◐	◐	◐	◐		○	○
RAROP ResidCat Cracking	?	?	?	?	◐	◐		?	?
Texaco Partial Oxidation	●	●	●	●				●	●
Shell Gasification Process	●	●	●	●				●	●
M-Gas	●	●	●	●				●	●
H-Oil/LC Fining	●	●	●	●	●	●	◐	●	●
Fixed Bed HDS	●	●	◐	◐	●	●	◐	●	●
Shell HDS/HDM	●	●	●	◐	●	●	◐	●	●
Aurabon	●	●	●	◐	◐	●	◐	●	●
HFC	●	●	●	◐	●	●	◐	●	●
CANMET	●	●	◐	◐	●	●	◐	●	●
Asphaltene Cracking	●	●	◐	●	●	●		●	●
FAROP Hydrocracker	?	?	?	?	●	●	?	●	●
Kombie Kammer	?	?	?	?	●	●	?	●	●
CATEGORY KEY:	● Characteristic acceptable in feedstock in any quantities; ◐ Characteristic acceptable in feedstock in limited quantities; ○ Characteristic generally unacceptable in feedstock				● High; ◐ Medium; ○ Low			● No emmissions; ◐ Minor emmissions; ○ External cleanup required	

[1] Although process may accept high contaminant levels in feed, product qualities or environmental concerns may necessitate substantial additional processing.

(continued)

Exhibit II-9: (continued)

BARRIERS TO CONVENTIONAL REFINING / PROCESS	RESOURCE BARRIERS				TECHNOLOGICAL BARRIERS	ECONOMIC BARRIERS	
	FUEL NEEDS	HYDROGEN	OXYGEN	CATALYST	DEMONSTRATED FEASIBILITY/ RELIABILITY	INVESTMENT	OPERATING COSTS
Solvent Deasphalting	○	◒	◒	◒	●	●	○
Visbreaking	◒	◒	◒	◒	●	●	●
Delayed Coking	◒	◒	◒	◒	●	◒	◒
Fluid Coking	●	◒	◒	◒	●	◒	◒
Eureka	◒	◒	◒	◒	●	◒	◒
Lurgi Ruhrgas	◒	◒	◒	◒	◒	●	◒
Taciuk	◒	◒	◒	◒	○	●	◒
MCC	●	◒	◒	◒	○	(?)	◒
COSMOS	◒	◒	◒	◒	○	(?)	○
Flexicoking	●	◒	◒	◒	●	◒	◒
Dynacracking	◒	◒	◒	◒	○	◒	◒
RAROP Catalytic Cracking	●	●	◒	○	○	(?)	(?)
Heavy Oil Cracking	◒	◒	◒	○	●	◒	○
RAROP ResidCat Cracking	(?)	◒	◒	(?)	○	(?)	(?)
Texaco Partial Oxidation	◒	●	○	◒	●	◒	◒
Shell Gasification Process	◒	●	○	◒	●	◒	◒
M-Gas	●	●	○	◒	○	◒	◒
H-OIL/LC Fining	○	○	◒	○	●	○	○
Fixed Bed HDS	○	○	◒	○	●	○	○
Shell HDS/HDM	○	○	◒	○	●	○	○
Aurabon	○	○	◒	◒	◒	○	○
HFC	○	○	◒	●	○	○	○
CANMET	○	○	◒	◒	◒	○	○
Asphaltene Cracking	○	○	◒	○	○	○	○
RAROP Hydrocracker	○	○	◒	○	○	(?)	(?)
Kombie Kammer	○	○	◒	○	◒	(?)	(?)
CATEGORY KEY:	● Resource Producer ◒ Internally provides required or need relatively little resources ○ Resource Consumer				● Commercially proven reliable ◒ Near term Commercial ○ Pilot plant or bench scale	● Low ◒ Medium ○ High	

III. Tar Sand Resources

1. BITUMEN PROPERTIES ARE NOT RIGIDLY DEFINED

Tar sands or oil sands are sedimentary rocks (consolidated or unconsolidated) which contain bitumen or other very heavy oils with exceptionally high viscosities.[1-3] These deposits are usually found on the fringes of sedimentary basins. The rocks are generally of a Lower Cretaceous Age (~120 million years old), whereas the bitumen are of the Upper Devonian Age (~360 million years old). Tar sand deposits vary in depth from outcrops to beds with overburden often more than 2,000 feet. Bitumen saturation is usually very high (up to 90 percent), but the maximum oil content is frequently only 8-15 percent by weight of saturated sand. The sand is wetted by water and not by bitumen, thus separation with hot water is not difficult if the tar sands can be surface mined. Exhibit III-1 depicts the typical tar sand particles' arrangement.

EXHIBIT III-1

Tar Sand Particle Arrangement

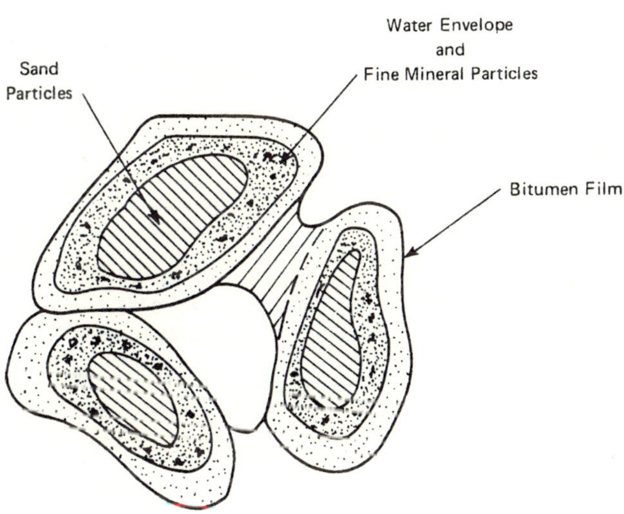

In its natural state, the bitumen is extremely viscous and cannot be recovered by conventional petroleum production methods. The bitumen extracted from tar sands is also high in asphaltenes, sulfur, heavy metals, oxygen, and nitrogen. The bitumen or very heavy oil in tar sands has very low gravity (<12° API), high viscosity (25,000-5,000,000 cp at 50°F) and high sulfur content (~5 wt-percent). The characteristics of tar sands deposits and properties of bitumen and heavy oils are not rigidly defined but rather range continuously from semi-solid bitumen through the very heavy oils to heavy oils that flow slowly under reservoir conditions. A distinction is often made between bitumen and heavy oil and some of these differences include:

Bitumen	Heavy Oil
. Semi-solid, no mobility under reservoir conditions	. Very viscous, some mobility under reservoir conditions
. No primary recovery under reservoir conditions possible	. Some primary recovery is possible though frequently uneconomical
. Viscosity greater than 25,000 cp at reservoir temperature	. Viscosity less than 25,000 cp at reservoir temperature
. Usually designated as tar sand, oil sand or bituminous sand deposits.	. Usually designated as heavy oil deposits.

For the purposes of this study, bitumen or very heavy oil is defined as having a gravity less than 10-12° API and viscosities >100,000 cp at reservoir temperature. The above classification is an arbitrary one on a continuous scale since there is no universally acceptable division point between heavy oil and bitumen.

Most of the viscous oil deposits located near the Athabasca River in Alberta, Canada are classified as bitumen contained within tar sand deposits. The primary exception is the deposits near Lloydminster in S.E. Saskatchewan, which are of the heavy oil type. In Venezuela, almost all of the reserves contain heavy oil of 8-18° API, rather than bitumen of the Athabasca type. Moreover, (because of high reservoir temperature and pressure) most of the Venezuelan fields have some primary production. In light of this, the above definition of very heavy oil was expanded somewhat so that Venezuelan heavy oil would be included in the study.

Generally, the viscosity of bitumen decreases rapidly as temperature increases. Exhibit III-2 shows the temperature, viscosity relationships for several different grades of oil. Production of bitumen from tar sands in deep formation usually entails the application of

EXHIBIT III-2

Viscosity/Temperature Relationship of Typical Very Heavy Oils

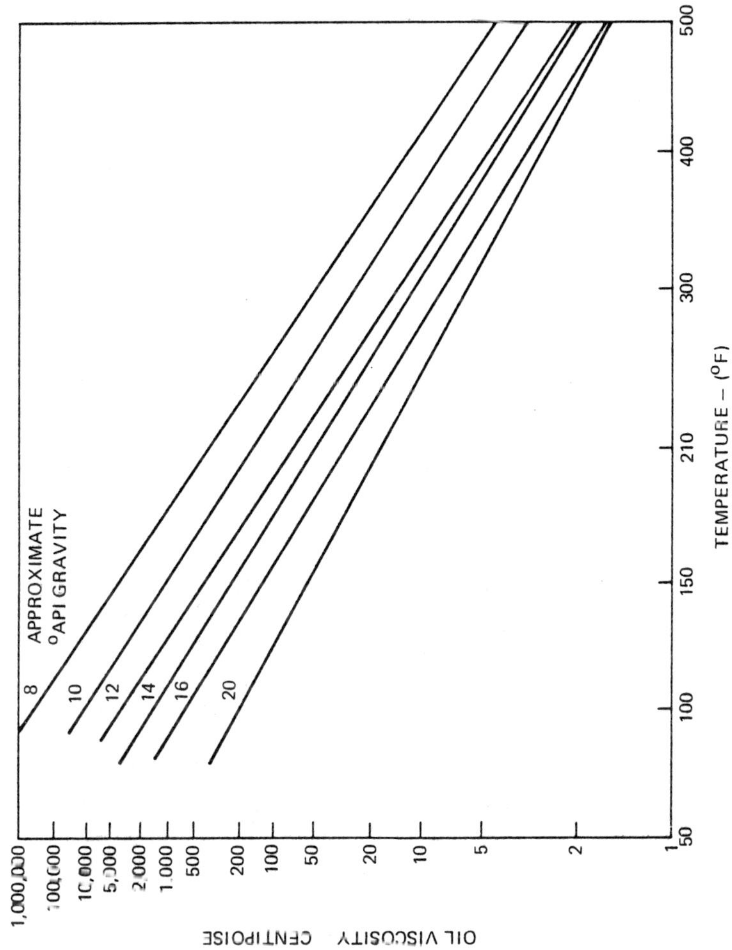

thermal energy or diluents to reduce the viscosity sufficiently so that the bitumen will flow to the production wells and can be pumped to the surface. Bitumen is frequently upgraded before pipelining and refining.

2. BITUMEN AND VERY HEAVY OIL RESOURCES ARE DISTRIBUTED WORLDWIDE

There are tar sand deposits on every continent except Australia. It is estimated that worldwide resources amount to 2.3 x 10^{12} barrels of in-place bitumen or very heavy oil.[4-8] The majority of tar sands resources are in Canada (1 x 10^{12} barrels in-place) and Venezuela (0.7 x 10^{12} barrels in-place). Other large to medium deposits are found in the United States, Malagasy, Albania, Trinidad, and U.S.S.R. Exhibit III-3 lists the major tar sand deposits of the world including the areal extent, pay thickness and bitumen characteristics.[9] Except for the Canadian deposits, statistical information on other areas is limited and based largely on estimates by experts. Following is a discussion of the world's major tar sand deposits in the various countries.

(1) Canadian Tar Sands Deposits Are Significant, But of Varying Properties

Canada has the largest known deposits of tar sand in the world.[10] These deposits total approximately 1 x 10^{12} barrels of heavy oil or bitumen in-place. Most of the deposits are located in the province of Alberta, forming a 600-mile arc which extends from Peace River in west central Alberta to Lloydminster in western Saskatchewan, an areal extent of 18,534 square miles. The four major deposits are Athabasca, Wabasca, Peace River, and Cold Lake. The Lloydminster deposit is not considered as tar sand because it contains conventional heavy oil. Huge amounts of bitumen also exist within an area known as the "Carbonate Triangle" in north central Alberta. The Carbonate Triangle includes Devonian carbonate formations, composed mainly of dolomite with porosities ranging from 19-28 percent and bitumen saturation averaging over 70 percent. It is estimated that these carbonate formations contain 1.35 x 10^{12} barrels of bitumen in-place, a potential giant oil supply for Canada. More exploratory drilling is required to confirm the amount of bitumen in the carbonate deposits. Exhibit III-4 is a map of Alberta displaying the location of tar sand and heavy oil deposits.[4]

The amount and properties of Alberta tar sands and heavy oil deposits are shown in Exhibit III-5 and III-6.[11] Most of the deposits have a high porosity and permeability but many are buried too deep for surface mining. Sulfur content of the bitumen is high, ranging from 4.4 to 6.1 percent by weight. Of the total bitumen reserves of 980 million barrels in Canada; only

EXHIBIT III-3 Major Tar Sand Deposits of the World

COUNTRY	NAME OF DEPOSIT	AREAL EXTENT acres	PAY THICKNESS ft	PAY THICKNESS average	BITUMEN SAT (wt%)	CHARACTER OF OIL °API at 60° F	OVERBURDEN THICKNESS (ft)	IN-PLACE RESERVES (10^6 bbl)
Canada	McMurray–Wabiskaw, Alberta	5,750,000	0– 300	(175)	–	–	0–1900	625,900
	Bluesky–Gething, Alberta	1,200,000	0– 400		–	–	700–2600	51,500
	Grand Rapids, Alberta	1,100,000	400	(280)	–	–	300–1400	33,400
	Total "Athabasca" tar sands	8,000,000		(150±)	2–18	10.5		710,800
	Melville Island, N.W.T.	?	60– 80		?– 16	10	0–2000*	?
Eastern Venezuela	Oficina–Temblador tar belt	5,750,000	3– 10			10	0–3000*	200,000
Malagasy	Bemolanga	96,000	80– 300	(100)	10		0– 100	1,750
U.S.A.	Asphalt Ridge, Utah	11,000*	11– 254	(98)	11	8.6–12	0–2000	900
			24– 200	(100)				
	Sunnyside, Utah	34,300*	10– 350		9	10 –12	0– 150	500
Albania	Selenizza	5,306	33– 330	(50)*	8–14	4.6–13.2	shallow	371*
U.S.A.	Whiterocks, Utah	1,900	900–1000		10	12	nil	250
	Edna, California	6,595	0–1200*	(250)	9–16	13	0– 600*	165
	Peor Springs, Utah	1,735	1– 250	(34)	9*		shallow	87
Eastern Venezuela	Guanoco	1,000	2– 9	(4)	64	8	nil	62
Trinidad	La Brea	126	0– 270	(135)	54	1 – 2	nil	60
U.S.A.	Santa Rosa, New Mexico	4,630+	0– 100	(20)	4– 8		0– 40	57
	Sisquoc, California	175	0– 185	(85)	14–18	4 – 8	15– 70	50
	Asphalt, Kentucky	7,000	5– 36	(15)	3–10		6– 30	48
Rumania	Derna	459	6– 25		15–22		shallow	25
U.S.S.R.	Cheildag, Kazakhstan	82*		(200)	5–13		shallow	24
U.S.A.	Davis–Dismal Creek, Kentucky	1,900	10– 50	(15)	5		15– 30	22
	Santa Cruz, California	1,200	5– 50	(11)	10–12		0– 100	20
	Kyrock, Kentucky	900*	15– 40	(20)	6– 8		15	18

*Statistics asterisked where estimated

EXHIBIT III-4

Location of Bitumen and Heavy Oil Deposits in Alberta

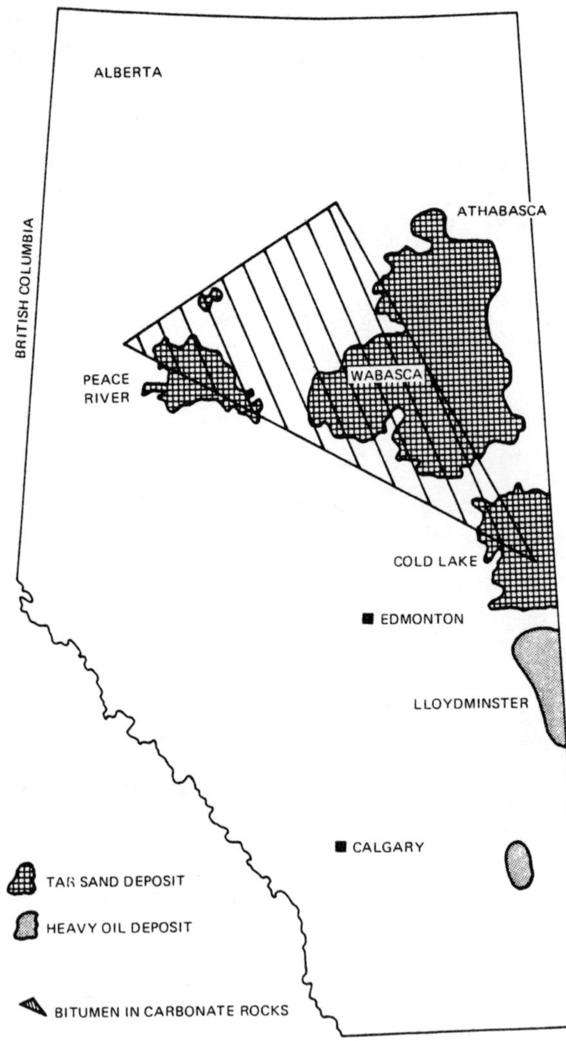

EXHIBIT III-5

Alberta Oil Sand Deposits

Field		ATHABASCA	PEACE RIVER	COLD LAKE B	WABASCA A	WABASCA B	CARBONATE TRIANGLE	LLOYDMINSTER H.O.	S.E. ALBERTA H.O.
Formation		Wabiskaw/McMurray	Bluesky-Gething	Clearwater	Grand Rapids	Webiskaw		Sparky	
Proved in Place Reserve	MMb	869,000	92,000	40,000	66,000	53,000	1,350,000	4000+	2000+
Depth	ft	0-2000	1500-2600	1350	300-1100	1000-2200		1000-2000	
Average Pay Thickness[1]	ft	88	47	36	40	24	No	45	
Reservoir Temperature	°F	50	62	55	50		Published Information	74	
Reservoir Pressure	psig		530	440			Available	445	
Initial G.O.R.	cu.ft/bbl							50	
Initial Oil Saturation	%	44-98	45-88	73	69	64			
Porosity	%	39	25-31	34-40	30-36	27-30		33	
Permeability	md	up to 5000	up to 1000	up to 3000	up to 3500			1000	

Source: Reference 11

[1] 3% Bitumen Cutoff

EXHIBIT III-6

Properties of Alberta Oil Sand Bitumens and Heavy Oils

Field			ATHABASCA	PEACE RIVER	COLD LAKE B	WABASCA A	CARBONATE	LLOYDMINSTER H.O.	S.E. ALBERTA H.O.
Formation (Segregation)			Wabiskaw/McMurray	Bluesky-Gething	Clearwater	Grand Rapids	Triangle		
Structure									
Hydrocarbon Properties									
Gravity		°API	8-10	8-9	10-12	8-13	7	10-18	12-18
Viscosity	- Reservoir Temp.	cps	5,000,000	100,000	100,000	8,000,000	1,000,000	3000	3000
	- 60°F								
	- 70°F								
	- 100°F								
	- 140°F		3700						
	- 212°F		1000	100	100	2000	700	20	20
	- 400°F			7					
Pour Point		°C	10				26	-22	
Elemental Analysis	- Carbon	% wt	83.1	82.2	83.7	83.0	83.7	82.8	
	- Hydrogen		10.6	10.0	10.5	10.3	10.3	11.8	
	- Sulphur		4.8	5.6	4.7	5.5	5.0	3.4	3.5
	- Nitrogen		0.4	0.1	0.2	0.4	0.4	0.3	
	- Oxygen		1.1	2.1	0.9	0.8		1.7	
	- C/H Ratio		7.8	8.2	7.9	8.1	8.1	7.0	
Molecular Weight			570-620	520	490	600		410	410
Hydrocarbon Type	- Saturates	% wt	22		33			47	43
	- Aromatics		21		29			24	20
	- Resins		39		23			17	25
	- Asphaltenes		18	19.8	15	18.6	20.0	12	12
Metals	- Varadium	ppm wt	250		240	210	400-700	100	
	- Nickel		100		70	75	80	40	
Conradson Carbon			13.5		12.6		14.4	8	12.2
Distillation Temperature	- IBP	°C	260		170	275			75
	- 5% vol		285		255	330		80	175
	- 10% vol		320		300	370	325	175	220
	- 30% vol		425		405		365	230	290
	- 50% vol		530		555		475	310	
							565		

about 8 percent is surface-mineable, 76 percent is reversable by in-situ extraction methods, and the recovery method for the remaining 16 percent is uncertain.

(2) Venezuelan Heavy Oil Deposits

The Orinoco Tar Belt is one of the largest untapped heavy oil accumulations in the world. The deposits are located in southeastern Venezuela in the states of Guarico, Anzoategui, Monagas, and Delta Amacuro and cover an areal extent of more than 16,000 square miles.[12] Exhibit III-7 shows the location of the Orinoco Belt. The deposits present in this area contain heavy oils with high viscosity and low API gravity (8-18°). Due to the high reservoir temperature, some of the heavy oil has mobility under reservoir conditions and limited amounts can be recovered by primary production methods. Recently, the Tar Belt name was officially changed to Orinoco Heavy Oil Belt to reflect the wide range of oil quality.

It is estimated that the heavy oil belt contains approximately 700 billion barrels of in-place heavy oil at depths of 600 to 7000 feet. With the discovery of the underlying Guiana Shield, this figure has been revised upward to double or triple the original amount and may total 2.1 X 10^{12} barrels of in-place heavy oil. The belt can be divided into the four following areas according to their location, principal developer and geological characteristics:

Area	Principal Assigned Developer	Sediments	Age(Yrs.)	Oil Accumulations
Cerro Negro	Lagoven	Tertiary	70MM	Oficina-Temblador
Hamaca	Meneven	Tertiary-Cretaceous	110MM	Oficina-Temblador
Zuata	Maraven	Tertiary-Cretaceous-Paleozoic	70-250MM	Oficina-Merecure-Temblador-Hato Viego
Machete	Corpoven	Tertiary-Cretaceous Paleozoic	70-250MM	Chaguaramas-Roblecito-La Pas Cua-Infante-La Cruz-Herto Viego

EXHIBIT III-7 Orinoco Heavy Oil Belt

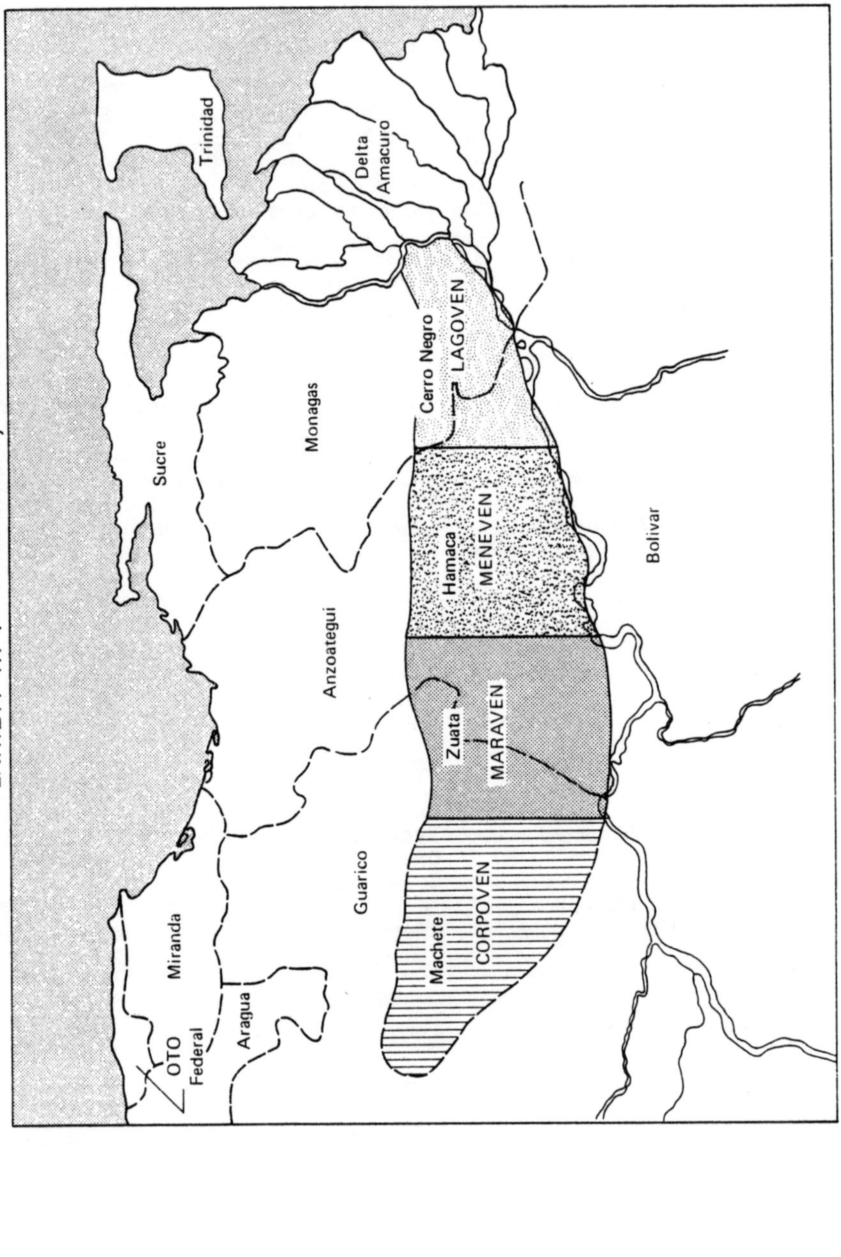

The main heavy oil belt formations are unconsolidated sandstones with porosities averaging between 20 and 35 percent. Exhibit III-8 outlines some of the oil properties and reservoir conditions in the heavy oil belt.

EXHIBIT III-8

Properties and Conditions in the Orinoco Heavy Oil Belt[13]

Depth:	600 feet - 7,000 feet
Reservoir Temperature:	95-150°F (function of depth)
Initial Pressure:	Hydrostatic
Sulphur Control:	2-4 Wt. percent
Metal Content (nickel and vanadium):	200-500 ppm
Gravity of the Crude:	7-18° API
Dead Oil Viscosity	100-20,000 cp at 130°F
Gas/Oil Ratio	50-170 scf/b

The heavy oil in Venezuela has a high metallic content, consisting of primarily nickel and vanadium. This has serious implications on the choice of upgrading process as heavy metals can reduce the selectivity of catalysts.

After the nationalization of its oil industry in 1976, Venezuela organized a state run oil company called Petroleos de Venezuela to carry out its petroleum operations. As indicated in Exhibit III-7, Petroleos de Venezuela is organized as four operating companies: Lagoven, formerly Exxon of Venezuela; Meneven, formerly Gulf of Venezuela; Maraven, formerly Shell; and Corpoven, a consortium of six smaller companies.

(3) United States Tar Sands Resources

The United States has the third largest tar sand resources in the world. There are a total of some 546 deposits located in 22 states.[14,15] The U.S. in-place bitumen resource is estimated to be thirty billion barrels. Most of the resource is found in five states: California, Kentucky, New Mexico, Texas, and Utah. Exhibit III-9 is a list of U.S. bitumen-bearing deposits and estimated resources. Within Utah, six major deposits contain over 90 percent of the total U.S. resource. These are:

EXHIBIT III-9

Deposits of Bitumen-Bearing Rocks in the United States With Resources Over 1,000,000 Barrels

STATE AND NAME OF DEPOSIT	ESTIMATED RESOURCES (MILLIONS OF BARRELS) LOW	HIGH
California:		
Edna	141.4	166.4
South Casmalia	46.4	. . . .
North Casmalia	40.0	. . . .
Sisquoc	26.0	50.0
Santa Cruz	10.0	. . . .
McKittrick	4.8	9.0
Point Arena	1.2	. . . .
California Total	269.8	323.0
Kentucky:		
Kyrock Area	18.4	. . . .
Davis-Dismal Area	7.5	11.3
Bee Spring Area	7.6	. . . .
Kentucky Total	33.5	37.3
New Mexico: Santa Rosa	57.2	. . . .
Texas: Uvalde (Recently revised)	124.1	3,000 (10,000)
Utah:		
Tar Sand Triangle	12,504.0	16,004.0
P. R. Spring	4,000.0	4,500.0
Sunnyside	3,500.0	4,000.0
Circle Cliffs	1,000.0	1,300.0
Asphalt Ridge	1,000.0	1,200.0
Hill Creek	300.0	1,160.0
San Rafael Swell	385.0	470.0
Asphalt Ridge, Northwest	100.0	125.0
Whiterocks	65.0	125.0
Raven Ridge	75.0	100.0
Argyle Canyon	50.0	75.0
Wickiup	60.0	75.0
Rimrock	25.0	30.0
Cottonwood-Jacks Canyon	20.0	25.0
Minnie Maud Creek	10.0	15.0
Willow Creek	10.0	15.0
Pariette	12.0	15.0
White Canyon	12.0	15.0
Littlewater Hills	10.0	12.0
Lake Fork	6.5	10.0
Nine Mile Canyon	5.0	10.0
Chapita Wells	7.5	8.0
Ten Mile Wash	1.5	6.0
Tabiona	1.3	4.6
Thistle	2.2	2.5
Spring Branch	1.5	2.0
Cow Wash	1.0	1.2
Utah Total	23,164.5	29,305.3
United States Total	23,650.1	32,722.8

	Utah Bitumen Resources (Billions of Barrels In-Place)
Tar Sand Triangle	16
Circle Cliffs	1.3
Sunnyside	3.5 - 4.0
P. R. Springs	4.0 1 4.5
Hill Creek	1.2
Asphalt Ridge	1.1
	28

(4) Malagasy Resources[16]

Aside from Canada, Venezuela and U.S.A., the next largest known tar sand deposit is just off the South African coast in the country of Malagasy (formerly Madagascar). The Bemolanga deposit covers an area of 150 square miles and has in-place resources of approximately two billion barrels, of which about 520 million barrels are mineable. Some of the Bemolanga deposit characteristics include:

- Pay thickness: 30 feet
- Bitumen content: 6 to 9 percent by weight
- Porosity: 19.5 percent
- Permeability: 446 md.

(5) Eastern Europe and U.S.S.R. Resources[17,18]

The Selenizza deposit in Albania covers an 8-square-mile area with an estimated in-place reserve of 370 million barrels. The bitumen is embedded in lenticular sands, and has a very high sulfur content of up to 7 weight percent. Most of the Albanian resource is marine deposits with uneven pay thickness and thin overburden.

The Derna deposit in Romania is adjacent to the Tataros asphalt deposit located in eastern Rumania. The formulation is composed of sedimentary rocks of the Tertiary type, and the oil-bearing strata is rather close to the surface. The in-place reserve is approximately 25 million barrels. Romania started to import crude oil from the U.S.S.R. in 1979 so it is conceivable that they will start developing the Derna deposit if world crude oil supply, especially that of the eastern bloc, continues to dwindle.

The Soviet Union has a rather small deposit of approximately 24 million barrels at Cheildag, located about 15 miles southwest of Baku. The bitumen is of the heavy oil type impregnated in carbonate or sandstone rocks. The reservoir properties vary greatly: porosity, from 3-5 percent to 40-45 percent; permeability, from a few units to 1,000-1,200 md; and pay thickness from 3 feet to 120 feet. Other deposits of heavy oil occur in the Asselian, Sakmerian, Ufimain, Lower Kazanian and Upper Kazanian regions of the Tartar Republic. The Russian heavy oils have widely varying properties: asphaltenes, 5-35 percent; resins, 12-60 percent; oils, 20-70 percent; and viscosities from hundreds to many thousand centipoise. Major heavy oil accumulations are present in the Tuman-Pechora bitumen-petroliferous basin near Tarega, Usa and Izhma, etc.

REFERENCES

III. TAR SAND RESOURCES

1. The 27th Canadian Chemical Engineering Conference, "Oil Sands," Oil Sands Symposium, Calgary, Alberta, October 23-27, 1977.

2. G. V. Chilingarian, T. F. Yen, Editors, "Bitumens, Asphalt and Tar Sands," Elsevier Scientific Publishing Company, Amsterdam, 1978.

3. Otto P. Strauss and Elizabeth M. Lown, "Oil Sand and Oil Shale Chemistry," Verlay Chemie, New York, 1978.

4. Alberta Oil Sands Technology and Research Authority, "Fourth Annual Report," March 31, 1979.

5. R. F. Meyer, Editor, "The Future Supply of Nature-Made Petroleum and Gas," First Unitar Conference on Energy and the Future, Luxenburg, Austria, July 5-16, 1976.

6. P. H. Phizackerley, L. O. Scott, "Major Tar Sand Deposits of the World," Seventh World Petroleum Congress, Mexico City, 1967.

7. "Big Alberta Carbonate Triangle Resource Seen" Oil and Gas Journal p. 40, June 9, 1980.

8. R. D. Humphreys and R. Schutte, "The Alberta Oil Sands: Energy for the Future," pp. 1-9, Verlay Chemie International, Inc. 1978.

9. Op. cit., Reference 2.

10. Op. cit., Reference 4.

11. John Starr, Jose M. Prats, "Chemical Properties and Reservoir Characteristics of Bitumen and Heavy Oil from Canada and Venezuela," presented at the First International Conference on the Future of Heavy Crude and Tar Sands, Edmonton, June 1979.

12. Edison Perozo, "Orinoco Oil Belt Exploration, Evaluation and Exploitation," presented at the Latin American Oil Congress, Rio de Janerio, July 1-4, 1980.

13. Carlos J. Borregarles, "Production Characteristics and Oil Recovery in the Orinoco Oil Belt," presented at the First International Conference on the Future of Heavy Crude and Tar Sands, Edmonton, June 1979.

14. C. Q. Cupps, L. C. Marchand and J. J. Stason, "Tapping the Vast Tar Sand Reserves of the U.S.," World Oil, September 1979.

15. Report No. HCP/79014-01, "The Economic Potential of Domestic Tar Sands," prepared for U.S. Department of Energy, Division of Oil, Gas and Shale Technology, 1978.

16. Colonel H. Andriansolo, M. D. Rakoto Andriantisilavo and E. Raveloson, "The Bitumen and Heavy Oil Deposits in Madagascar," presented at the First Unitar International Conference on the Future of Heavy Crude and Tar Sands, Edmonton, June 1979.

17. Op. cit., Reference 4.

18. E. M. Khalimov, R. Kh. Muslimov and G. T. Yudin, "Bitumen Deposits of the U.S.S.R. and the Ways of Their Development."

IV. Tar Sands Production Methods

1. PROCESS SELECTION OVERVIEW

Key characteristics of bitumen deposits which make their recovery particularly difficult include:

- Lack of mobility of bitumen at reservoir conditions
- Lack of primary energy in the formation
- Extreme heterogeneity of the deposits
- Low effective permeability of reservoir rocks.

Generally tar sand deposits must contain at least 8-10 percent by weight of bitumen or very heavy oil before large-scale development can be considered. In addition, the amount of recoverable reserves must be large enough to supply the project at the design capacity for at least 20-25 years. Tar sands deposits can be classified into three categories according to their overburden thickness which is critical in selecting amongst applicable recovery methods:

Overburden Depth	Applicable Recovery Method	Estimated Percent of Total World Tar Sands Resources
0 - 150 feet	Surface mining and extraction	5
≥ 500 feet	In-situ recovery	70
150-500 feet	Uncertain, no proven technology	25

Surface mining and extraction methods are suitable primarily for shallow deposits or outcrops, because the removal of overburden usually becomes uneconomic for overburden depths of greater than 150 feet. In-situ recovery methods are usually applied to deposits deeper than 500 feet since a thick overburden is required to contain the pressure and prevent the loss of gas and liquid products. Currently there is no proven recovery technology for deposits at intermediate depths which are too deep for surface mining and too shallow for in-situ methods. Underground mining of tar sands is being evaluated in Russia and United States for production of deposits at intermediate depths. Exhibit IV-1 shows how the deposit depth determines the application of various bitumen recovery processes. Surface mining methods are, by far, more developed than their In-situ counterparts. In addition, there is a class of "novel" recovery methods which are conceptually feasible but will not be ready for large-scale application for at least 10 years.

EXHIBIT IV-1

Classification of Bitumen and Very Heavy Oil Production Processes

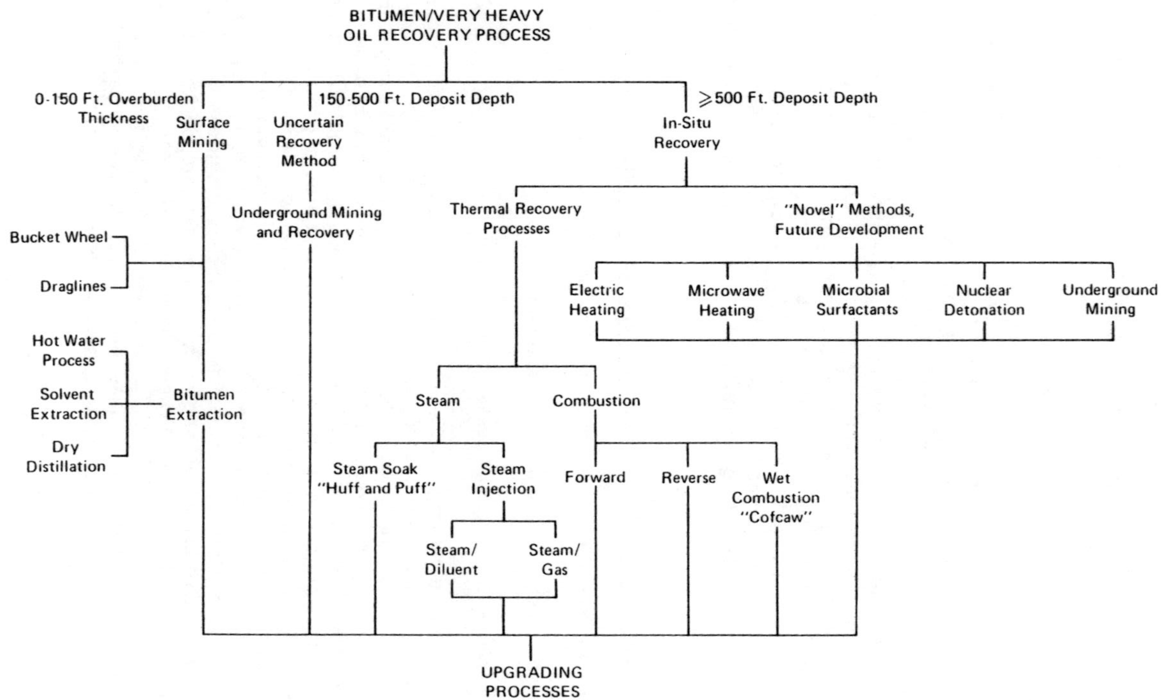

To date there is no single universally applicable bitumen recovery process, because the operability of a particular process is highly dependent on overburden depth, reservoir characteristics, and bitumen properties. Extensive field testing is required on each deposit to obtain information such as well life, percent recovery, production rates, steam/oil ratio, and operating costs. This is especially true for in-situ recovery processes. It takes at least 5 years for field testing to provide any meaningful operating information, and even then reproducibility is not guaranteed. In view of this, it is difficult to generalize on the applicability of a particular process.

The status of development for the different processes is shown in Exhibit IV-2. The level of detail of process evaluation in this report is a direct function of its scale of development and likelihood of near-term applicability. As such, following are general discussions of production methods; commercial processes are evaluated in greatest detail, followed by pilot projects. "Novel" recovery methods which appear promising and have progressed beyond bench scale are also evaluated.

2. SURFACE RECOVERY PROCESSES

The general steps involved in recovering bitumen from surface-mined tar sands are:

- Tar sands mining
- Bitumen extraction and diluent recovery
- Bitumen upgrading.

Surface mining is conducted in two stages: overburden and muskeg removal, followed by excavation of tar sands. The two commercially proven surface mining methods are discussed subsequently under the Suncor (formerly GCOS) and Syncrude sections.

(1) Hot Water Extraction of Surface Mined Tar Sands

There are three extraction processes for surface-mined tar sands, but only one, the Hot Water Process shown in Exhibit IV-3, is proven commercially.[3] The fact that tar sands are wetted by water and not by bitumen makes separation by a hot water process feasible.

The hot water process, based on K. A. Clark's method, is the most common extraction process to recover bitumen from surface-mined tar sands.[1-3] It is employed by Suncor (formerly GCOS), Syncrude, and the proposed Alsands plant. In this process, there are four fundamental steps:

- Feed conditioning
- Bitumen separation
- Waste disposal
- Final cleaning of bitumen concentrate.

EXHIBIT IV-2

Bitumen and Very Heavy Oil Production Methods
(Status of Foreign Development)

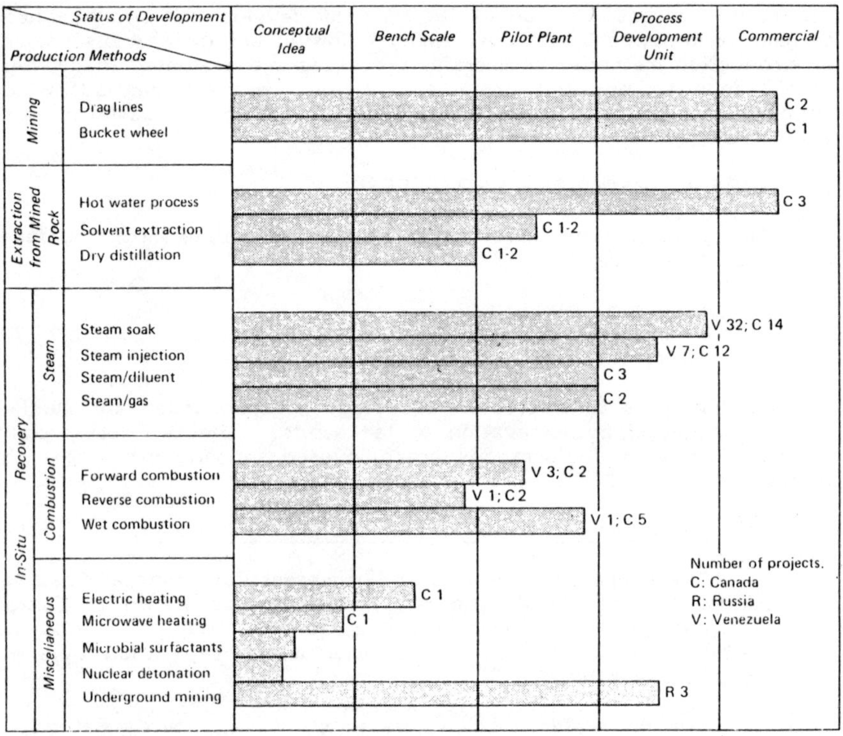

EXHIBIT IV-3

Hot Water Extraction Process

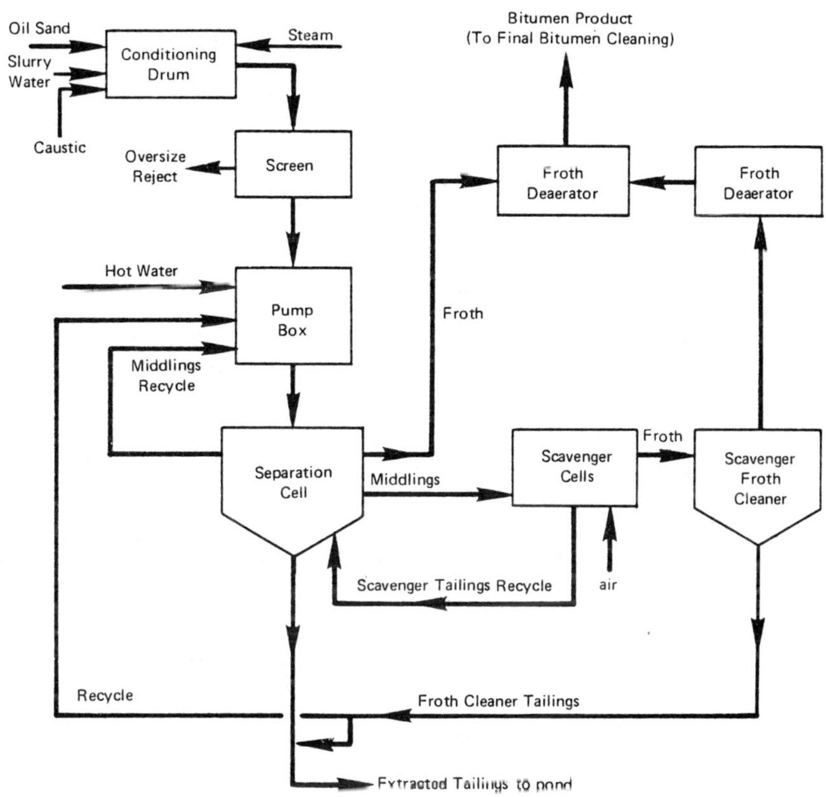

Tar sands conditioning is accomplished by slurring the sand with water and caustic to form a pulp of 60 to 85 percent solids at 175-200°F. The conditioning step is accomplished in a horizontal rotating drum into which steam is injected below the slurry surface. Maximum density difference between water and bitumen occurs within this temperature range; therefore, separation is made easier. Chemical conditioners, such as caustic soda (NaOH), are added at this stage to improve separation of the bitumen from the sand. A recent study shows that the sodium hydroxide neutralizes organic bitumen acid to form salts which are surface-active.[4] These surfactants are the primary agents responsible for improved bitumen recovery.

The effluent from the conditioning drum is screened to remove coarse particles, stones, or frozen lumps. It is then mixed with additional hot water and sent to the separation cell. Since the sand is wet by water, the sand particles settle quickly to the conical bottom of the cell. The bitumen floats to the top of the cell and is skimmed off and pumped to the froth degenerator. Effluent from the center portion of the cell, called middlings, is pumped to a scavenger cell to complete the bitumen separation from the mineral particles and water by air flotation. The tailings from these cells are delivered to a tailings pond where water clarification and treatment take place.

The final step before upgrading involves the cleaning of raw bitumen which is in the form of froth containing water, air and fine particulate. Basically, there are three methods of bitumen cleanup. The first method is used commercially and involves blending the froth with naphtha obtained in the upgrading process and centrifuging the resultant mixture. The second method requires that the mineral matter first be removed by washing with water, then water is removed by thermal dehydration. The third method involves distilling the raw bitumen at atmospheric pressure up to about 850°F. Some of the feed is converted to coke which remains in the still, together with the mineral matter and the nondistillate portion of the bitumen. The filtered and ash-free bitumen product is then subjected to further upgrading.

Although commercially proven, the Hot Water Process has some disadvantages which are significant:

- The tailings stream contains sand and fine clay mixture which has very poor settling characteristics. This necessitates the construction of large tailings ponds.

- Approximately 10 percent of the bitumen in the feed is lost to the tailings stream.

- Large amounts of water and steam are required in the conditioning step.
- Large amounts of energy are required for the production of hot water and low pressure steam.

(2) Solvent Extraction of Surface Mined Tar Sands

Solvent extraction is being evaluated for extracting bitumen or heavy oil from tar sands or diatomaceous earth.[5] The Dravo Company recently patented a solvent extraction method which is similar in principle to the solvent extraction processes used in extracting vegetable oil from oil seeds. As outlined in Exhibit IV-4,[5] solvent extraction of tar sands involves three basic steps:

- Tar sands preparation
- Bitumen extraction
- Recovery and separation of bitumen product and solvent.

EXHIBIT IV-4

Basic Steps of Solvent Extraction

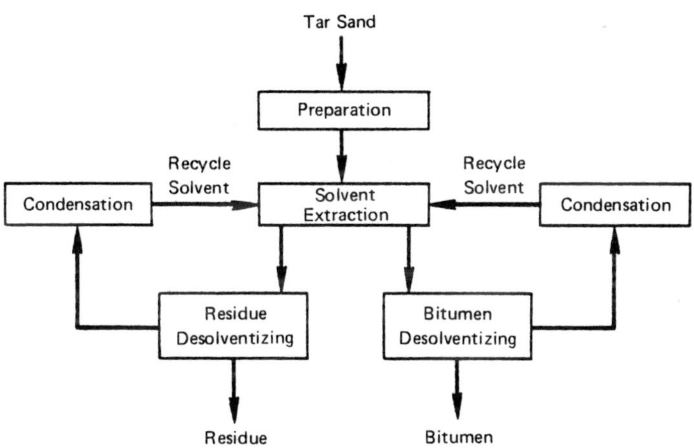

Cyclic or aromatic hydrocarbons in the boiling range of 170-210°F are very good solvents for the extraction of tar sands since they easily dissolve the asphaltenes (15-25 percent) contained in bitumen. Cyclohexane is superior to benzene or toluene, which is undesirable because of environmental reasons. The extracted miscella (oil or bitumen in solvent) usually contains 25 percent bitumen by weight. Dravo indicates that 96 percent of the bitumen can be recovered, and solvent recovery is as high as 99 percent.

The preparation step for Utah-type tar sands consists of grinding and mixing the feed tar sands with recycle solvent. After the mixture is heated to the extraction temperature, it is fed into a percolation extractor. Solvent is recovered from spent sand by countercurrent contact with steam while the bitumen is concentrated and stripped free of solvent by multiple effect evaporation and steam stripping. Exhibit IV-5[5] is the flow diagram for the solvent extraction process. For diatomaceous sands, a countercurrent decantation system is used because the fine particles cannot be easily separated from the solvent mixture. The system has six stages, and the miscella overflowing from each stage is increasingly clarified until it reaches the distillation system where concentration and stripping take place.

Detailed information and process economics on the solvent extraction are proprietary. Recently, an agreement was reached between Amoco and Dravo for the joint development of this process. Getty Oil is also evaluating solvent extraction for recovery of heavy oil from diatomaceous earth in McKittrick, California.

(3) Lurgi-Ruhrgas Dry Distillation Process

The Lurgi-Ruhrgas Process represents a suitable alternative to the Hot Water Process, and does not have some of its disadvantages.[6-8] L-R distillation is a dry process, thus circumventing use of hot water in the extraction stage and bitumen coking in the upgrading stage. This versatile dry distillation process was originally developed by Lurgi Mineralöltechnick GmbH in 1949-55, and can be applied for retorting coal, oil shale, or tar sand. A schematic diagram of the L-R Process as applied to extracting bitumen from tar sands is shown in Exhibit IV-6.[7]

The mined tar sand is fed into a mixer from a feed bin and contacted with hot recirculated sand at 1,200°F. The sand is heated in the lift heater by burning off the coke that is deposited on sand in the mixer. At a final retorting temperature of 900 to 1,000°F, the bitumen is broken down resulting in liquid, vapor, and entrained solids. The product stream is sent to

EXHIBIT IV-5

Process Flow for Solvent Extraction of Tar Sands

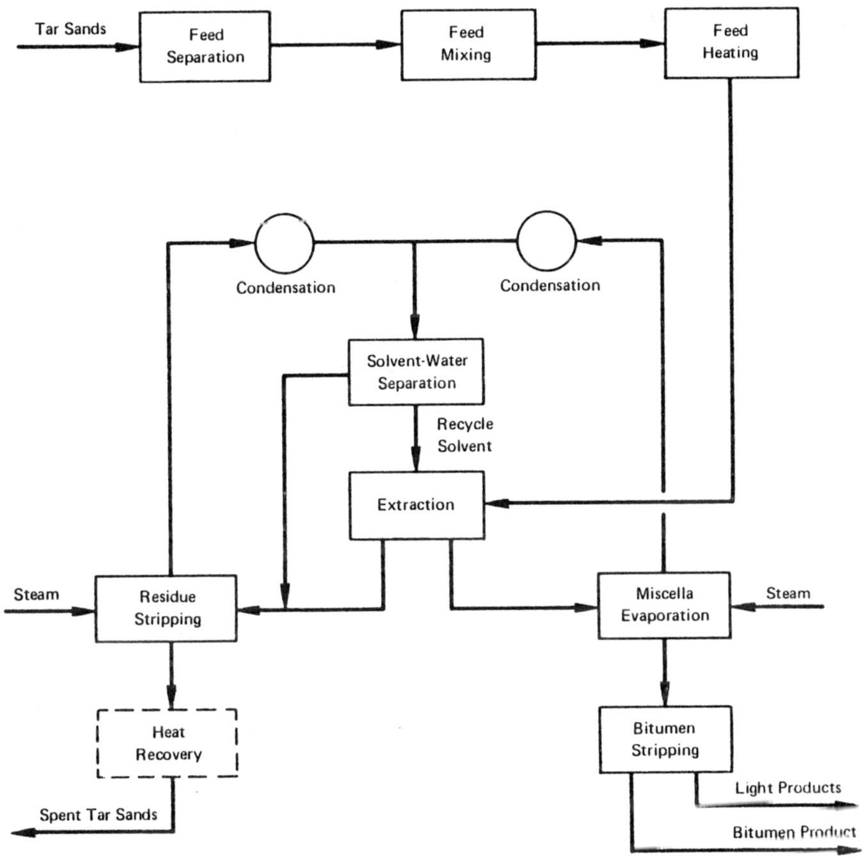

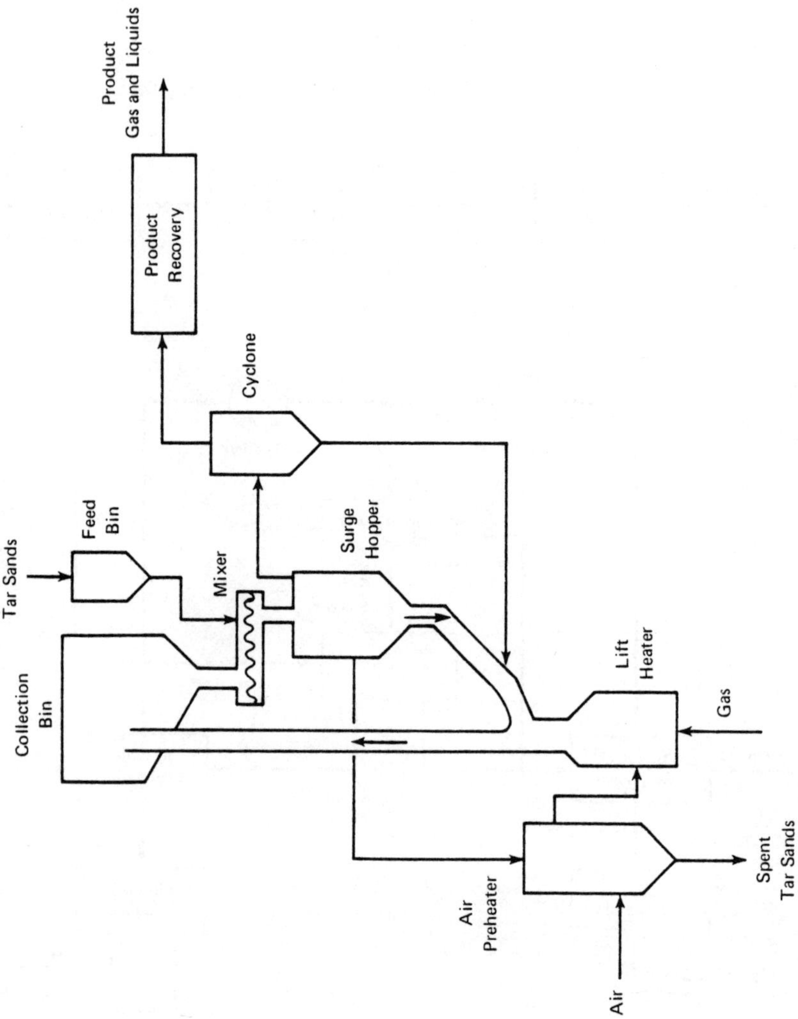

EXHIBIT IV-6 Lurgi-Ruhrgas Plant for Recovery of Tar Sands

cyclone to remove particulate followed by scrubbers to recover the light gases and product liquids. The hot sand from the mixer drops to a surge hopper where it is divided into two streams: part of it is recirculated to the mixer through the lift heater, and the rest is sent to the air preheater. The product liquids from the L-R Unit must be upgraded before they can be piped and distributed as synthetic crude oil.

The overall bitumen recovery in the Lurgi-Ruhrgas dry distillation process is estimated at 70 percent which compares favorably to approximately 65 percent for the Hot Water Process used by both Suncor and Syncrude Canada. The L-R Process does have some drawbacks, the main ones being:

- Severe heat loss in the spent sand
- Small unit throughput and poor scale-up economics
- Expensive construction and replication of retorting equipment
- Preparation and handling of tar sand feed.

AOSTRA (Alberta Oil Sand Technology and Research Authority) is sponsoring a study on the comparative economics of the L-R Process and the Hot Water Process. Furthermore, Lurgi is testing a pilot unit in Essen, Germany. This test is funded by AOSTRA at a cost of approximately $600,000, and is expected to be completed soon.

3. IN-SITU TAR SANDS RECOVERY PROCESSES

Selection of a particular in-situ recovery method depends heavily on the reservoir characteristics and bitumen properties.[9] A low bitumen gravity usually means a high viscosity which limits the bitumen fluidity. Reservoir porosity is a measure of voids and pore spaces in the rock formation, and its ability to hold fluids. Athabascan tar sand deposits have porosities ranging from 20 to 40 percent and the bitumen in-place is equal 1,000 to 1,500 barrels per acre-foot. Reservoir permeability is a measure of the amount of fluid that can be forced through the rock formation given pressure drop driving force. Formation permeability is usually expressed in millidarcies, and depends on the size of the pores and interconnecting passages. A derived quantity called transmissibility measures the rate of oil flow through the rock formation. It is defined as the product of permeability and formation thickness divided by oil viscosity.

A selection guide for preliminary screening of various in-situ recovery methods is presented in Exhibit IV-7.[9] For example, the heterogeneity of a formation is an additional site-specific factor which must be considered. Reservoirs with massive faulting and fracturing will generally produce poor results because of heat and product loss.

EXHIBIT IV-7

Selection Guide for In-Situ Thermal Recovery Methods

IN SITU RECOVERY PROCESS	GRAVITY °API	VISCOSITY AT RES. CONDITIONS (cp)	POROSITY %	PERMEABILITY md	TRANSMISSIBILITY	NET PAY FT	DEPTH FT	ROCK TYPE
Continuous Steam Injection	8-20	150- 5,000	⩾30	⩾1,000	⩾100	>30	500-4,000	Sandstone or Carbonate
Steam Soak, 'Huff n Puff'	10-40	1,500-10,000	⩾30	⩾ 800	⩾100	>50	500-2,000	Sandstone or Carbonate
Forward Combustion	<10	300- 2,000	⩾20	> 300	⩾ 40	10-70	500	Standstone
Reverse Combustion	<10	No upper limit	⩾20	> 300	⩾ 40	10-150	500	Standstone
Wet Combustion	<10	100,000	⩾20	> 400	⩾ 70	10	500	Standstone
'Novel' Methods	Not Advanced Enough to Allow Generalization. Should be Combined with Some Thermal Methods, Usually Steam Injection							
Miscible and Immiscible Displacement, e.g., CO_2, Polymer, Surfactant	Not Applicable if Used Alone Because of the Low API Gravity and High Viscosity of Bitumen and Very Heavy Oil							

Carbonate formations are not suitable for in-situ combustion because these rocks decompose at high temperatures, thus reducing the heat efficiency.

(1) Thermal Methods

Thermal methods can be broken down into two classes: steam injection and combustion, examples of which are discussed below.[10,11] The purpose of all thermal recovery processes is to raise the bitumen (or heavy oil) temperature so they become mobile and will flow to the production wells.

1. Cyclic Steam Injection

Cyclic steam injection is also known as "steam soak," "steam stimulation," and "huff-and-puff." There are three phases in this process; and the same well is used for steam injection and oil production. Initially, high-pressure steam or hot water are injected into the well for a period of days or weeks. The well is then capped and allowed to soak for a few days during which the heat flows throughout the formation. In the last phase, the oil flows to the well and is pumped to the surface. Pressure in the well is controlled, resulting in vaporization of some of the water that condensed during the injection phase. This regenerated steam drives the heated oil and some water toward the producing well. Exhibit IV-8 depicts the sequence of events in cyclic steam stimulation process.[10,11]

Cyclic steam stimulation is not suitable for reservoirs with little or no primary energy because, as the name implies, it is primarily a stimulation process with some elements of displacement. Thus, after several "huff-and-puff" cycles, most wells are converted to steam drive or other methods of continuous production.

2. Steam-Drive Recovery

This process is also known as "steam displacement" and "steam flooding". Steam-drive involves injecting high quality steam into injection wells to displace the bitumen (or heavy oil) into nearby production wells. This method generally recovers a greater percentage of the in-place oil over a shorter period of time than cyclic steam injection. As steam-drive is a displacement process, it requires drilling separate injection and production wells, and establishing flow paths between them. Exhibit IV-9 shows the steam-drive method. It is by no means problem free; some of the inherent disadvantages and their proposed solutions are:

EXHIBIT IV-8

Cyclic Steam Injection

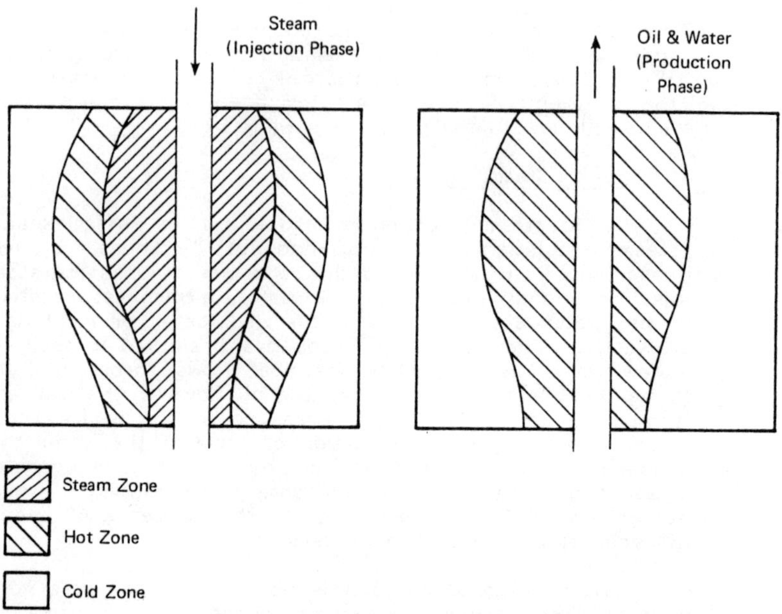

EXHIBIT IV-9

Steam Drive Recovery

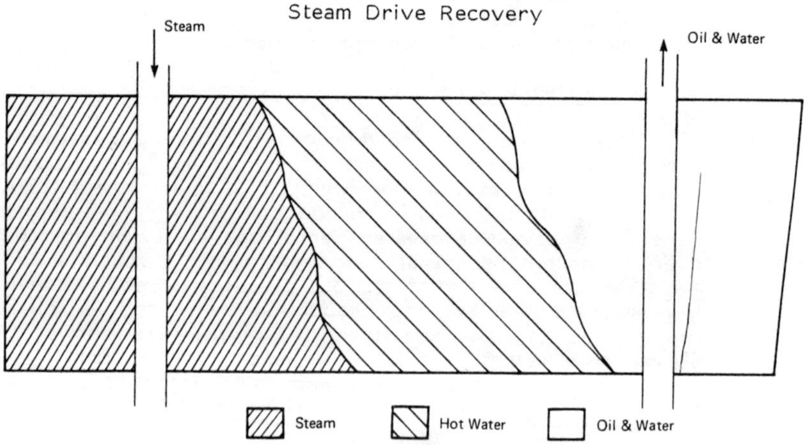

Drawbacks	Solutions
. Long duration of process	. Optimize by alternating injection and production of a large number of wells
. High energy requirement for steam generation	. More efficient steam generators, and coupled with steam and air or solvent injections
. Low efficiency due to heat loss	. Better insulated steam distribution lines and wells; e.g., thermal packers
. Low recovery because of directional permeability, uneven fracture, and bottom-water zones (thief zones).	. Better screening and fracturing techniques. Oil zones can be heated by conduction from bottom water zones.

Steam is often used in conjunction with gas, emulsifiers, or solvents. The choice additive depends on the degree of continuity and oil saturation of the reservoirs, and the homogeneity of the oil zones. When additives are used, their chemical alteration and deposition must be studied.

(1) Steam/Alkaline Additives

Alkaline solution can be injected with the steam to reduce interfacial tension. The resultant emulsion usually improves the vertical continuity of steam displacement.

(2) Steam/Solvent Drive

Solvents can be used to extract bitumen from tar sands, but they must be miscible with bitumen and must not cause asphaltene precipitation. Aromatic solvents like benzene and toluene are good candidates, except for environmental drawbacks. It has been found that naphtha, one of the products from the hydro-treaters of commercial tar sand plants, can also be used as a solvent or diluent.

(3) Steam/Gas Drive

When used in conjunction with steam, gases such as air, methane, and carbon dioxide frequently improve oil recovery. It has been found that combining natural gas and steam reduces the steam/oil ratio significantly.

3. In-Situ Combustion Recovery Processes[12]

In-situ combustion processes include a broad class of thermal recovery processes in which part of the bitumen or heavy oil is burned in the reservoir to generate the required heat. In-situ combustion is also known as "underground combustion," "underground combustion drive," and "fireflooding." Aside from the conventional dry forward combustion other variations include reverse combustion and wet combustion, the so-called COFCAW process (combination of forward combustion and water flooding).

(1) Forward Combustion

In forward combustion, ignition is initiated at the air injection well and the fire front moves forward towards the production well. Since bitumen or heavy oil has very little mobility under reservoir conditions, it is necessary to fracture the formation before beginning forward combustion. The forward combustion process is shown in Exhibit IV-10.[10] After fracturing the formation, the oil is ignited at the injection well, and the flame front propagates. The products of combustion and the displaced fluids then move through the system to the production well. Since the flame front and the injected air move in the same direction, the process is appropriately termed forward combustion.

In-situ combustion causes some cracking of the bitumen resulting in coke deposition and partial upgrading of the bitumen. The burned zone consists of sand and coke, which acts as fuel source for further combustion. The temperature of the combustion front usually ranges from 800 to 1,500°F.

(2) Reverse Combustion

In this process, ignition takes place at the production well and propagates towards the air injection well. Thus, the combustion front is traveling in opposite (reverse) direction to the air and oil flow resulting in better displacement of viscous bitumen. Reverse combustion is particularly adaptable for recovering very viscous low gravity bitumen that cannot be moved through the cold zone by forward combustion. Disadvantages of reverse combustion include:

- Lower thermal efficiency because of heat lost in the combustion zone

EXHIBIT IV-10

Forward Combustion Process

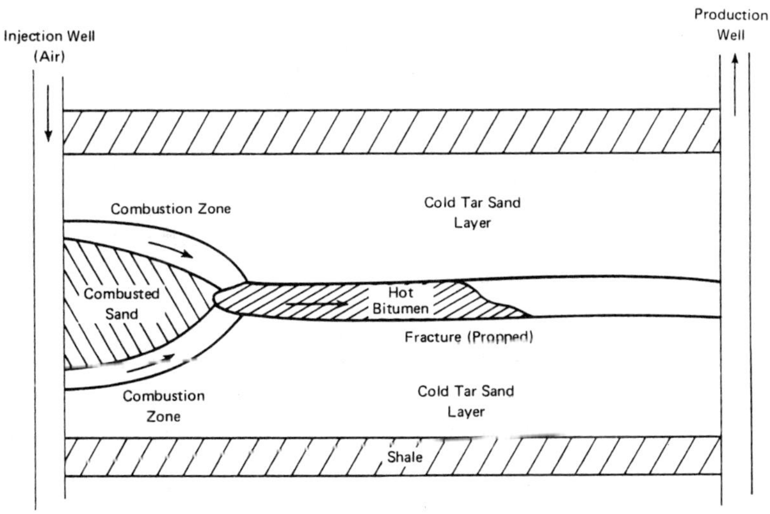

EXHIBIT IV-11

Wet Combustion

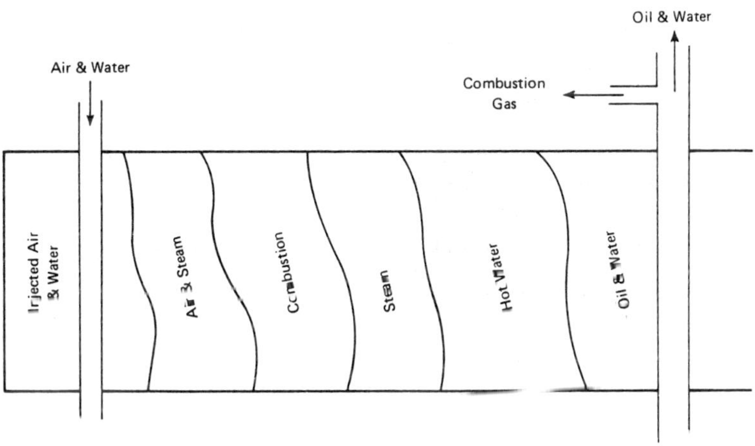

- Hot product causing pump and well problems
- Partial combustion of light liquid products.

(3) Wet Combustion

This process is also called fire-water flooding or "combination of forward combustion and water flooding" (COFCAW). It was originally developed by Amoco Canada, aimed at producing bitumen from Athabasca tar sand deposits. Exhibit IV-11 is a schematic of the wet-combustion process." The process involves three phases:

- Heat-up phase (8 months) - air is injected to promote forward in-situ combustion. Production of bitumen is controlled to ensure even heat distribution in the formation.
- Blow-down phase (4 months) - air injection is suspended and production is throttled to lower the formation pressure.
- Displacement phase (6 months) - fire/water flooding is carried out to supply heat and driving fluids for the residual in-place bitumen. The air/water ratio used should be 500-3,000 scf/bbl.

Wet combustion is usually superior to the dry combustion process because it improves oil recovery and increases heat efficiency. The injected water vaporizes in the hot reservoir, transporting heat ahead of the combustion zone. This results in higher sweep efficiency and lower fuel usage.

Some of the advantages and disadvantages of the wet in-situ combustion process are:

Advantages	Disadvantages
. Less heat loss than steam-related processes.	
. Spontaneous ignition can be achieved if the formation temperature is high enough.	. Inherent stability problems
. Low-power requirement.	. Lower percentage recovery because part of the oil is used as fuel

Coking or pyrolysis of the bitumen, this results in partial product upgrading.	Corrosion and thermal damage to well casings.

(2) Novel In-Situ Recovery Methods

1. Use of Electric Power to Recover Bitumen[13-15]

Electric heating is a novel method aimed at recovering bitumen from deeply buried tar sands. Electric resistive heating and inductive heating are being investigated as methods for increasing the temperature of the bitumen. Both methods are designed to supply heat to the tar sand formation, thereby reducing the viscosity of the impregnated bitumen so it can be recovered by more conventional production methods. Unlike other in-situ recovery processes, electric heating does not require initial fracturing of the formation, thus it has better contact and directional control.

Electric resistance heating was first investigated by Atlantic Richfield Canada and is relatively simple. It utilizes resistance heating in the same way as an ordinary electric heater, with the tar sands acting as the resistive element. Current flows from one electrode well through the formation to a second electrode well, thereby converting electric energy to thermal energy, which dissipated throughout the formation. Brine is often pumped into the formations to improve electrical conductivity of the tar sands. Exhibit IV-12 is a simplified schematic diagram of the electric resistance heating method.

Electrical induction heating involves generating thermal energy within an induction coil when an alternating current is passed through it. The heat then propagates radially outward, heating the deposit contained by the coil. The induction coil is usually toroidal in shape and is formed by drilling an inner and outer set of shafts and tunnels encompassing the deposit. Electrical conductors are threaded through these openings, and the ends of the coil are brought to the surface through the inner set of shafts. Exhibit IV-13 is a conceptual diagram of the induction heating process.[13]

At present, electric heating has progressed from the conceptual stage to pilot-scale experiments. Extensive field tests are still needed to test the feasibility of electric resistive and inductive heating as applied to in-situ recovery of bitumen from tar sands. The advantages and disadvantages associated with this process are listed below.

EXHIBIT IV-12

Electric Resistive Heating

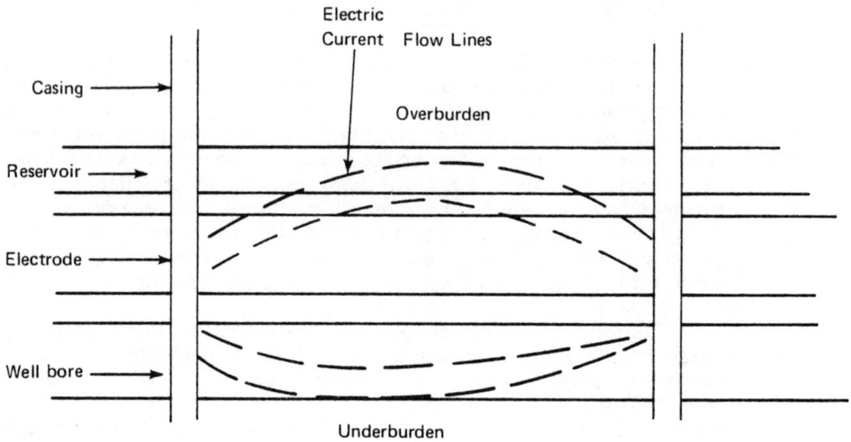

EXHIBIT IV-13

Electric Induction Heating

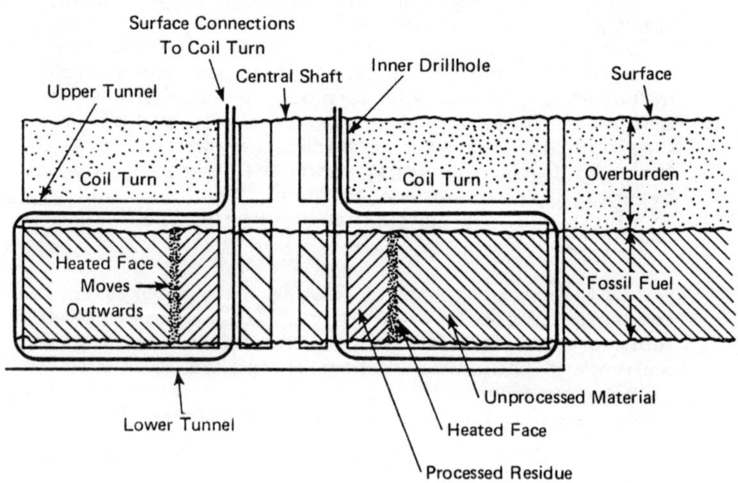

Advantages	Disadvantages
. Less environmental problems	. Poor thermal conductivity of bitumen
. Little surface and formation disturbance	. Localized heating often leads to evaporation of water (loss of electrical medium for resistive heating only)
. Better control of process.	. Prevalent uncertainties in design and fabrications
	. Expensive form of energy
	. Uneven heating
	. Unproven technology.

2. Electromagnetic Heating[16,17]

This is a broad class of methods which induce heating by electromagnetic radiation. These effects are displayed from the lowest frequencies (e.g., 60 cycles/second) to the highest (X-rays, gamma ray, etc.). Two methods which have been found to show some promise are:

- Radio frequency heating (10^4 - 10^7 cycles/second)
- Microwave heating (10^9 - 10^{11} cycles/second).

Resistive heating has the disadvantage of localized heating around the electrodes or induction coils, and, as a result, is relatively inefficient. Electromagnetic heating is based on dielectric heating, and does not depend on the conductivity of heating media; e.g., water or brine, thus more even heating and penetration can be achieved.

If the correct frequency and placement of electrodes are selected, radio frequency heating can be propagated up to a distance of ~300 ft. within a volume of 3×10^7 - 3×10^8 feet^3. The Illinois Institute of Technology Research Institute has recently devised a method of electrode placement which achieves high efficiency and minimum radiation leakage. This method involves a three-parallel-plate configuration which can be approximated in actual field tests by parallel rows of buried tubular conductors. After heating, the mobile bitumen can be pumped out by conventional methods like surfactant flooding or by gravity drainage coupled with gas drive if the electrodes are placed horizontally.

Microwave heating entails placement of antennas inside wellbores. The antennas are connected to an aboveground power supply by coaxial cables. See Exhibit IV-14.[16] While this is a volumetric heating method, it suffers from the disadvantage of overheating near the antennas, and underheating at a distance. Moreover, the size of the antenna required is directly proportional to the wavelength. For most deposits which require lower frequencies and thus longer wavelengths, the size of the antenna may become unwieldly and impractical inside typical wellbores. Other difficulties include complicated well completions and precise control required for optimization of power transfer.

IITRI will soon carry out actual field testing of the radio frequency heating concept at the Tar Sand Triangle in Utah. This project is funded by the U.S. Department of Energy and is aimed at gathering operating information and determining process parameters.

3. Microbial Recovery Processes[18]

This radically new concept calls for using the metabolic processes of bacteria for bitumen extraction or heavy oil recovery. Certain micro-organisms are known to feed on the heavy fractions of the oil, biodegrading them, and producing gases and surface-active agents. As a result of these metabolic activities, the crude oil is made lighter and less viscous. The gaseous products also act to increase the formation pressure and drive the bitumen toward the producing well.

An alternate but similar process is to grow the bacteria on surface tanks and extract enzymes from them which are then injected with water into the wells. These enzymes "digest" the heavy fractions of the oil, thus reducing its viscosity, and changing the surface tension. This method eliminates the problem of formation "plugging" caused by the accumulation of dead biomass.

The two most serious problems with the microbial process are: (1) no universal enzyme has been found which can act on a wide range of hydrocarbons, and (2) the uneven distribution of agents injected into the formation. Microbial recovery can be classified as an esoteric process which shows some promise, but is still a long way from any large-scale development. Currently, microbial field tests are being carried out in Poland, Hungary, and Czechoslovakia, but empirical data are extremely limited.

EXHIBIT IV-14

Microwave Heating System

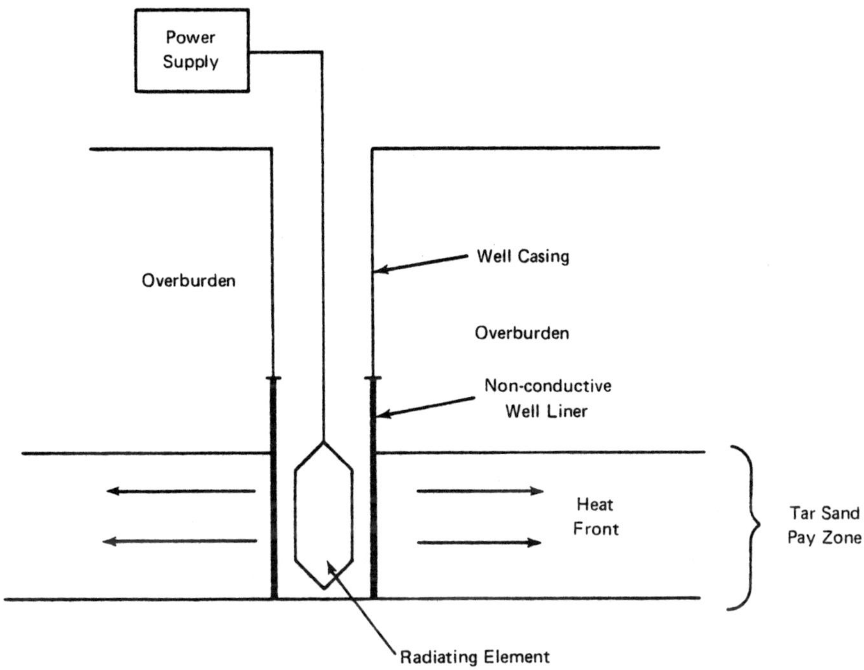

4. Nuclear Detonation[19]

Underground nuclear detonation has been proposed as a scheme to extract bitumen from deep formations (>1,000 feet). Detonation of the nuclear device would fracture the formation, creating a rubble-filled cavity into which the heated, fluid bitumen would drain. The gravity drainage phase could be accelerated by injecting air or steam.

The main drawbacks of this method are: (1) uncontrollable and sometimes severe fracturing of the formation, (2) loss of bitumen to adjacent permeable formations if the original deposit is not continuous, and (3) containment of the nuclear explosion radioactive substances and contamination of the oil and gas products. The last two problems require that the deposits be deep, thick, and continuous. The many serious problems associated with this process may prevent its future development.

(3) Underground Mining of Tar Sands[20-23]

Russia is continuing to develop underground mining methods for the recovery of bitumen from moderately deep tar sand deposits. Underground oil mining was previously practiced in Germany from 1900 to the late 1930s in the Weitz, Pechelbroun and Heide fields, but these operations later became uneconomic. At present the only commercial underground tar sands mining now operating is located in the Yarega field near Ukhta in the Komi Autonomous Republic. The Yarega field is about 750 miles northeast of Moscow. A second tar sands mining project is being developed in the 100-year old Balakany field located near Baku in Azerbijan.

Underground mining of tar sands is accomplished by sinking large diameter access shafts from the ground surface to the tar sand deposit. Horizontal access tunnels are then driven at the bottom of the deposit to connect the vertical shafts. Small diameter holes are drilled from the horizontal tunnel upward at various angles into the tar sand pay zone. The viscous oil flows by gravity into the tunnel where it is collected and pumped to the ground surface. The oil flow rate can be increased significantly by injecting steam into the formation. A second mining method that has been attempted with less success is the Uhtinsk system in which the horizontal tunnels are driven at the top of the tar sands deposit and several vertical holes are then drilled down into the underlying pay section.

Underground tar sands mining was first started at Yarega in 1940 when it was found that only 1 to 2 percent of the oil-in-place could be produced by conventional drilling and production methods. The Yarega deposit is at a depth of 600 to 650 feet and has a pay thickness of some 50 feet. Bitumen gravity is 16° API and its viscosity is 15,000-20,000 cp at the reservoir temperature of 43°F. Raising the bitumen temperature to 200°F reduces its viscosity to 30 cp. Hard rock formation overlies the sandstone reservoir. Reservoir porosity is 24 percent and its permeability varies from 500 to 5,000 md. Jet turbine drills were used to sink two vertical shafts up to 12 feet diameter in order to access the Yarega deposit. One shaft is used for transporting the miners, equipment, and mine tailings; and the other shaft is used to provide mine ventilation. The horizontal tunnels are driven to form blocks in the deposit about 500 feet on a side. Oil production holes (3 to 4 inch diameter) are drilled upward at various angles, several hundred feet into the deposit. Spacing between the oil production holes is 40 to 60 feet.

Steam injection was not used in the Yarega mine until 1968 and oil production declined from a peak of 3,500 b/d in 1950 to about 1,000 b/d in 1968. Oil production increased rapidly after steam injection was started and is now estimated to total 15,000 b/d from the three Yarega mines. The Russians have estimated that to produce 10,000 bbl of bitumen from the deposit requires 150 feet of tunnel, 2,300 feet of oil production wells and the injection of 20,000 bbl of steam. Oil recovery before steam injection was used and was estimated at only 10 percent of the original oil-in-place.

The second underground oil mining project that is being developed at the Balakhany field near Baku has encountered problems which have postponed the start of production. Production was originally planned to start in 1978, but has been delayed to 1981 because the shafts and tunnels must penetrate unstable sands and clays which are susceptible to cave-ins and instability problems. It was thought that the mining technology developed for Yarega could be used, but the Balakhany bitumen contains more natural gas which creates a safety problem and requires automatic gas detection and alarm equipment. Two vertical shafts, 26 feet in diameter, have been sunk to the bottom of the Balakhany oil sand formation and work is now underway on driving the horizontal tunnels. Plans call for recovering a cumulative total of 500,000 bbl of heavy oil from the field by the late 1980s. The recovery of heavy oil by primary and secondary production from the Azerbaijan fields near Baku has averaged in the range of only 5 to 20 percent. The goal for bitumen recovery by underground mining is 40 to 50 percent of the original in-place oil. Surface strip mining is also being considered for the shallow bitumen deposits near Baku. It is estimated that the area has 25 fields with deposits at depths of less than 1,500 feet.

4. FACTORS AFFECTING TAR SANDS DEVELOPMENT

Many factors will affect the future development of tar sands resources, and these factors strongly influence the selection of both recovery and upgrading processes (discussed in Chapter V) for specific tar sands deposits. The important selection factors can be grouped into three general areas:

- Site-specific factors
- Process-related factors
- External factors.

Some factors impact more than one of the above areas and are particularly important as process selection criteria. For example, environmental regulations are determined by external factors, but environmental controls are strongly dependent on site-specific and process-specific factors. Another example is in-situ recovery processes which require large amounts of fuel that can be supplied from either the upgrading process or from outside sources, such as coal or natural gas.

(1) Site Specific Factors

Site-specific factors are fixed by the deposit characteristics and location, bitumen quality, and availability of other resources, including water and fuel supplies. The following site-specific factors must be carefully evaluated when selecting a recovery and upgrading process for a specific tar sands deposit:

- Deposit characteristics
  - Overburden depth, pay thickness, and areal extent of deposit
  - Reservoir porosity, permeability, and transmissibility
  - Deposit porosity, percent oil saturation, and amount of oil-in-place
  - Reservoir pressure, temperature, gas/oil ratio, and aquifer location
  - Site location, climate and accessibility
  - Deposit heterogenity, rock type, and faulting
- Bitumen quality factors
  - Bitumen gravity, viscosity, sulfur, nitrogen, and metals content

Infrastructure factors

- Availability of water and electric power
- Transportation facilities to and from the project site
- Housing facilities at the site.

Selection guides, such as the one shown in Exhibit IV-15, aid in preliminary process screening. However, as indicated by the number of factors above, the process selection is complicated by many interrelated variables. Most importantly, the variable characteristics of tar sands deposits require that extensive, long-term field tests be conducted. Field testing can take five years, and even then extrapolating the results to a commercial project involves many risks and uncertainties. By testing a prospective process in the field, real operating problems can be uncovered and corrected. Unfortunately, there is no shortcut to obtaining field operating information. To date the operating experiences of commercial tar sands plants are inadequate for optimizing process selection. It is, therefore, reasonable to assume that future tar sand projects will continue to require extensive techno-economic analysis and field testing of alternative recovery and upgrading processes for each specific deposit.

(2) Process-Related Factors

Process-related factors include fixed costs, operating costs, energy efficiency, emissions, process reliability and development status, and time required to reach full production.

Fixed costs are items such as mine investment, process plant facilities investment, utilities investment, and working capital. Operating costs include raw materials, supplies, maintenance labor, and operating labor. Energy efficiency is important because it often affects both fixed and operating costs, and government regulatory agencies usually require maximum resource recovery. Process emissions of solid, liquid, and gaseous pollutants must be controlled to comply with local regulations. Thermal-based processes have a common disadvantage in that they generate SOx, NOx, and particulates. Disposal of spent tar sands tailings from the hot-water extraction process is complicated by the low settling rates of fine solids. This necessitates the construction of large tailings storage ponds. Steam injection processes commonly use process-produced fuel, which generates pollutants such as sulfur and nitrogen oxides, and particulates. To comply with emission standards usually means the construction of scrubbers to remove the pollutants from stack gas. Generally, this costs an additional $1-2 per barrel of oil produced.

EXHIBIT IV-15

Effect of Site Specific Factors on Tar Sands Recovery Process

			SITE SPECIFIC FACTORS			
			FORMATION CHARACTERISTICS	HETEROGENEITY OF DEPOSITS	AMOUNT OF RESERVES	AVAILABILITY OF FEED WATER
SURFACE RECOVERY	MINING	DRAGLINES	◒	◒	◒	◒
		BUCKET WHEEL	◒	◒	◒	◒
	EXTRACTION	HOT WATER PROCESS	●	●	◒	○
		SOLVENT EXTRACTION	◒	◒	○	◒
		DRY DISTILLATION	◒	◒	◒	●
IN-SITU RECOVERY		STEAM SOAK	○	○	◒	○
		STEAM DRIVE	○	○	◒	○
		STEAM/DILUENT	○	○	○	○
		STEAM/GAS	○	○	○	○
		FORWARD COMBUSTION	◒	◒	◒	●
		REVERSE COMBUSTION	◒	◒	○	●
		WET COMBUSTION	◒	◒	◒	◒
		ELECTRIC HEATING	◒	○	◒	◒
		MICROWAVE HEATING	◒	●	○	◒
		MICROBIAL SURFACTANTS	○	○	○	◒
		NUCLEAR DETONATION	◒	◒	○	○
		UNDERGROUND MINING	○	○	○	◒

● DEPENDENT
◒ MODERATELY DEPENDENT
○ HIGHLY DEPENDENT

The status of development of a process is a measure of its feasibility and near-term applicability. It takes at least five years for completion of pilot projects and nine years for the planning, procurement, and construction of a commercial tar sands project. Thus, barring unforeseen economic and technical difficulties, it takes a minimum of 14 years for a project to progress from the conceptual to the commercial stage. With the vast amount of deeply buried tar sands still untapped, future efforts will be directed more towards the improvement and development of in-situ recovery processes. Steam injection processes are proven technologies, but they suffer from emissions problems and low thermal efficiencies. Other in-situ processes, with the exception of the in-situ combustion processes, are not well enough developed to be used in large-scale projects during the next 10 years.

(3) External Factors

Assuming the site-specific and process-related factors have been solved, a project's ultimate success will often depend on a set of factors which are external to the taxonomy of tar sands technology. These external factors are related to the overall business and economic climate, which may or may not be conducive to the development of tar sands resources. Specifically, the business and economic factors include:

- Cost inflation
- Taxes and royalties
- World energy prices
- Interfuel competition
- Environmental constraints
- Business financial risks.

The first three external factors will have important impact on the development of tar sands as an alternate source of energy. Cost inflation is a potential major obstacle because of the long lead time for a plant to be built. It disrupts project planning and can very often render a prospective project uneconomical. An inflationary environment also increases business financial risks and stifles development.

Unfavorable tax and royalty policies reduce the profitability of any business venture. The Alsands project in Alberta has temporarily been suspended pending the outcome of pricing negotiations between the provincial government of Alberta and the federal government of Canada. This is but one of the examples of how restrictive climate can hamper the progress of a developing industry. It is imperative that government spell out such policies early to spur development. The current Canadian situation is counterproductive, as government policy on royalties is negotiated after a company has committed tremendous resources to a large-scale development.

World energy prices will ultimately determine the competitiveness of a particular source of energy. Current OPEC light crude prices of $30-38/barrel have made many alternate sources of energy attractive. The results of a recent study by Bechtel,[64] Inc., which compares the prices of various alternative sources of energy, is reproduced in Exhibit IV-16. This indicates that oil from tar sands compares favorably with other synfuels. However, it should be pointed out that it is difficult and risky to generalize on the cost of tar sands plants.

Interfuel competition, environmental constraints, and business financial risks affect specific projects in varying degrees. Environmental constraints typically affect surface-mining processes more than in-situ ones. Business financial risks are usually high for processes that are in their initial state of development and that are prone to early obsolescence.

Based on the three sets of factors described, the different tar sands production processes can be compared qualitatively. These three classes of factors affect each process differently, and the magnitude is often difficult to measure. Processes should only be compared within their own classifications, as it is meaningless to compare surface extraction with in-situ recovery methods. Exhibits IV-15, 17, and 18 show how the processes compare qualitatively. It should be noted that the list is by no means exhaustive, and some of the "novel" production methods on which only limited information is available are based largely on intuitive evaluation.

In the sections that follow, we can readily see how these factors have come into play as the various technologies are applied to commercial and pilot projects.

5. COMMERCIAL TAR SANDS PROJECTS

At present there are four commercial scale tar sand projects in various stages of operation, construction, and planning. Three of these are surface mining projects located along a 20-mile stretch of the Athabasca River and the fourth is an in-situ recovery project located at Cold Lake. Exhibit IV-19 is a summary of the four commercial projects.[38] The Alsands Consortium project and the Esso Cold Lake project are in the design and planning stage. Exhibit IV-20 is a plot showing the estimated project construction costs on the basis of current dollars per daily barrel of bitumen production. The results indicate that construction costs are estimated to increase by a factor of 8 over an 18-year period which is equal to a compounded annual rate of 12 percent.

EXHIBIT IV-16

Comparison of Fuel Production Costs

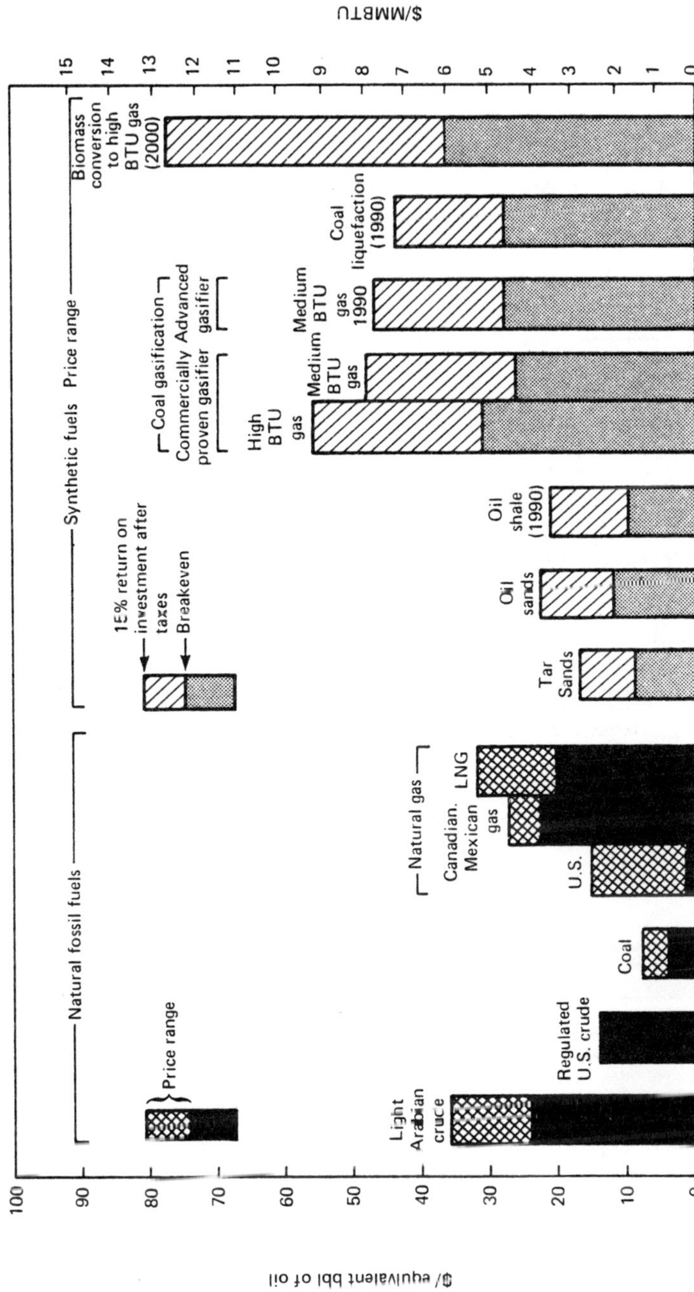

Source: Bechtel Inc. In constant 1979 dollars

EXHIBIT IV-17

Effect of Process Specific Factors on Tar Sands Recovery Processes

			PROCESS SPECIFIC FACTORS				
		PROCESS	STATUS OF DEVELOPMENT	FIXED COSTS	OPERATING COSTS	ENERGY EFFICIENCY	EMISSIONS/ TAILINGS
SURFACE RECOVERY	MINING	DRAGLINES	●	●	◒	◒	◒
		BUCKET WHEEL	●	◒	●	◒	◒
	EXTRACTION	HOT WATER PROCESS	●	◒	◒	○	○
		SOLVENT EXTRACTION	◒	○	○	◒	◒
		DRY DISTILLATION	◒	○	◒	○	●
IN-SITU RECOVERY		STEAM SOAK	●	●	◒	●	◒
		STEAM DRIVE	●	◒	●	◒	○
		STEAM/DILUENT	◒	◒	◒	◒	◒
		STEAM/GAS	◒	◒	◒	◒	◒
		FORWARD COMBUSTION	◒	◒	○	◒	◒
		REVERSE COMBUSTION	○	◒	○	◒	◒
		WET COMBUSTION	◒	◒	◒	●	◒
		ELECTRIC HEATING	○	◒	○	◒	●
		MICROWAVE HEATING	○	◒	◒	◒	●
		MICROBIAL SURFACTANTS	○	●	◒	●	◒
		NUCLEAR DETONATION	○	○	○	●	○
		UNDERGROUND MINING	◒	◒	◒	◒	◒

● FAVORABLE
◒ NEUTRAL
○ UNFAVORABLE

EXHIBIT IV-18

Effect of External Factors of Tar Sands Recovery Processes

			External Factors		
		Process	Degree of Competition	Environmental Constraints	Business Risks
Surface Recovery	Mining	Draglines	Medium	High	Low
		Bucket Wheel	Medium	High	Low
	Extraction	Hot Water Process	High	High	Low
		Solvent Extraction	Medium	Medium	High
		Dry Distillation	Medium	Medium	High
In-Situ Recovery		Steam Soak	High	Medium	Low
		Steam Drive	High	Medium	Low
		Steam/Diluent	Medium	Medium	Medium
		Steam/Gas	Medium	Medium	Medium
		Forward Combustion	Medium	Medium	Medium
		Reverse Combustion	Low	Medium	High
		Wet Combustion	Medium	Medium	Medium
		Electric Heating	Medium	Low	Medium
		Microwave Heating	Medium	Medium	High
		Microbial Surfactants	Low	Medium	High
		Nuclear Detonation	Low	High	High
		Underground Mining	Medium	Medium	Medium

● LOW
◒ MEDIUM
○ HIGH

Exhibit IV-19: Commercial Tar Sand Projects in Alberta, Canada

Operator	Location	Recovery Method	Upgrading	Geological Formation	O.I.P. Barrels	Anticipated Recovery Barrels	Oil Gravity (API)	Project Status
Suncor (Great Canadian Oil Sands)	Suncor Project, Ft. McMurray, Athabasca Region	Open pit mining using dragline, bucketwheel excavators to conveyors	Delayed coking followed by hydrogenation	Wabiskaw-McMurray	1.03×10^9	6.3×10^6	9-10	1962-ERCB approval 1964-ERCB approval for 31,500 B/D construction production; 1967-1972-ERCB appl.-45,000 B/D for 50-60,000 B/D 1973-ERCB approval for 65,000 B/D
Syncrude-Esso Resources Ltd. (31.25%) Canada-Cities Service Ltd. (22%); Gulf Canada Res. Inc. (16.75%); PetroCanada Expl. Inc. (15%); Alta. Energy (10%); Pan Canadian Petroleum Inc. (5%)	Syncrude Project, Mildred Lake, Ft. McMurray, Athabasca Region	Surface mining-use of dragline to bucketwheel reclaimers to conveyor belt to extraction plant	Two-stage fluid coking followed by hydrotreating	–	9.4×10^9	–	6-10	Production mid-1986 (140,000 B/D). First oil sand mining June 1977. April 1978-first oil sand processing in extraction plant. 1st fluid coker up-July 1978; 2nd opening ceremonies Sept. 1978 coker on stream, Oct. 78-1st synthetic crude pumped to pipeline, July 1978 (planned prod. 130,000 B/D by 1981)
Alsands-Shell Canada Res. Ltd. (25%); Shell Expl. Ltd. (20%); Amoco Can. Ret. Ltd. (10%); Chevron Standard Ltd. (8%); Dome Pet. Ltd. (4%); Gulf Can. Ltd. (8%); Hudsons Bay Oil & Gas Ltd. (8%); PetroCanada (9%); Petrofina Can. Ltd. (8%)	Aslands Project, Athabasca Region	Open pit mining using dragline, bucketwheel reclaimers and conveyor belt	Fluid coking followed by hydrotreating	–	8.2×10^9	44×10^9	6-10	Dec. 1978-ERCB application full production 140,000 B/D 1986-1987. Preliminary construction suspended.
Esso Resources Canada Ltd.	Esso Projects, Grand Centre, Cold Lake Region	Cyclic steam stimulation	Flexicoking followed by hydrotreating	Lower Cretaceous Mannville Group (Clearwater)	8.2×10^9	20% O.I.P.	10	ERCB Application Nov. 1977; planning suspended

EXHIBIT IV-20

Estimated Investment Required for Commercial Tar Sands Projects

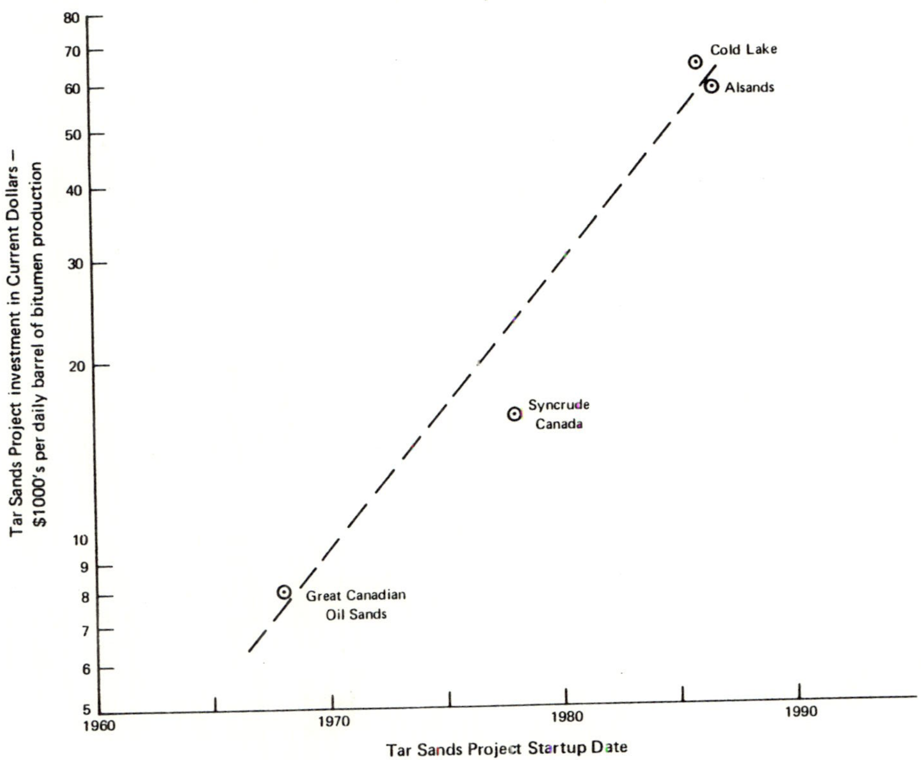

(1) Great Canadian Oil Sands - Suncor[24-26]

GCOS has extensive experience in tar sands. In 1960 GCOS applied to the Alberta Oil and Gas Conservation Board (now the Energy Resources Conservation Board) for construction of a synthetic crude oil plant with a capacity of 31,500 b/d. In 1962 Sun Oil Company gained controlling interest of GCOS and revised its application to a 45,000 b/d of synthetic crude oil. Construction began in 1964, and the plant was brought on-stream in 1967. Recently GCOS has been officially renamed Suncor.

The GCOS plant is the first commercial scale tar sand mining and processing plant. The operation involves four different stages:

Stage	Operation
1	Overburden removal and surface tar sand mining
2	Tar sand transport and preparation
3	Bitumen extraction
4	Bitumen upgrading

Exhibit IV-21 is a schematic diagram of the GCOS Plant.[2] The first stage consists of removing the overburden by 15-cubic yard front-end loaders. A bucket wheel excavator is also utilized for removing the overburden when required. Mining is carried out by a two-bench system, employing two giant bucket-wheel excavators, each with a design capacity of up to 100,000 tons/day.

The mined tar sands are transported to the extraction plant by a series of conveyor belts, moving at a speed of 1,000 feet/minute. The load from the conveyor belts is then dumped into surge bins capable of handling 10,000 tons.

GCOS uses the hot water process described previously for extracting raw bitumen from tar sands. The final bitumen cleanup stage of the extraction process entails mixing the froth with naphtha to reduce its viscosity. The mixture is then pumped to the separation plant where water and mineral particles are separated from raw bitumen by two centrifuges connected in series.

In the upgrading process, the bitumen is first preheated to remove diluents. Bitumen upgrading is achieved by passing the oil through a series of six delayed-coking drums, each 26 feet in diameter and 95 feet in height. The overhead products consist of three streams: naphtha, kerosene, and gas oil. Each stream is hydrotreated separately, and their products are blended to a low-sulfur synthetic crude oil of approximately 30° API.

EXHIBIT IV-21

Schematic Diagram of GCOS Plant

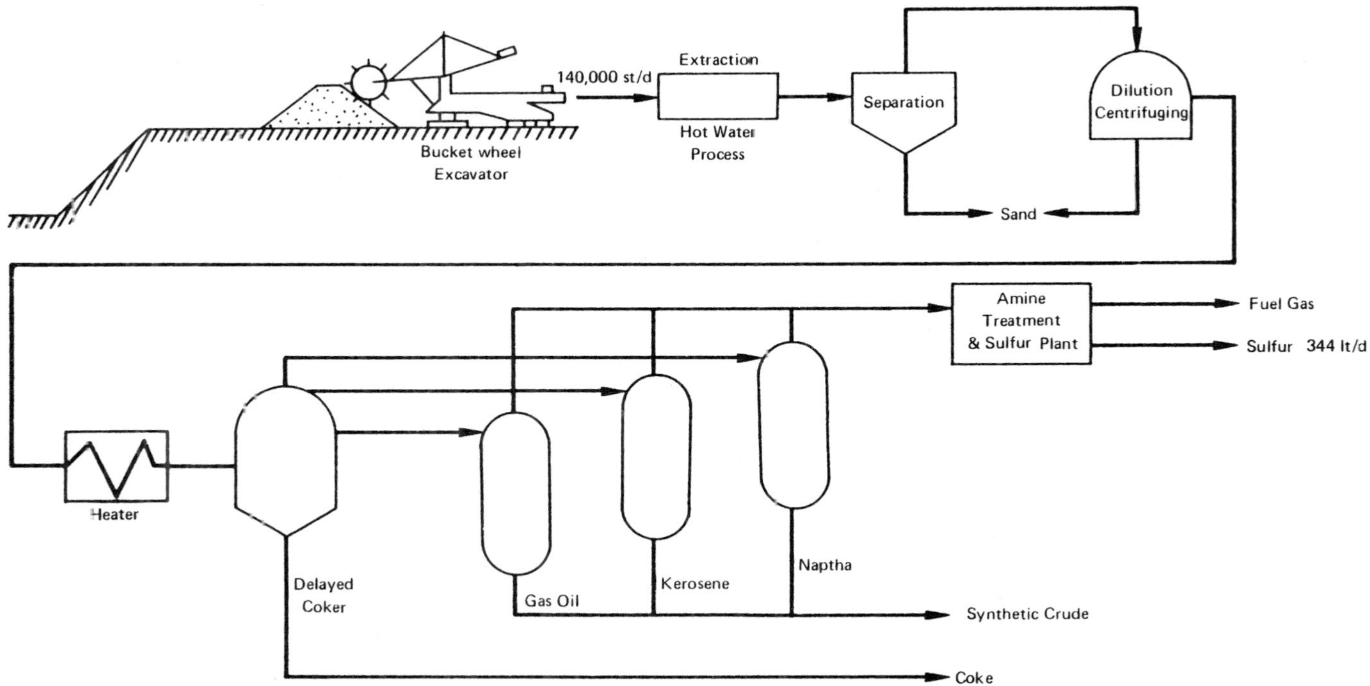

In 1978 GCOS applied to the Alberta Energy Resources Conservation Board for expansion of the plant to 65,000 b/d. This expansion included adding a third mining train; expanding the bitumen extraction facilities by adding a fifth extraction line and adding two coke drums and a coke heater for the upgrading process. The estimated total cost of the expansion is $186 million. Permission to expand was granted in early 1979, and construction is currently underway, with Canadian Bechtel Limited being retained as the prime contractor.

Overall synthetic crude oil recovery after expansion would be 48.1 weight percent of the original in-place bitumen. Exhibit IV-22 shows the recovery efficiencies on a weight percent basis of the three stages of operation.[3]

EXHIBIT IV-22

Recovery Efficiency CGOS Project
(Based on wt Percent of Original Bitumen In-Place)

STAGE	STAGE PERCENT RECOVERY 1973	STAGE PERCENT RECOVERY 1982/87	PRODUCT DESCRIPTION	OVERALL PERCENT RECOVERY 1973	OVERALL PERCENT RECOVERY 1982/87
Mining	80	80	Plant fuel	80	80
Primary and final extraction	90	91.4	Crude bitumen	72	73
Upgrading (delayed coking)	65	65.8	Synthetic crude oil	47	48

Source: Reference 3.

The GCOS tar sand plant was a pioneer venture and as a result experienced material handling problems which had never been confronted before. For example, for every two tons of ore mined, a ton of overburden must be removed. After extraction the sand must be removed to the tailings area. It is estimated that for every ton of oil produced, it is necessary to move approximately 12 tons of solids. For the size of GCOS, this adds up to roughly 130 thousand tons of solid removal every day.

Weather conditions, which can vary from above 90°F in the summer to 55°F below zero in winter, create major operating problems and constraints. During winter equipment wears out faster, and the oil sands quickly freeze on contact with cold equipment. In the warm and humid months, the tar sands are transformed into a pulpy mass and bog down all transportation equipment.

Except for 1974, GCOS incurred a financial loss every year until 1978 when the world price of crude oil rose enough for the synthetic crude oil from tar sands to become economically competitive and profitable. The pessimistic economics, coupled with technological and environmental difficulties have delayed the rapid development of synthetic crude oil from tar sands.

(2) Syncrude Canada

The Syncrude project began construction in late 1973 on a site (covering 7,000 acres) 25 miles north of Fort McMurray.[28,29] The plant came on-stream in August 1978 costing $2 billion to complete. The great significance of this plant is its immense size, 129,000 bbl/calendar day, more than two times the size of GCOS. Approximately two tons of solids have to be moved for every barrel of synthetic crude oil produced.

After the withdrawal of Atlantic Richfield Canada from the original consortium, an infusion of funds by the governments of Canada, Alberta, and Ontario was required. In 1979 Pan Canadian Petroleum purchased the Ontario government share. The present consortium is made up of:

	Percentage
Imperial Oil	31.25
Canada-Cities Service	22.00
Gulf Oil Canada	16.75
Petro-Canada (Government-owned)	15.00
Alberta Energy Company	10.00
Pan Canadian Petroleum, Ltd.	5.00
	100.00

The Syncrude Plant uses two electrically operated draglines for overburden removal and tar sand mining. Each dragline operates from a bench about midway in the open pit mining face. The company is currently applying to the Energy Conservation Board to build an out-of-pit permanent overburden discard site as the mined-out area cannot accommodate all the mined material after it has been removed and processed.

The extraction stage of the Syncrude Plant is quite similar to that employed by GCOS. There are essentially two stages of bitumen recovery, followed by a final bitumen cleanup and recovery of raw bitumen from a dilution centrifuge process.

In the upgrading stage, the two product streams, naphtha and gas oil are hydrotreated separately to remove sulfur, nitrogen, and oxygen. They are then blended to give low-sulfur synthetic crude oil which can be pipelined and is for further refining. The required hydrogen is produced by stream reforming purchased natural gas. By limiting the air through the coker burner, CO is produced, collected, and used as boiler fuel. Exhibit IV-23[2] is a schematic diagram of the Syncrude Plant.

The province of Alberta provided Syncrude Canada Ltd. with some financial incentives to make this mammoth project a reality. In 1973 the Alberta government declared Syncrude as a joint venture and waived the royalty payment. Instead, a 50 percent share of the net profits would be paid by the company after production officially began--the date when five million barrels of synthetic crude oil had been delivered. This milestone was reached in February 1979. If no profits were realized after a few years of operation, Alberta could then elect to receive a given production royalty of seven and a half percent in the form of either oil or current dollar value of oil.

By the end of this year Syncrude will have moved about 170 million tons of overburden plus tar sand and produced about 48 million barrels of synthetic crude. The present and projected status of the Syncrude Plant is summarized in Exhibit IV-24.

As expected, a project of this scale and complexity will have problems during the initial phases of its operation. Compared with the GCOS plant, Syncrude's start-up has been relatively smooth partly because the company has benefited from GCOS' experience. Some of the problems and solutions in tar sands mining, extraction, and upgrading in the Syncrude plant are presented in Exhibit IV-25.[30]

(3) Alsands Consortium[31-33]

In late 1979, the Alberta Energy Conservation Board granted permission to the Alsands Project Group to build a synthetic crude oil plant with a design capacity of 140,000 b/cd.

The plant will be located on the east side of the Athabasca River (40 miles north of Fort McMurray) and cover 42,000 acres. It is estimated that the project area has sufficient reserves to sustain continuous mining for over 100 years at the design capacity.

The Alsands Project Group is a consortium headed by Shell Canada and is comprised of:

- Amoco Canada Petroleum Company Ltd. (10 percent)
- Chevron Standard Limited (8 percent)
- Dome Petroleum Limited (4 percent)
- Gulf Canada Resources Inc. (8 percent)
- Hudson Bay Oil & Gas Company Ltd. (8 percent)

EXHIBIT IV-23

Schematic Diagram of Syncrude Canada Project

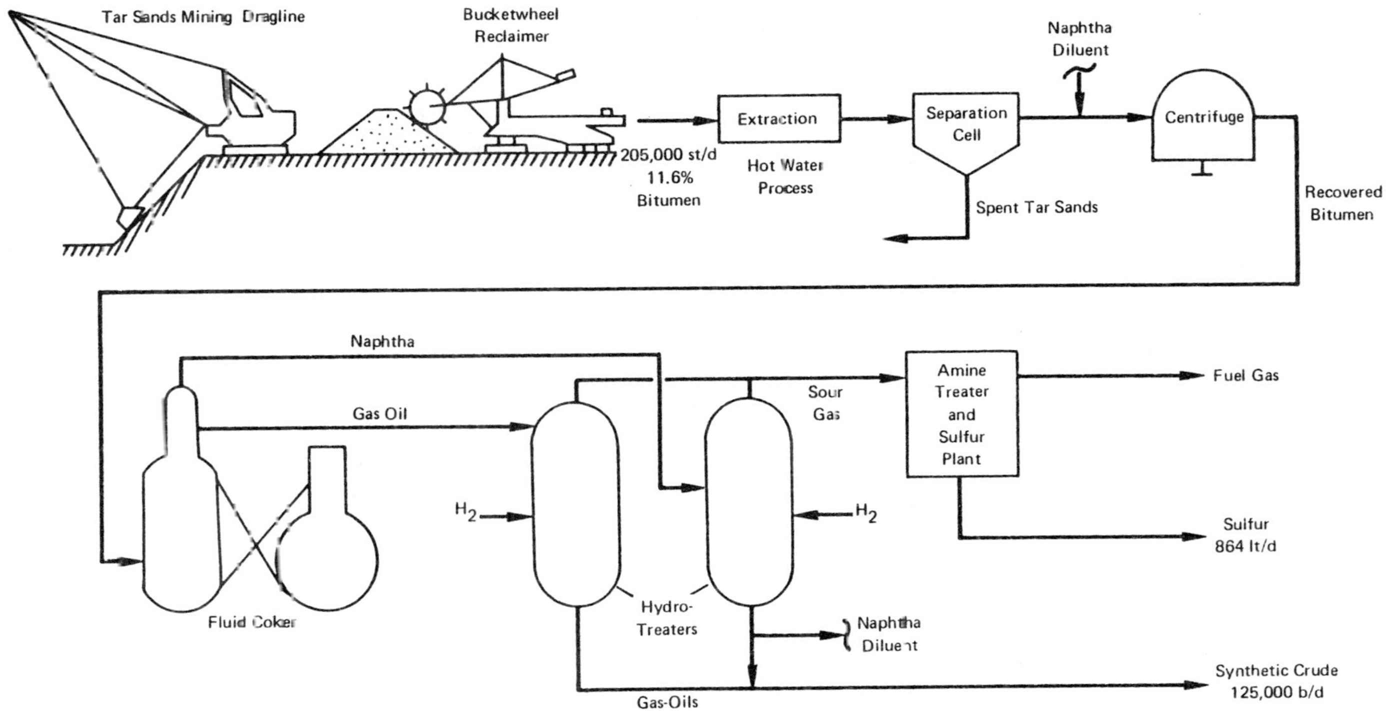

EXHIBIT IV-24

Syncrude Canada Project Production Record

Exhibit 20

SYNCRUDE CANADA PROJECT PRODUCTION RECORD

	AVERAGE DAILY PRODUCTION b/cd		SYNTHETIC CRUDE SHIPPED MM b/y
1978	9,700		3.55
1979	49,400		18.026
1980	90,400		33.00*
		Total	54.576

* Projected for 1980.

EXHIBIT IV-25

Selected Major Problems and Solutions of Syncrude

PROBLEM	SOLUTION
MINING	
. Low equipment productivity	- Equipment modifications - Improved maintenance - Overburden pre-stripping
. Geotechnical - Clearwater strata - Siltstones	 - Overburden pre-stripping - Overburden blasting
. Greater oil sand dilution	- Overburden pre-stripping
. Frozen tar sand	- Lower windrow inventory - Equipment modifications
. Reclamation	- Improve tailings sand properties - Select appropriate plant species
EXTRACTION	
. Low extraction recovery	- Equipment modifications - Improved operating and maintenance procedures
. High water content froth	- Reduce bitumen in tailings - Additional settling equipment
. Centrifuge component life	- Redesign and modify equipment components
. Tailings pond contamination	- Reduce bitumen in tailings - Recover bitumen from pond - Bind deterrent program
UPGRADING	
. Coker carryover	- Modifications of cyclones - Improvements to piping runs
. Coke formation and deposition	- Modifications to fuel nozzles - High reactor temperature
. Coke utilization	- Increased capacity of storage ponds - Study utilization alternatives and economics
. Electrostatic Precipitator plugging	- Modifications to transport system - Modification to hopper heat tracing

Source: Reference 15.

- Pacific Petroleum Ltd. (9 percent)
- Petrofina Canada Ltd. (8 percent)
- Shell Canada Resources Ltd. (25 percent)
- Shell Explorer Limited (20 percent).

The $6.7-8 billion plant is scheduled to begin construction in early 1981 and go on stream in late 1986. The consortium has picked a joint venture construction group, headed by Bechtel Canada Ltd. as prime contractor.

The Alsands project will draw heavily on the experience of Syncrude Canada and GCOS projects. Alsands plans to use the commercially proven technology of open-pit mining, hot water extraction, and fluid coking with minor modifications.

Tar sand ore will be mined and stockpiled by four large draglines, using both single and double bench mining configurations. Four bucket-wheel reclaimers will then transfer the tar sands from the stockpiles to the conveyor transportation system.

The commercial proven hot water process and froth flotation will be used to extract the raw bitumen. However, there will be minor modifications in the treatment of tailings. Dr. R. A. Ritter, the former dean of chemical engineering at Calgary University, has developed an electrocoagulation cell which can agglomerate the clay particles in the tailings' water, thus promoting much more rapid settling. This will minimize the construction of huge tailings ponds dikes that are normally required for setting the fine particulate.

Originally, an innovative bitumen upgrading process was to be used. This is the Taciuk process, designed by William Taciuk and developed by UMATAC Industrial Process Ltd. The Taciuk process uses a kiln which acts as a coker as well as a retort. The bitumen feed is partially cracked in the kiln, producing a low grade product stream (10-20° API) as well as depositing coke on the sand particles. The deposited coke then undergoes combustion to give clean, dry sand. The partially cracked bitumen can be upgraded further or mixed with diluents and shipped as heavy oil. Due to uncertainties in scale up plus technical difficulties in excessive energy loss and sulfur dioxide emission, the Taciuk process was abandoned in favor of the more conventional fluid coking process.

The fluid coking scheme selected for Alsands departs slightly from the one used by Syncrude. The hydrogen feed for the hydrotreaters will eventually be produced by gasification of residual coke; although initially, hydrogen will be produced by partial oxidation of coker gas, oil, or by steam reforming natural gas. A process flow sheet of the Alsands project is shown in Exhibit IV-26.[31]

EXHIBIT IV-26

Schematic Diagram of the Alsands Plant

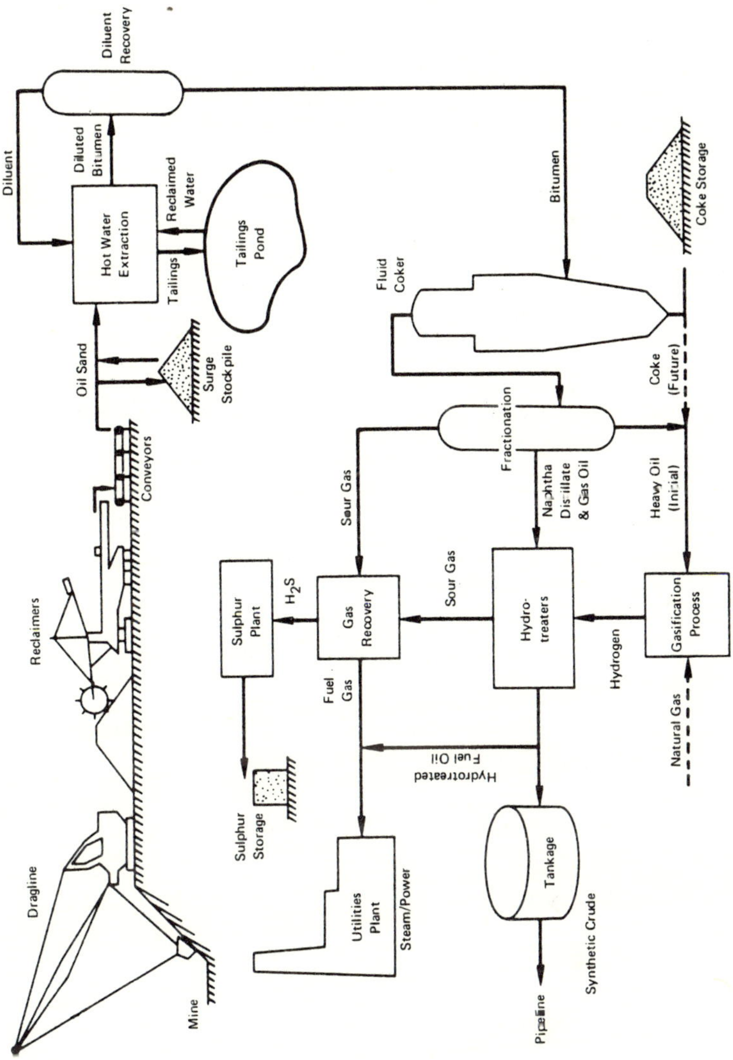

(4) Esso - Cold Lake [34,35]

Esso Resources Canada's in-situ oil sands project at Cold Lake will be the first of its kind. The plant will be located next to Cold Lake on a 200,000 acre site of the Clearwater formation, approximately 140 miles northeast of Edmonton. Esso Resources Canada Limited is a wholly owned subsidiary of Imperial Oil Limited which, in turn, is controlled by Exxon Corporation of the United States (69 percent interest).

Based on a design capacity of 160,000 b/d, 20 percent bitumen recovery and an in-place reserve of 8.9 billion barrels in the Clearwater formation, it is estimated that the site can sustain continuous operation for over 25 years. Esso is aiming at a bitumen recovery rate of 50 percent by 1993.

The Cold Lake Plant has already been approved by the Energy Conservation Board subject to some minor modifications. Like the Alsands Plant, it is waiting for a final agreement between the provincial and federal government on crude prices and royalties. Fluor Canada will head up construction of the plant that is scheduled to begin in late 1981 and be completed in 1986. The cost is estimated at $7-8.8 billion, up from the original $5 billion because of inflation and costly modifications.

The Cold Lake project will be the first commercial scale in-situ tar sand project. If operationally and economically successful, it could pave the way for a generation of in-situ recovery projects.

Basically, the process can be divided into four stages:

- In-situ bitumen production by cyclic steam injection
- Surface bitumen cleaning and separation
- Surface bitumen upgrading
- Waste treatment and disposal.

Through operation of the adjacent Leming pilot plant, Esso has gained extensive experience in in-situ recovery of bitumen. Specifically, it has been determined that bitumen recovery is a strong function of:

- Reservoir sand grain size
- Bitumen saturation levels
- Well completion techniques
- Well spacing and orientation
- Well production operating practice
- Economic considerations.

Bitumen recovery of the Cold Lake Plant involves an in-situ steam stimulation process. The latent heat of steam mobilizes the highly viscous bitumen, creating a bitumen water emulsion which can be produced by conventional means. Production of the bitumen is achieved through blowdown of the reservoir pressure, then by pumping the well. The cycle will be terminated at low production rates, temperatures, or high water/bitumen ratios. The entire cycle can be repeated several times.

The wells will be drilled directionally in clusters of twenty, with an average well spacing of 4 acres. Over the life of the project, 8,000 injection and production wells will be drilled with 2,000 wells in various stages of production at any time. Initially, Esso will drill 1,400 wells, and new ones will be added at a rate of 320 wells/year. Exhibit IV-27 is a typical well configuration.[34]

Each well cluster will have a station monitoring and controlling the flow rates of steam and bitumen. The bitumen will be pumped through a network of surface piping to one of two central production plants. In these plants, water and gas will be removed in separators and the residual water will be removed by heat and electrostatic dehydration.

The upgrading process will include flexicoking and hydrotreating. Raw bitumen from the cleaning and separation step will first be desalted and distilled in atmospheric and vacuum units. The heavy ends will be fed to the Flexicoker while the light ends will be sent directly to the hydrotreaters. Exhibit IV-28 is a flowsheet of the Cold Lake Project.[34] Flexicoking is an Exxon proprietary process which was developed at the Baytown, Texas, pilot facility. It combines fluid coking and coke gasification.

The light ends from primary distillation and coker liquid products will be hydrotreated and blended to produce synthetic crude oil. Steam reforming of natural gas will be used to produce the necessary hydrogen.

Hydrogen sulfide will be removed from the hydrotreater gas stream in a DEA gas purification unit. The resultant acid gas will then go to a sulfur recovery plant. Overall sulfur recovery efficiency is expected to be 99.8 percent during normal operation.

The synthetic crude oil conversion is estimated to be 75.5 weight percent based on the bitumen entering the upgrading unit.

Total water required would amount to 900 thousand b/d consisting of both fresh and recycled water. The breakdown of water usage is as follows:

EXHIBIT IV-27

Well Configuration In Esso's Cold Lake Project

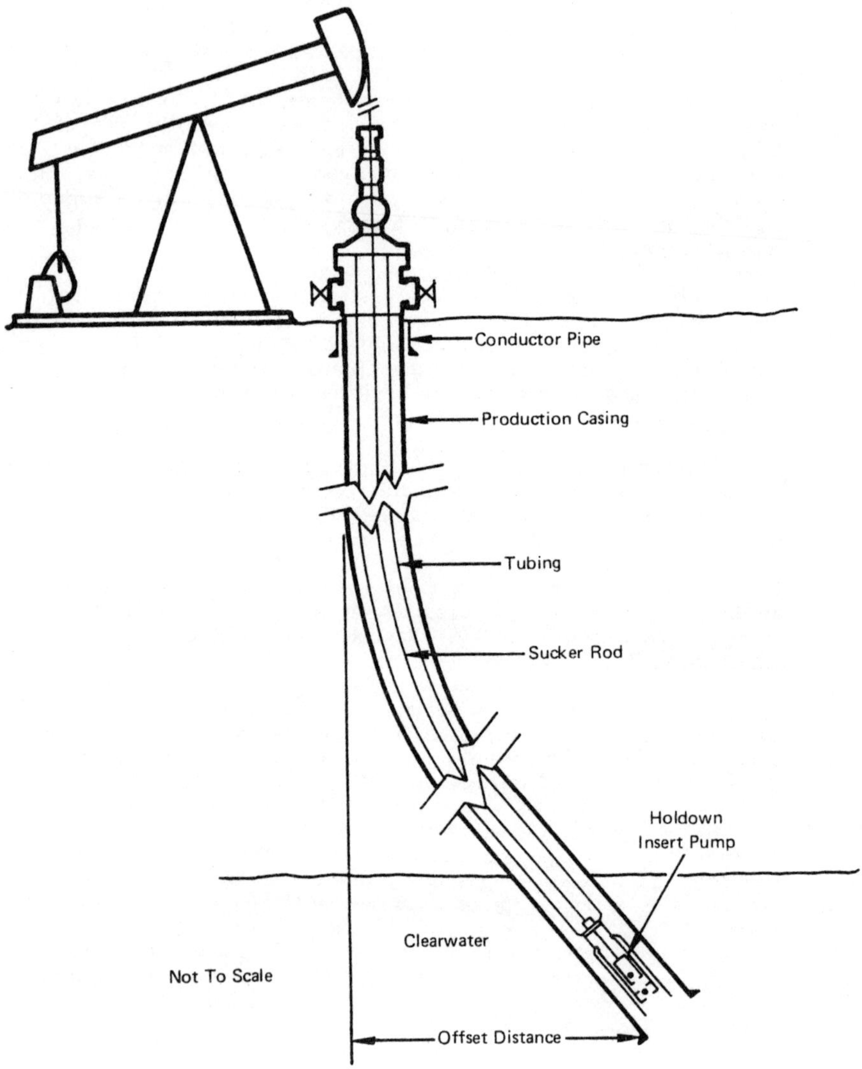

EXHIBIT IV-28

Schematic Diagram of Esso's Cold Lake Project

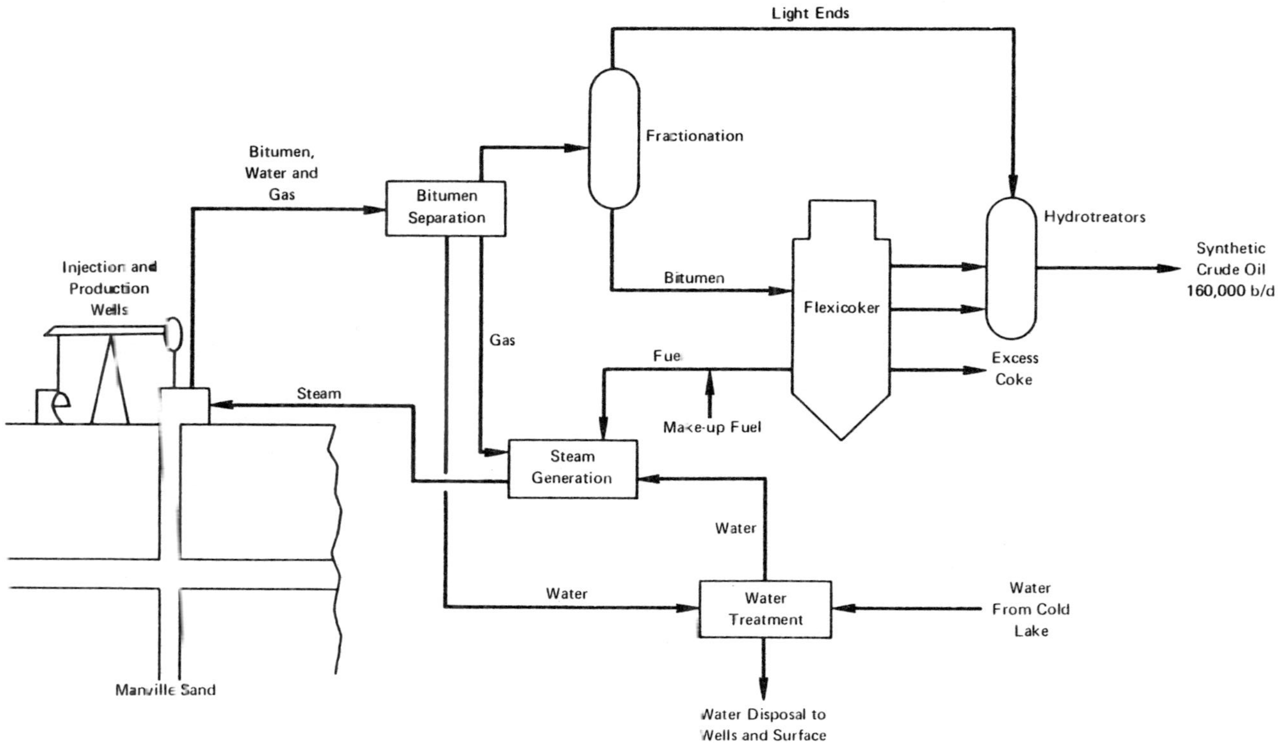

	Water, 10^3 b/d
Water consumed by project	333
Water sent to disposal	252
Water recycled to process	315
	900

Esso's original proposal was to obtain the entire 585 thousand barrels per day of makeup water from Cold Lake but this was rejected by ERCB; and fresh water will instead be piped from the North Saskatchewan River about 48 miles away. A central water treatment plant will handle all requirements of the project. Water required for steam generation will be clarified, filtered, softened, and degassed. Waste water from the process will require treatment because of its oil content. Part of the highly saline water which cannot be recycled will be disposed in deep wells (83 thousand b/d), and the remainder discharged to the nearby Beaver River (168 thousand b/d).

The estimated amount, source, and treatment of various pollutants from the Cold Lake project are:

Pollutants	Amount	Source	Upstream Treatment
Sulfur dioxide	128 st/d	Upgrading Steam Generation	Hydrogenation/ Absorption using DEA
Particulates	0.1 lb/1,000 lb of flue gas	Utility Plant	Electrostatic Precipitation
Fly Ash	18 to 23 st/d	Utility Plant	"
Ammonia	122 st/d	Stripping Facilities	--

6. PILOT PROJECTS FOR BITUMEN RECOVERY

(1) Canadian Pilot Projects

With about one trillion barrels of oil-in-place, the Canadian tar sand deposits represent the largest known deposit of the world. However, only 10 percent of the deposits are surface-mineable, and the rest has to be recovered by in-situ methods. Using current technology, an estimated about 27 billion bbl of bitumen is recoverable, compared with proven Canadian reserves of 6 billion barrels of the lighter conventional crudes.

There is no doubt that tar sands/heavy oil production will play a key role in Canada's future oil supply. Tapping the vast Alberta-Saskatchewan tar sands resource is a formidable challenge demanding technological expertise as well as compromises between Canadian Federal and provincial governments, and industry on pricing, taxes, royalties, environmental standards, and a wide range of other issues.

There are three areas where Canadian development efforts are being concentrated:

- In-situ technology--especially in gathering field tests information and developing viable methods of production
- Surface mining technology aimed at improving existing proven technology in separation and upgrading, as well as reducing waste disposal problems
- Basic research aimed at developing solid theoretical foundations for tar sand technology and inventing new processes.

These development efforts are being carried out through a series of programs which involve government, industry, and academic institutions. One of the more eminent agencies is the Alberta Oil Sands Technology and Research Authority (AOSTRA).[36-38] It was established in June 1974 to fund research and development programs directed towards the recovery and processing of bitumen from the tar sand and heavy oil deposits. It has a total funding of $178 million from the Alberta government and has been negotiating international agreements with the United States, Venezuela, and Madagascar.

At present, there are four commercial tar sand plants in various stages of construction, operation, and expansion. These were described in sections above and in a summary provided in Exhibit IV-19. For pilot projects, there are more than 35 active and proposed ones. Some of these are partially funded by the Canadian government, but most of them are undertaken by individual companies and limited partnerships. A summary of the tar sand and heavy oil pilot projects in Alberta is provided in Exhibit IV-29 and IV-30.[38] Most of the in-situ pilot projects use a combination of thermal recovery methods, steam, and/or combustion. The technology of these processes is fairly well developed but extensive field test data need to be gathered before any scaled up plants can be planned. The representative pilot projects are described in detail in the following section.

In addition to commercial and pilot projects, there is a wide range of academic research projects. These are aimed at gaining a better understanding of tar sand/heavy-oil mining, production, and upgrading phenomena.

EXHIBIT IV-29 In-Situ Projects in Alberta, Canada

OPERATOR	LOCATION AND PROJECT	RECOVERY METHOD	GEOLOGICAL FORMATION	ANTICIPATED RECOVERY %	OIL GRAVITY (°API)	PROJECT STATUS
Amoco Canada Petroleum Co. Ltd. (12½%, operator); AOSTRA (50%); Petro-Canada Ltd. (12½%); Shell Can. Res. (12½%); Suncor Inc. (12½%)	Gregoire Lake, Ft. McMurray Athabasca, Gregoire Lake Block I Pilot	Combination of forward combustion and waterflooding (COFCAW)	Lower McMurray Unit	50% O.I.P.	6-8	Operation began 1977 - 18 months - preheat and blowdown; 5 yrs. COFCAW; Mar./80 preheat
Amoco	St. Lina, Cold Lake		Upper Grand Rapids			Lease 1980/81
BP Exploration Can. Ltd. (50%); AOSTRA, (50%), Hudsons Bay Oil & Gas Ltd. (17½%); Pan Canadian (12½%)	Cold Lake, Bonnyville, Marguerite Lake Project	Sequential steam stimulation and wet forward combustion drive	Clearwater-Manville "C" Unit	31-40% O.I.P.	11-12	1974-pre-pilot study construction complete Oct. 1979. Operating phase began Oct. 31/79. Jan./80-steam soak/stimulation phase of producing wells
CDC/Tenneco Oil	Athabasca		Wabiskaw-McMurray			Lease 1980/84
Chevron Standard	Beaver Crossing, Cold Lake	Steam injection, huff & puff, and cycling	Lower Grand Rapids			Start 1977 - Completion 1981
Esso Resources	Bourque, Cold Lake	Steam injection, huff & puff and cycling	Clearwater			Start 1977 - Completion 1982
Esso Resources	Ethel, Cold Lake	Steam	Clearwater			Start 1964 - Completion 1984
Esso Resources	May, Cold Lake	Steam	Clearwater			Start 1971 - Completion 1981
Esso Resources	Leming, Cold Lake	Steam	Clearwater			Start 1973
General American	Joli Fou, Athabasca	Steam	Wabiskaw-McMurray			Lease 1980/84
Gulf Canada	Cold Lake	Huff & puff steam injection	Clearwater		11	Start 1974 - Completion 1980
Gulf Canada	Desmerais, Athabasca		Wabiskaw-McMurray			Lease 1980/84
Gulf Canada	Pelican, Wabasca	Steam	Wabiskaw-McMurray			Start 1979 - Completion 1979

(continued)

EXHIBIT IV-29 (Cont.)

OPERATOR	LOCATION AND PROJECT	RECOVERY METHOD	GEOLOGICAL FORMATION	ANTICIPATED RECOVERY %	OIL GRAVITY (°API)	PROJECT STATUS
Gulf Canada Res. Ltd. (25%); Numac Oil & Gas Ltd. (25%); AOSTRA (50%)	Surmont, Athabasca	Fracturing				1977 - Initial fracturing
Gulf Canada Res.	Wood River, Wabasca	Interfacial burn along oil-water interface and steam-induced fracture from injection well to production wells	Upper Grand Rapids		5-6	Start 74 July 78 - air injection on interfacial burn; Feb 79 steam injection to induce fracture; Mar. 79-air injection resumed.
Norcen Energy Resources Ltd.	Primrose Lake, Cold Lake	Huff & puff steam injection	Clearwater		10	Start 1975 - Completion 1980
PetroCanada	Mildred Lake, Athabasca	Horizontal drilling	Wabiskaw-McMurray			Start 1979 - Completion 1980
PetroCanada	Gregoire Lake (NW)	Electric/steam	Wabiskaw-McMurray			Start 1980 - Completion 1984
PetroCanada	Muriel Lake, Cold Lake		Lower Grand Rapids			
PetroCanada	Primrose Lake, Cold Lake	Steam injection huff & puff, and cycling	Clearwater			Start 1977 - Completion 1982
RTR Oil Sands Mildred Lake	Mildred Lake, Athabasca		Wabiskaw-McMurray			Completion 1981
Shell Canada Resources Ltd. (18.75%); Shell Explorer (18.75%); AOSTRA (50%); Amoco Canada Petroleum Co. Ltd. (12.5%)	Peace River	Steam Drive	Bullhead		9	Start October 1979 - Completion 1980
Texaco Expl. Canada	Ft. McMurray, Athabasca	Steam	Wabiskaw-McMurray	39% O.I.P.		Start 1972 - Completion 1980

Exhibit IV-30: Heavy Oil Projects in Alberta, Canada

Operator	Location and Project	Recovery Method	Geological Formation	Anticipated Recovery (% O.I.P.)	Oil Gravity (°API)	Project Status
Alberta Energy Co. Ltd. (25%) (Operator); AOSTRA (50%); Westcoast Petroleum Ltd. (12.5%); Kaiser Oil Ltd. (12.5%)	Suffield Military Range, (Medicine Hat), Suffield Heavy Oil Project (SHOP)	In-situ combustion recovery	Upper Manville J. (Glauconitic sand)	30-35	19	Construction began spring/summer 1980-air injection Fall 1980 Phase A-4 yr
Dome Petroleum	Morgan	–	Lloydminster A	–	–	–
General Crude, (Mobil-SGC), Husky Oil	Silverdale, Lloydminster	Wet combustion (combustion thermal drive process)	Sparky Q/K	30-35	12 (Kitscoty) 15-16 (Silverdale)	Start 1975-completion 1980
Husky Oil	Lloydminster	Wet combustion and steam stimulation	Sparky C	–	12-14.5 (Waseca) 23 (Wainwright)	1967/69 steam 1969 fire-flood 1976-77 shutdown
Murphy Oil	Celtic, Lloydminster	Wet combustion (steam stimulation)	–	–	–	July 1969-1st injection
Murphy Oil	Cold Lake	–	–	–	–	–
Murphy Oil	Lindbergh	Steam	Lower Grand Rapids	–	–	Start 1974-completion 1980
PetroCanada Expl. Inc. (50%); AOSTRA (50%)	Wainwright, Lloydminster, Kinsella Heavy Oil project (KHOP)	Steamflood and firefloods	Viking, Kinsella Wainwright D Pool	30	15	1979/80-Primary production 1980-81-steamflood and dry/wet combustion Phase B-extension to 1985
Tesoro Petroleum Corp.	Provost	Combustion	–	–	–	Start 1974-completion 1980
Union Oil Co. of Canada Ltd. (25%-Operator); AOSTRA (50%); Canadian Superior Oil (25%)	Buffalo Creek, Athabasca, Grosmont Thermal Recovery Project	–	Devonian Carbonate Triangle (Grosmont)	–	–	Start 1974
Worldwide Energy (WECO); Sun Oil	Fort Kent, Bonnyville, Cold Lake	Steam stimulation	Upper Grand Rapids	–	–	Start 1976-completion 1983
Union Texas of Canada	Ardmore, Cold Lake	Huff and puff injection	Clearwater	–	11	Start 1974-completion 1980
Worldwide Energy; Pacific Petroleum; Pan Ocean Oil	Muriel Lake, Cold Lake	Steam injection, huff and puff, and cycling	Clearwater	–	–	Start 1980

1. Amoco Canada's, Gregorie Lake Pilot Project[39]

This is an in-situ combustion project that has been operating since 1977. Amoco Canada Petroleum Company Ltd. is the field operator and has a 12½ percent working interest. The remaining partners are AOSTRA (50 percent), Petro-Canada, Shell Canada, and Suncor (12½ percent each).

The pilot project is located south of Gregorie Lake, approximately 25 miles southwest of Fort McMurray. It consists of nine, inverted 5-spot well patterns in a grid 3 by 3 as shown in Exhibit IV-31. Some of the reservoir and production data are listed below in Exhibit IV-32.

EXHIBIT IV-31

Well Pattern for Block 1, Amoco Canada's, Gregorie Lake Pilot Project

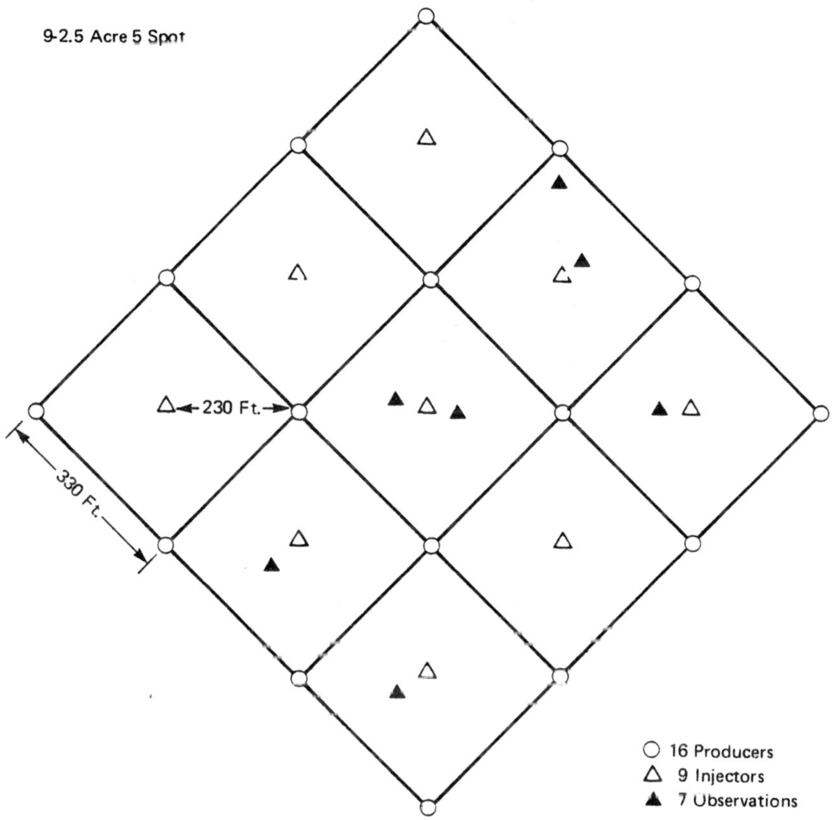

EXHIBIT IV-32

Reservoir and Production Data of Amoco Canada, Gregorie Lake Pilot Project

Rock Formation	Unconsolidated Sandstone
Age	Lower Cretaceous
Overburden Thickness	300 m (984 ft.)
Pay Thickness	27 m (89 ft.)
Reservoir Pressure	1130 kpa (164 psi)
Original Oil in Place (Pattern Volume)	14,308 m^3 (90,000 bbl)
Recovery OOIP (Expected)	50%
Viscosity at 30°C	100,000 cp
°API Gravity	8
Porosity	30%
Permeability	4,700 md
Peak Production (Expected)	125 m^3/d (786 b/d)

Amoco Canada employs its patented COFCAW (combination of foward combustion and water flooding) recovery method. It is essentially a wet combustion method consisting of three phases:

- Preheat - reservoir is heated by forward combustion, and the in-place bitumen is made mobile
- Blowdown - pressure is reduced and temperature stabilized
- Displacement - wet combustion is carried out. A steam pocket precedes the combustion front, displacing the bitumen towards the production well.

It is estimated that the preheat and blowdown phase will require 18 months, while the displacement phase will last up to 5 years. The Gregorie Lake project schedule is shown in Exhibit IV-33.[39]

EXHIBIT IV-33

AMOCO Canada's Gregorie Lake Project Schedule

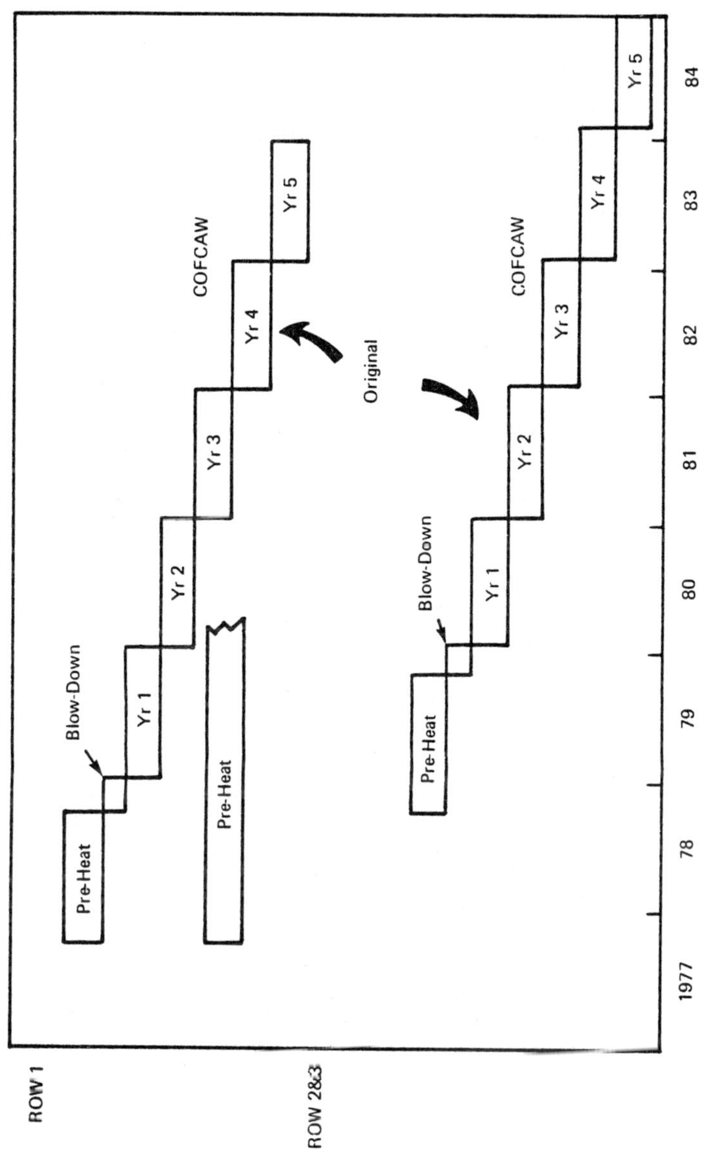

No on-site bitumen upgrading is planned for Gregorie Lake. Since this pilot employs a combustion process, some in-situ bitumen upgrading will occur. After the bitumen is recovered, it will be diluted and/or emulsified so it can be trucked or pipelined.

Operational and detailed technical data on the project are limited, but Amoco Canada is optimistic about the feasibility of this process. By the end of 1980, $46 million will have been invested in the project and an estimated $25 million (1977 dollars) will be needed for completion. This project is being closely monitored by Amoco Canada as it represents an important part of the company's synthetic fuel program.

2. British Petroleum's, Marguerite Lake Pilot Project[40]

"Phase A" of this project uses a sequential process of cyclic steam stimulation and in-situ combustion. British Petroleum Exploration Canada is the field operator with a 20 percent interest. The remainder of participants include: AOSTRA (50 percent), Hudson's Bay Oil and Gas Company Limited (17½ percent), and Pan Canadian Petroleum Limited (12½ percent). The estimated project cost is $18.4 million for 7 years of operation.

The project site is located in the Cold Lake region, approximately 45 km (28 miles) north of Bonnyville, Alberta. British Petroleum has been developing in-situ thermal recovery methods since the 1960s, and has also obtained information through participation in other pilot projects.

The well pattern includes four five-spot patterns with five acre well spacings (four injection wells and nine production wells). Four extra wells are drilled adjacent to the pattern in order to obtain data on in-situ combustion. The well pattern and the location of related facilities are shown in Exhibit IV-34. Some of the reservoir characteristics are listed in Exhibit IV-35.

The original plan calls for operating the main-pattern wells for 3 years using steam injection, after which the center wells will be switched to air/water injection. The four special test wells will provide operating information for the main wells on in-situ combustion drive. Exhibit IV-36 is the Marguerite Lake project schedule.

EXHIBIT IV-34

Marguerite Lake Project Phase A Well Locations

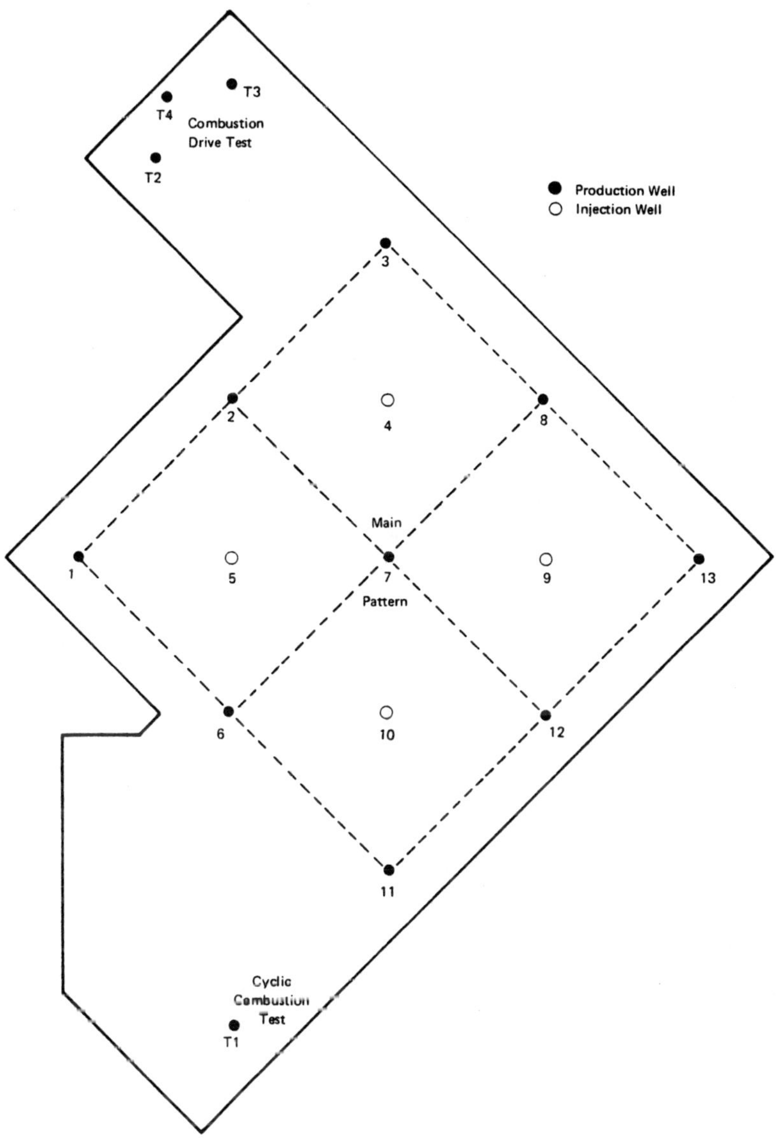

EXHIBIT IV-36

Marguerite Lake Project Schedule

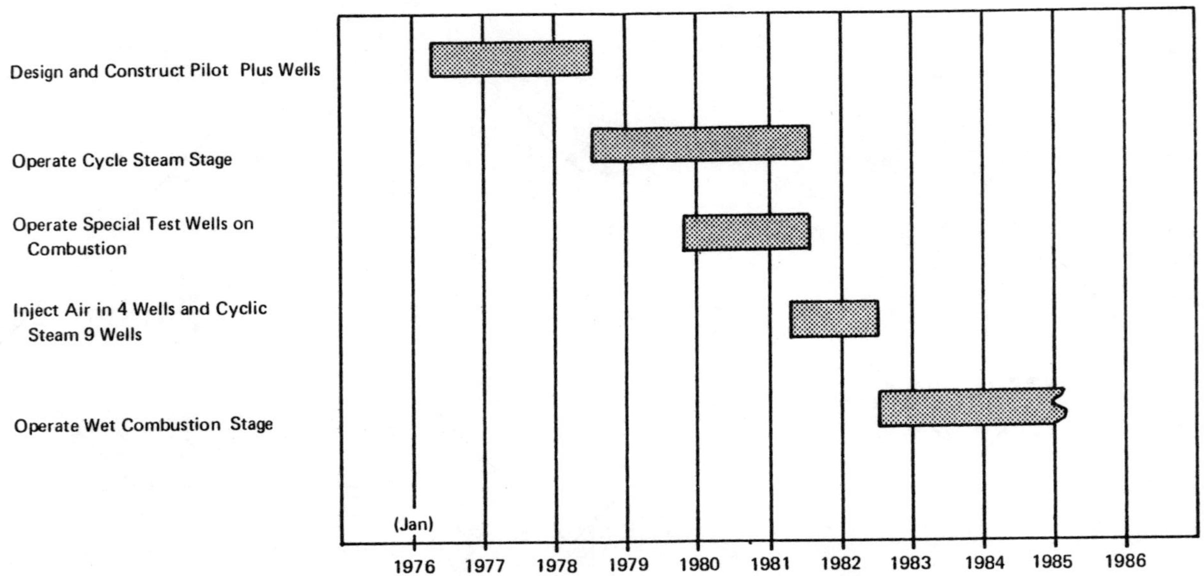

EXHIBIT IV-35

BP Marguerite Lake "Phase A" Project
Reservoir Characteristics

Formation	460 m (1,510 ft.)
Pay Thickness	35 m (115 ft.)
Reservoir Pressure	2,700 kPa (392 psi)
Reservoir Temperature	15°C (59°F)
Original Oil in Place	21 MM m^3 (130 MM bbl)
Viscosity at 38°C	6,800 mPa · S (6,800 cp)
66°C	680 mPa · S (680 cp)
96°C	130 mPa · S (130 cp)
Porosity	30%
Sulfur Content	4.3 wt. % of bitumen
Well Orientation	Northeast/Southwest

After the bitumen is recovered, it is dehydrated by an on-site dehydration system consisting of heat exchangers, a knockout drum, and a heater treater. It is then stored in three 400 m^3 (14,000 $feet^3$) oil tanks prior to transportation by trucks to terminals at Lloydminster or Wainwright.

Project steam requirements are provided by two 14-megawatt steam generators capable of producing 80 percent quality steam at approximately 13,000 kpa (1,900 psi). Air for in-situ combustion purposes is provided by screw compressor connected to a radial reciprocating compressor. The final maximum discharge air pressure is 12,000 kpa (1,740 psi). The entire project is highly instrumented and automated to record temperature, pressure, and injection/production data. The site chosen is representative of that portion of the lease which would support a commercial project. It is possible that BP will further expand its scale of development after enough operating experience has been gained.

3. Shell Canada's, Peace River Pilot Project[41]

The primary objective of this pilot project is to obtain reservoir performance data to establish the feasibility of a commercial scale in-situ bitumen recovery project at Peace River during the 1990s.

Shell Canada's present pilot project is based on 16 years of field test experience at Peace River. Four previous pilot projects include two steam soak tests, one combustion test, and one steam drive test. Past field test results led Shell to believe that steam drive recovery offered the highest probability of success for this particular formation.

Shell Canada Resources is the field operator and has an 18.75 percent interest in the project. The AOSTRA (50 percent), Amoco Canada Petroleum Company Ltd. (12.5 percent), and Shell Explorer Limited (18.75 percent). Construction of the $70 million pilot project was completed in October 1979 and operation has since begun. Project operating costs are estimated to be about $10 million (1979 dollars) per year for an expected project life of 9 years.

The pilot project includes a total of 44 wells:

24	Production
7	Injection
11	Observation
2	Disposal
44	

These wells are arranged in 7 seven-spot patterns on 7-acre spacings, with three production wells to each injection well as shown in Exhibit IV-37.

The pilot project is located on a leased site 30 km (19 miles) northeast of Peace River adjacent to Cadotte Lake. Some of the reservoir data are given in Exhibit IV-38:

The basic bitumen recovery process used by Shell in the Peace River pilot project involves steam soak and steam drive. It has seven phases:

- Cold productivity tests - short term
- Steam soak
- Steam drive

EXHIBIT IV-37

Well Pattern at Shell Canada's Peace River Plat Project

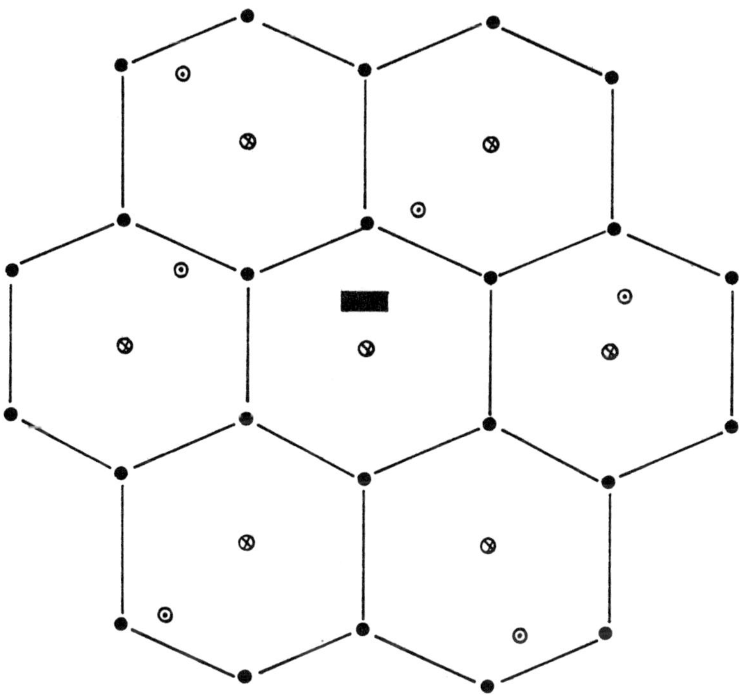

7-7 Spot Pattern

⊗ Injection Well

● Production Well

⊙ Observation Well

EXHIBIT IV-38

Shell Canada Peace River Reservoir Data

Formation	Bullhead
Overburden Thickness	550 m (1,805 ft.)
Pay Thickness	27 m (89 ft.)
Original Oil in Place	15,650 MM m^3 (51,500 MM bbl)
Bitumen Recovery	
Minimum	50%
Likely	60%
Maximum	65+%
Reservoir Temperature	17°C (63°F)
Porosity (average)	28%
Bitumen Saturation	77%
Permeability (average)	440 md
Oil/Steam Ratio	
Minimum	0.2
Likely	0.25
Maximum	0.30+
Viscosity at 17°C	200,000 cp
at 204°C	7 cp

- Back pressure production wells - after heat breakthrough
- Controlled steam injection - balances production
- One or more pressure buildup and blowdown cycles
- Final stage - gravity drainage, heat scavenging, and stripper operations.

The pilot project is designed primarily to obtain reservoir performance data such as bitumen/steam ratio, percent bitumen recovery, bitumen properties, and production rates versus time. It is not meant to test specific separation and upgrading processes.

After the bitumen is brought above ground, it is dehydrated by a series of steps which involved diluent addition, free water knock-out, and gas separation. It is then desalted, fractionated into three streams, and then stored or trucked to sales points. Some of the project capacities are:

- Source water - 3,957 m^3/d (25,000 b/d)
- Steam generation - 2,860 m^3/d (18,000 b/d, 80 percent quality) water equivalent @ 16,200 kpa (2,350 psi)
- Bitumen handling -550 m^3/d (3,500 b/d)
- Produced water - 2,460 m^3/d (15,500 b/d)
- Produced gas - 28 x 10^3 m^3/d (1 MMCF/d).

To supplement and further define the results obtained from the pilot project, laboratory work and numerical simulation are being carried out simultaneously. Conceptual design of a full-scale commercial plant has already begun. However, since it takes 5 year to generate sufficient pilot plant data, commercial production is not expected to commence until the early 1990s. The time schedule for the pilot and full-scale plants is included in Exhibit IV-39.

4. Petro-Canada's Heavy Oil Pilot Project at Kinsella, Lloydminster[42]

This pilot project is one of several heavy oil projects currently underway at the Lloydminster deposit. It is operated by Petro Canada Exploration and jointly owned by Petro-Canada and AOSTRA.

The Viking Kinsella project site is located roughly 80 km (50 miles) southwest of Lloydminster, and 50 km (30 miles) northwest of Wainwright. The wells were originally drilled in five-spot patterns, but later converted to a series of two-spots because of geological irregularities. The upper portion of the oil reservoir contains unconsolidated sand; but the lower portion has a high clay content with low permeability. Some of the reservoir parameters are listed in Exhibit IV-40.

EXHIBIT IV-39

Shell Canada Peace River Project Schedule

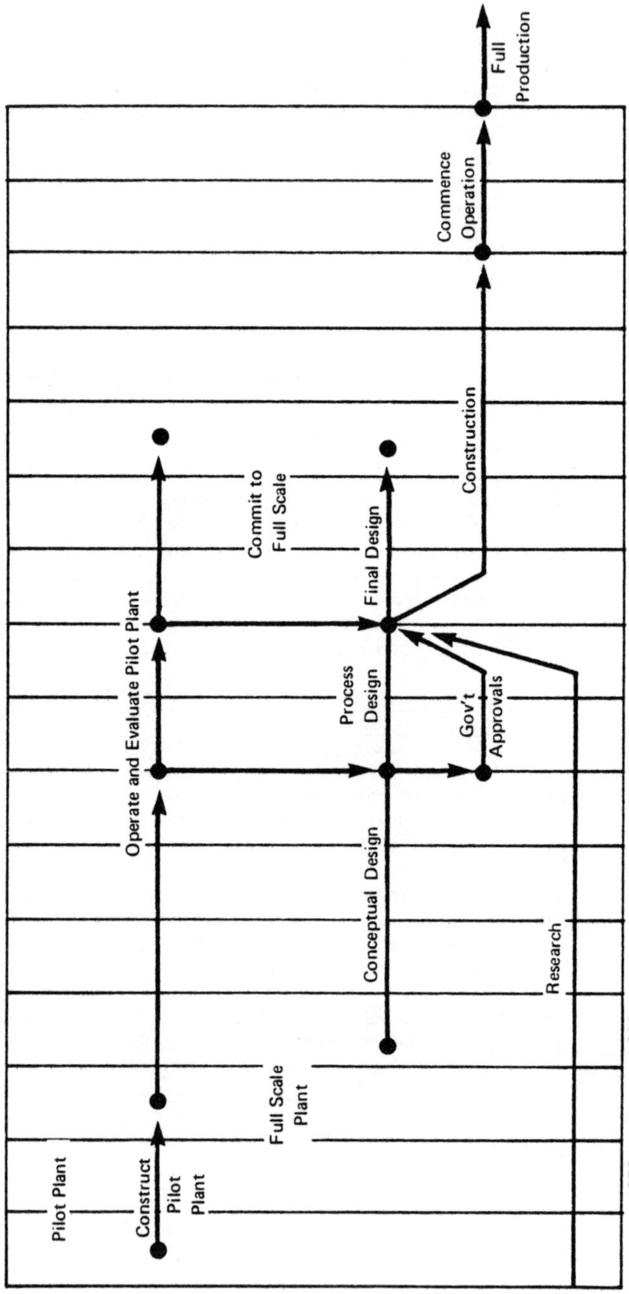

EXHIBIT IV-40

Reservoir Parameters of Kinsella
Wainwright "D" Pool

Well Spacing	2 ha
Overburden Thickness	680 m (2,230 ft.)
Gross Pay Thickness	5.5 m (18 ft.)
Net Pay Thickness	4.7 m (15 ft.)
Recovery (Expected Percent)	30% OIP
Porosity (Average)	30%
Permeability (Average)	360 md
Oil Saturation	70%
Reservoir Pressure	5,100 kPa (740 psi)
Oil Density	965 kg/m^3 (.198 lb/ft^3)
Oil Viscosity at 28±c	400 m pa · S (400 cp)
°API Gravity	15

Two different recovery techniques are being evaluated and compared: steam flooding and wet combustion. Seven injection wells and eight production wells are drilled directionally into the deposit. The eight production wells have two types of well completions: perforated casing and open-hole gravel pack. Three of the injection wells are used for steam flooding while the remainder are used for air/water injection.

The liquid products pass through on-site separators and are transported to the adjacent Kinsella "B" pool where they are blended and treated. The product crude oil is pipelined to Hardisty, Alberta, for storage.

Like most pilot projects, operating data of the Kinsella pilot are largely proprietary. Primary production has begun, and it is expected that field tests will continue into 1985 if the results are favorable. An approximate schedule of the Viking Kinsella project is shown in Exhibit IV-41.

EXHIBIT IV-41

Petro Canada's Kinsella Project Schedule

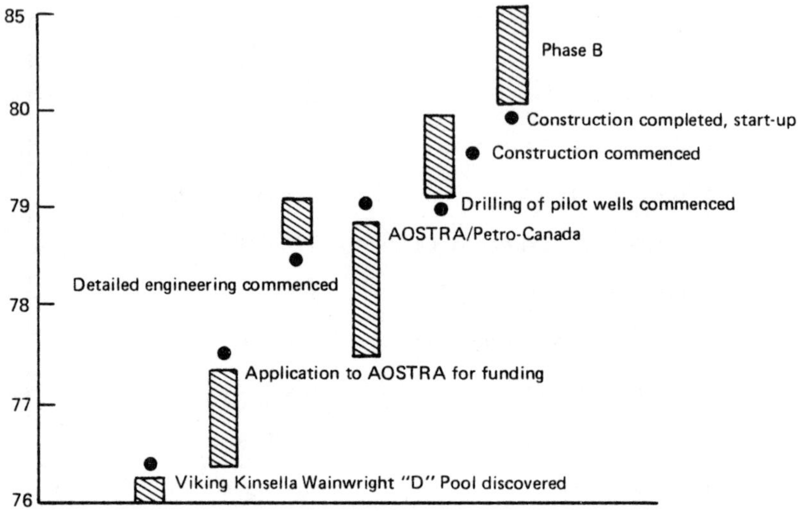

5. Union Oil's Heavy Oil Project at Buffalo Creek, Athabasca[43]

The "Carbonate Triangle" in northcentral Alberta contains an estimated 1.35 x 10^{12} barrels of heavy oil in place, possibly one of the largest oil deposits in the world. These are not tar-sand deposits, but instead are carbonate formations composed mainly of dolomite of the Upper Devonian Age (350 million years).

Recognizing the potential of this huge heavy oil pool, Union Oil Company of Canada began field testing in 1975, concentrating its efforts on the Upper Grosmont Carbonate formation. The present partnership consists of Union Oil Company of Canada (25 percent), AOSTRA (50 percent), and Canadian Superior Oil (25 percent).

The heavy oil pilot project is located at Buffalo Creek, along the east side of the Carbonate Triangle. Buffalo Creek is 70 km (43 miles) from the nearest all-weather access road. As part of the program, Union has constructed a road which provides access during the winter months.

The Upper Grosmont reservoir can be divided into three units, according to their heavy oil saturation and dolomite thickness. Exhibit IV-42 is a partial list of the Upper Grosmont two and three reservoir characteristics.

EXHIBIT IV-42

Upper Grosmont 2 & 3 Reservoir Characteristics

	Grosmont 2	Grosmont 3
Overburden thickness, metres (ft.)	285 (935)	257 (843)
Net pay thickness, metres (ft.)	23 (75)	16 (53)
Zone thickness, metres (ft.)	34 (112)	28 (92)
Average porosity, %	19	28
Average water saturation, %	25	35
Reservoir pressure kPa (psi)	1,400 (203)	1,130 (164)
Reservoir temperature °C (°F)	11 (52)	11 (52)
Bitumen gravity °API	7	7

The Grosmont formation combines thin overburden, high porosity, and heavy oil saturation, making it favorable for thermal recovery methods. Until recently, Union has been using primarily cyclic steam stimulation. In 1979, Union combined steam soak and in-situ combustion processes on its five-spot pattern. A summary of field test results is displayed in Exhibit IV-43.

Problems encountered during the course of the project and their solutions include:

Problems	Solutions
. Collapse of well casings due to high temperature steaming	. Improved cementing and casing technology
. Contaminated fuel gas supply.	. Better control of combustion gases
. Lost circulation problems due to communication between zones.	. Better casing and cementing procedures.

EXHIBIT IV-43

Union Oil's Buffalo Creek Field Test Results

YEAR	1975	1977	1977	1978-79	1980	1981*
Process	Steam Soak	Steam Soak	Steam Soak & Drive	Steam Soak & Combustion	Steam Soak	Steam/ Combustion
Wells						
Injection/Production	1	1	2	5 (Inverted 5-Spot)	1	5-16
Observation	–	1	1	2	4	–
Results						
Peak Daily Production per Well						
M^3	3	12	80	56	–	–
(bbls)	(20)	(78)	(500)	(350)	–	–
*Projections						

Concurrent with its heavy oil recovery pilot projects, Union Oil carried out several laboratory studies to form a theoretical backup for its pilot project results. Some of the more unique conclusions from the laboratory studies are:

- 70 percent oil recoveries are possible

- Carbon dioxide produced in the formation at high temperatures may result in viscosity reduction and act as a source of internal energy

- Carbonates appear to improve in-situ upgrading of bitumen during combustion

- Bitumen from the Grosmont formation is similar to Athabasca bitumen.

The high production rates of its pilot projects and postulated recovery rates have prompted Union Oil to press forward with its heavy oil development in the Carbonate Triangle. It has identified three areas where definitive answers or solutions are required:

- Final choice of a recovery process, either steam soak, combustion or a combination thereof

- Improved well casing and cementing procedures

- Identification of an area than can sustain potential commercial facility.

It is conceivable that the "Carbonate Triangle" will receive increasing attention in the future because of its excellent potential, and Canada's search for alternate sources of crude oil.

(2) Venezuela's Heavy Oil Development

The vast heavy oil resources of Venezuela have often been categorized as Athabasca-type tar sands. This is actually a misnomer as Venezuelan heavy oil is similar to the heavy oils of the Lloydminster and southeastern Alberta fields. Venezuelan heavy oil is lighter than the bitumen of Alberta tar sands, has a higher API gravity and lower viscosity. Venezuelan heavy oil gravity ranges from 8° API at the Jobo field to 18°API at the Merey field, and almost all heavy oil fields have some primary production. In light of this fact, the original study definition of very heavy oil (<10-12° API and no primary production) must be expanded somewhat to include Venezuelan heavy oils. Exhibit IV-44 is a detailed tabulation of the properties of the active heavy oil fields of Venezuela.[44]

Exhibit IV-44: Properties of Active Venezuelan Heavy Oil Fields

Region	. . . . Western Venezuela-Maracaibo Basin								Northern Edge of Orinoco Heavy Oil Belt in the Cerro Negro Region*					
Field	West of the Lake Boscan**	Bolivar Coast Tiajuana	. . . . . . Lagunillas			. . . . Bachaquero				Merey ***	Morichal	Pilon	Jobo	
Formation†			Laguna		Lagunillas		B-13	B-17						Jobo II
Age	Eocene	Miocene		Miocene		Miocene					. . . . Oligocene			
OOIR, MMbbl	17,000	19,000	—	23,500	—	20,500	—	—	6,600	—	5,000	2,300	8,600	—
Depth, ft	8-10,000	1,600-2,900	—	1,800-3,000	—	3,000-4,300	—	—	1,900-4,000	—	3,500-4,000	3,300	3,400-3,850	—
Average pay thickness, ft	74-405	51-162	—	41-332	—	45-122	—	—	15-40	—	224	80	45-169	—
Reservoir temp. °F	170-185	110-130	—	110-130	—	130-140	—	—	130-150	—	133	138	—	—
Reservoir pressure, psig	—	780	—	—	—	—	—	—	—	—	1,364	—	68-160	—
Initial G.O.R., ft³/bbl	15-150	90-200	—	90	—	145	—	—	60-250	—	70	141	76-88	—
Initial oil saturation, %	65-70	72-85	—	75-87	—	84-89	—	—	80-90	—	83	85	30-31	—
Porosity, %	22-26	31-38	—	34-35	—	31-33	—	—	22-37	—	19-32	32	1-5	—
Permeability, D	0.04-5	0.7-2.9	—	0.7-3.0	—	0.8-1.8	—	—	1-5	—	0.5-4.0	5	—	—
Specific heat, Btu/lb-°F	—	0.314	—	0.314	—	0.314	—	—	0.375	—	—	—	—	—
Thermal conductivity, Btu/ft² (day)	—	35	—	35	—	35	—	—	—	—	—	—	—	—
Viscosity at 100°F	26,200	3,700	6,600	—	498	—	1,120	294	—	205	3,200	1,110	—	62,265
140°F	2,950	707	1,000	—	136	—	290	90	—	66	590	237	—	5,535
Pour point, °F	63	32	48	—	-6	—	-4	-13	—	-22	32	14	—	68
Sulfur, % wt	5.2	2.7	3.0	—	2.2	—	2.9	2.4	—	2.3	2.1	1.9	—	3.7
Nitrogen, % wt	0.65	—	—	—	—	—	—	0.42	—	—	0.53	—	—	—
Asphaltenes, % wt	9-17	5.8	7.3	—	5.4	—	5.7	4.9	—	12.5	10.8	8.4	—	8.6
Vanadium, ppm wt	1,200	284	390	—	260	—	270	320	—	200	250	105	—	390
Nickel, ppm wt	150	38	64	—	40	—	20	42	—	60	65	60	—	106
Conradson carbon	15.0	11.2	12.8	—	10.0	—	11.2	10.0	—	10.0	14.0	9.5	—	14.1
Distillation, IBP °F	226	223	248	—	165	—	228	158	—	212	280	158	—	—
Temperature 5% volume	556	—	—	—	—	—	—	—	—	302	—	—	—	—
10% volume	—	536	545	—	446	—	513	392	—	392	—	500	—	—

*Does not include Machete, Zuata, Hamaca and part of Cerro Negro of the Orinoco Oil Belt.
**Includes Boscan, Los Claros, Garcia, Urdaneta West and East fields.
***Merey includes Miga, Melones and Oveja fields.
†Segregation.

The Venezuelan petroleum industry, which was nationalized in 1976, is presently organized under a single state-owned company Petroleos de Venezuela, S.A. (PDVSA) with three departments, (see Exhibit IV-45). Four competing companies (Lagoven, Meneven, Maraven and Corpoven) now undertake the exploration and production of domestic petroleum resources. In addition, they contract with foreign companies like Phillips Petroleum and Amoco to operate pilot and large-scale projects.

Since nationalization, PDVSA has disclosed plans for heavy capital investment to develop the country's oil resources. PDVSA's plans for investing $25 billion in the oil industry over the next 10 years are as follows:[45]

Planned Investment 1980-1989
(In Billions of Dollars)

Production	$11
Exploration	5.1
Refineries Improvement	5
Oronoco Heavy Oil Belt	2.3
Infrastructure	1.6
	$25

Venzuela began commercial oil production 60 years ago. Currently, it has 18 billion barrels of proven oil reserves, enough to last until 2000 at the current rate of increase in consumption. Recognizing the rapid depletion of its light crude reserves and the potential of its heavy oils, Venzuela has instituted a policy which focuses on the conservation of light crude reserves and expansion of heavy oil supplies. This plan calls for the rapid development of the Orinoco Heavy Oil Belt, achieving one million b/d production by the turn of the century.

Currently, about one third of Venezuela's daily production of 2.3 million barrels is in heavy oil with an average gravity of 15° API.[46-48] The bulk of this heavy oil (570,000 b/d) is produced from fields located on the eastern shore of Lake Maracaibo in western Venezuela. These fields, Tia Juana, Lagunillas and Bacherquero, have been in production for about 40 years and have combined resources of about 34 billion barrels. Conventional primary and secondary production techniques recover about 11 percent of the oil-in-place. Subsequent tertiary recovery using steam soak followed by steam injection would ultimately recover about 32 percent. At present thermal recovery methods account for about 171,000 b/d.

EXHIBIT IV-45

Petroleas de Venezuela Organization Chart

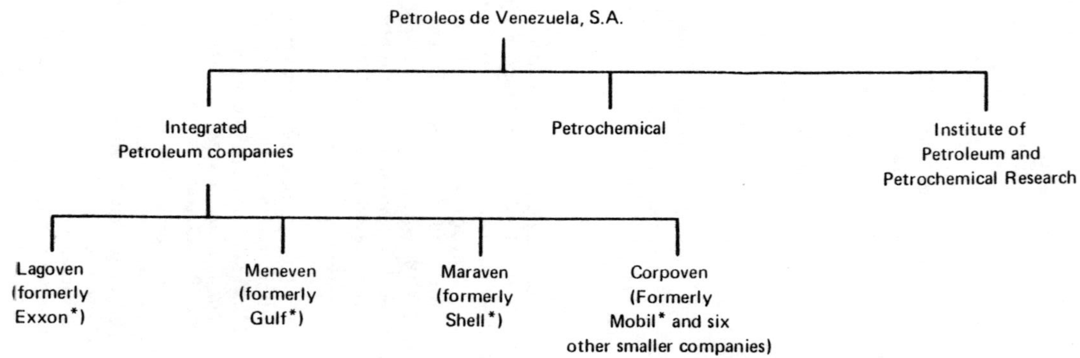

*Includes other smaller operators

Steam soak or "huff-n-puff" has been found to be an extremely useful method for enhanced oil recovery in Venezuelan heavy oil reservoirs, which are unconsolidated and are relatively shallow (<4,000 feet). Coupled with reservoir compaction, steam soak serves to maintain reservoir pressure and increase oil mobility. Some features of the steam-soak processes in the Maracaibo Basin include:

- Steam "slug" size of 5,000 tons
- Production cycle - 1 to 2 years
- Thermal efficiency - 4 bbls of oil per bbl of steam or 40 bbls of oil produced per barrel of oil burned.

The second stage of the thermal recovery process is usually continuous steam injection or steam drive. This is usually initiated when reservoir compaction is complete and no additional oil can be recovered by steam soak. The thermal efficiency of steam injection is lower (3 bbls of oil produced per barrel of oil burned) because of the large amount of steam required. In summary, the total oil recovery from the Maracaibo field at each recovery stage is as follows:[49]

	Average Percent OOIP
Primary Production	11
Primary + Steam Soak	18
Primary + Steam Soak + Steam Drive	32

To date, thermal recovery techniques have been applied to some 1,650 wells in the Maracaibo Basin, and 20 million tons of steam have been injected. The largest of the heavy oil projects is the M-6 steam-drive project in the Tia Juana Field in western Venezuela. It covers 1,400 acres and is arranged in a 19 six-spot pattern with a total of 150 wells. The original oil-in-place is about 525 million bbls. of 12° API crude. Primary production aided by steam soak is expected to recover 105 MM bbls or 20 percent. Final recovery would be approximately 220 MM bbls or 40 percent after steam drive is applied.

1. Orinoco Heavy Oil Belt Development

The Orinoco Heavy Oil Belt is located north of the Orinoco River in the southern part of the eastern Venezuelan basin, covering an area of about 42,000 km^2 (16,200 mi^2). It is estimated that the oil belt contains 700 billion to 2 trillion barrels of oil-in-place. The higher figure results from the discovery of the bordering Guyana Shield which lies at a depth of about 20,000 feet.

Reservoir characteristics of this vast area are not well known because of complexity of structural features and lack of geophysical surveys. A summary of the reservoir and crude characteristics is listed in Exhibit IV-46.[50] Much of the in-place oil is less than 10° API, although a significant amount ranges between 10-18° API. The heavy oil does have some mobility under the high reservoir temperature (~130°F) and pressure (~1,800 psig). Hence, primary production is possible even though production rates are low, ranging from 50 bbl/day to 500 bbl/day, per well depending on reservoir conditions. Venezuelan heavy oil has a high sulfur and metal content. These metals (vanadium and nickel) present special problems in refining the crude to salable products because they reduce the selectivity of catalysts used in refining. Upgrading processes are discussed in Chapter V.

EXHIBIT IV-46

Orinoco Heavy Oil Belt General Reservoir and Crude Characteristics

Rock Type	Unconsolidated
Porosity	20-35%
Reservoir Temperature	95-150°F
Sulfur Content	2-4 wt-%
Vanadium Content	200-500 ppm
Nickel Content	200-500 ppm
Gravity of Crude	7-18% °API
Viscosity	100-20,000 cp at 130°F, depending on API gravity
Solution GOR	50-170 scf/stb

Thus far, most of the development has been limited to the fields located in eastern Venezuela on the northern fringe of the oil belt. The fields involved (Merey, Morichal, Pilon, and Jobo I and II) in the Cerro Negro region are currently producing about 120,000 bbl/day of crude between 8-15° API.

Extensive heavy oil recovery programs have been carried out in the Jobo Field. The Jobo field contains rocks of the sandstone type, and the reservoir is divided into two parts, based on the deposit characteristics. Exhibit IV-47 lists reservoir data for the Jobo field.[51]

EXHIBIT IV-47

Jobo Field Reservoir Data

	Jobo Group I	Jobo Group II
Oil Gravity, °API	13.5	9.0
Depth, feet	3,650	3,950
Net pay, feet	60	210
Porosity, percent	30	31
Productive area, acres	6,677	9,922
Reservoir pressure, psig	1,400	1,000
OOIP, MM bbls	540	1,561
Primary Recovery percent	7.7	2.1
Ultimate primary reserves, MM bbls	41.7	33.1

The Jobo field was discovered in 1958. Primary production was very low, so steam soak or cyclic steam injection was initiated in 1974. The intermittent injection of steam substantially increased the mobility of the heavy oil, causing it to flow towards the wellbore. A portable 30 MM-Btu per hour steam generator was used for injecting 80 percent quality steam at 2,500 psi. Previous steam injection in some of the old wells had failed because the wells were not designed to withstand the tremendous amount of thermal and mechanical stress. In preparation for thermal recovery processes, the Jobo well completion techniques were specially designed. Well completion consisted of pre-stretched casing, two-stage cementing, bottom-hole thermal packer with sliding mandrel and as a result, the wells were able to withstand the high steam pressure and temperature; and, more importantly, heat loss was minimized. The wells were soaked for one week to two months with heat ratios of between 150 and 200 MM btu per foot of pay thickness. Exhibit IV-48 shows a schematic diagram of the Jobo field on-site facilities.[51] Steam stimulation in the Jobo field has increased the oil production rate by three-fold as indicated below:

Average Production Per Well, b/d

	Before Steam Soak	After Steam Soak
Jobo I	360	960
Jobo II	80	300

EXHIBIT IV-48 Jobo Field Schematic Diagram

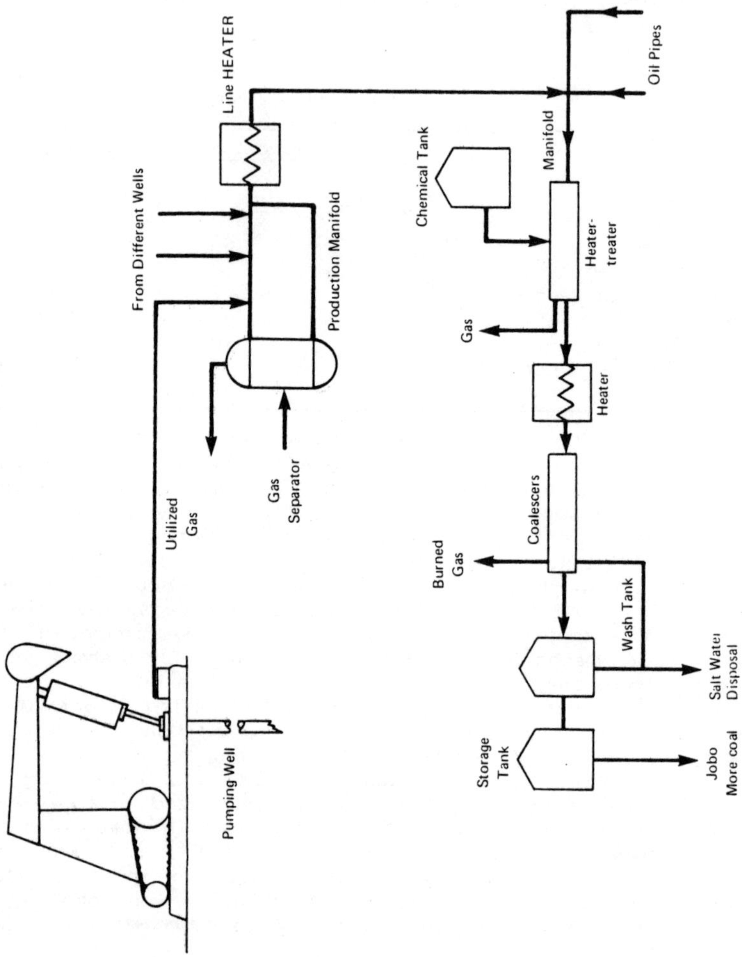

It is apparent from the results that steam soak is effective for both fields. However, because of the lower gravity and greater depth of Jobo II oil, its increase in production from steam soak was not as pronounced.

Phillips Petroleum Company conducted extensive development in Venezuelan oil fields before the country's nationalization of its petroleum industry in 1976.[52] From 1965 to 1969, Phillips operated three in-situ combustion pilot projects in the Morichal field. The first pilot project involving reverse combustion in a reservoir containing 8° API oil, failed because of uncontrollable ignition at the air-injection well. A second pilot project was carried out in the same reservoir using forward combustion but production was extremely low due to plugging of the gravel-packer liner in the production wells. The third pilot project was relatively successful except for difficult pumping and low production rates, resulting from the intrusion of large volumes of combustion gas from second-line wells. The problem was partially solved by installing a down-hole gas separator which prevented the gas from entering the producing well. The pilot projects by Phillips demonstrated some of the difficulties that can occur with in-situ combustion. In-situ combustion has not gained wide acceptance in Venezuela, although future development plans do include further testing and evaluation of this process.

Part of Phillips' program in the Morichal field included blending of the heavy oil with a diluent to produce a mixture that could be desalted, dehydrated, and transported through conventional pipelines. In order to reduce the amount of diluent required, a light oil (30-40° API) had to be used. The process is simple but become prohibitively costly and unattractive due to the high price and world shortage of light crude.

2. Large-Scale and Pilot Projects for Heavy Oil Recovery in Venezuela

There are a total of 53 pilot projects and large-scale projects in Venezuela for the recovery of heavy oil. Most of these projects are in the fields of the Maracaibo Basin, and approximately one fourth of the projects are in the Eastern Venezuelan fields of Merey, Morichal, Pilon, and Jobo.[53,54]

Large Scale Projects:		
. Recovery processes		
	- Gas and water injection	1
. Thermal recovery processes		
	- Steam soak (1,650 wells)	17
		18
Pilot Projects:		
. Non-thermal recovery processes		
	- Water injection	3
	- Caustic injection	1
	- Condensate injection	1
	- Polymers injection	1
		6
. Thermal recovery processes		
	- Steamflood	7
	- Steam soak	15
	- In-situ combustion (direct and dry)	3
	- In-situ combustion (reverse and dry)	1
	- In-situ combustion (wet)	1
	- Cyclic hot water	1
	- Thermal sandwich	1
	Total	29
		53

Three pilot projects that have just started production in the Orinoco Heavy Oil Belt are:

- Cerro Negro Steam Drive Project
- Jobo Steam Drive Pilot
- Melones Wet Combustion Project.

The combined capacity of these pilots are 17,000 b/d, costing $61 million dollars. The operating experience from these pilots and the extensive information obtained from some of the western and eastern heavy oil fields will better define a future path of development.

At present, development plans of the Orinoco Heavy Oil Belt can be divided into three categories:[55]

. Evaluation and exploration

- An exploration program aimed at the complete evaluation of the belt by 1983. This would entail shooting 16,000 kms (10,000 miles) of seismic lines and drilling 520 exploratory wells

- Drilling 380 wells between 1980-1984 to identify near-term areas for development

- The total cost of this program will be $430 million (1979 dollars)

. Production

- Lagoven (formerly Exxon) will complete a 125,000 b/d plant by 1988 in southern Monagas state to produce upgraded crude. The total cost of $3 billion will have the following breakdown: production (35 percent), upgrading (50 percent), infrastructure (15 percent).

- Meneven (formerly Gulf) will build a 75,000 b/d plant by 1988 to produce nonupgraded crude in southern Anzoategu Ei state. This project is currently in the design phase.

Cooperative agreements

- Agreement between the Venezuelan Ministry of Energy and Mines and the German Ministry for the Research and Technology aimed at the evaluation and research on:

  .. steam generation and distribution
  .. water treatment
  .. upgrade of Venezuelan heavy crude

- Agreement between AOSTRA and the Venezuelan Technological Petroleum Research Institute (INTEVEP). The specific projects are:

.. Study of the generation of H_2S and other contaminants in thermal recovery processes

.. Scientific exchange between INTEVEP and the Alberta Research Council.

It has been estimated that the development of the Orinoco Heavy Oil Belt will require an investment in the range of $3,000-$10,000 per b/d, depending upon which production and upgrading techniques will be used. The present focus on production technology is aimed toward improving the steam soak and steam injection methods while simultaneously increasing the viability of combustion methods, and other novel recovery methods. The projected percent oil recovery from the Orinoco heavy oil belt is:

	Percent Oil Recovery	
	Range Percent OOIP	Average Percent OOIP
Primary Production	0-7	5
Primary + Steam Soak	2-12	8
Primary + Steam Soak + Steam Drive	10-35	20

Based on an estimated in-place resource of 1 trillion barrels, the belt can potentially produce 200 billion barrels if thermal recovery methods are used. This can truly be classified as a super giant reserve.

(3) United States Tar Sands Development

This section is not intended as an exhaustive evaluation of domestic tar sand and very heavy oil developments, but rather is included to provide useful background information. Problems unique to U.S. tar sand development are discussed, and processes which are being developed in the United States and have implications on future worldwide tar sand development are surveyed.

With the exception of Utah, information on U.S. tar-sand resources and reservoir characteristics in other states is limited. Exhibit IV-49[56] lists the tar sand deposit characteristics in Utah. Properties of the Athabasca deposit are included for comparison. The following conclusions can be drawn on Utah's tar sand deposits:

EXHIBIT IV-49

Tar Sand Deposit Characteristics in Utah

	OVERBURDEN FT	POROSITY %	PERMEABILITY MD	OIL SATURATION %	OIL CONTENT % WT	NET PAY FT	BARRELS PER AVG FT OOIP	BITUMEN VISCOSITY 10^6 CP	SULFUR CONTENT WT %	OIL GRAVITY °API
Tar Sand Triangle	1517	23	–	50	6	125	1200	0.3	3.6	11
P.R. Spring	370	24	121	44	[illegible]	35	1000	–	0.3	10
Sunnyside	300-1000	25-30	–	60	[illegible]	50-100	1250	–	0.55	9
Asphalt Ridge	700	21	132	50	5	20-80	810	–	0.4	10
Hill Creek	300	20	125	30	3	34	450	–	0.4	9.5
(Athbasca– Included for Comparison)	0-2000	30-40	Up to 5000	44-98	10	88	–	1.5	4.5	7-8

- The bitumen is not exceptionally viscous
- Formation porosity and permeability are low
- Average oil saturation is low
- Utah deposits are too deep for surface mining.

Additionally, problems which are unique to U.S. tar sands and their implications are:

Problems	Implications
. Discontinuous reservoirs, both vertically and horizontally	. Complications in exploration and recovery
. Divergent oil saturations	. Inefficient use of recovery energy - Poor economics
. Numerous thief zones and fractures	. Loss of injected fluid, recovery aids
. Small conventional mining potential.	. Need for development of special recovery techniques - Long lead time to commercialization.

A recent study carried out by the Lewin Associates shows that over half of the 26 billion tar sands resource in Utah can possibly be recovered by in-situ thermal recovery methods. However, as indicated in Exhibit IV-50[57] economic constraints probably limit total tar sand reserves to only 8 billion barrels.

EXHIBIT IV-50

Utah Sands Resource

	AREA (SQ. MILES)	TOTAL SIZE (BILLION BARRELS)	IN-SITU THERMAL RECOVERY (BILLION BARRELS)	ECONOMICALLY FEASIBLE (BILLION BARRELS)
Tar-Sand Triangle	230	16	12	6
P.R. Spring	270	4	Negligible	None
Sunnyside	90	4	3	2
Asphalt Ridge	25	1	1.4	None
Hill Creek	125	1	Negligible	None
Total	740	26	16.4	8

Applying a 50 percent recovery factor and subtracting the energy consumed in the process, the potential net amount is approximately two billion barrels. If this is indeed the case, bitumen from tar sands as an alternate source of crude oil represents only a miniscule amount at the current U.S. crude demand of 15 million b/d. Of course, the economically feasible amount will increase if world crude oil price continues to spiral upwards. Breakthrough or improvement in in-situ thermal recovery methods will also serve the same purpose. A list of active and proposed U.S. tar sand projects is shown in Exhibit IV-51.[59]

Current development of U.S. tar sand resources is concentrated in three areas:

. Field tests
. Basic research
. Resource evaluation.

All but one of the U.S. pilot projects employ in-situ thermal recovery methods including steam soak, steam injection, in-situ combustion, and wet combustion. The only open-pit mining pilot project involves Getty's project in the McKittrick field of Kern County, California.[58,59] This area contains approximately one billion barrels of viscous oil impregnated in diatomaceous earth. The original objective was to mine the diatomaceous earth for use as filter aid but high petroleum prices have prompted Getty to focus on recovering heavy oil instead. Getty proposes to use either the Dravo solvent extraction process or the Lurgi retorting process to extract very heavy oil from stripped-mined diatomaceous sand. The Lurgi retorting process is similar to a dry distillation process: After the oil sand is dried in a rotary heater, it is heated to approximately 650°C (1202°F) to vaporize the hydrocarbons which are recovered and upgraded to produce synthetic crude oil. The spent sand is recycled to heat the feed oil sand. A pilot plant is currently being built and will be in operation by 1986. The plant will have a capacity of 20,000 b/d and is expected to require an investment ranging from $200 to $260 million.

The method of producing heavy oil using underground mining has also gained much attention recently. At present, Continental Oil is evaluating underground mining in Wyoming and Getty is planning a similar operation in California. A recent study by the Energy Development Consultants estimated that 300 billion barrels of oil could be recovered through this method.[60,61] This estimate includes many targets ranging from remaining oil-in-place to shale and tar sands. Selling price was pegged at $12.50/bbl for a 20 percent rate of return, and $15.25/bbl for a 50 percent return. Feasibility information and operating data are sketchy. Moreover, there are many structural and environmental problems yet to be resolved. In the U.S., this process is not anticipated to reach commercial scale in the next 10 years.

EXHIBIT IV-51

U.S. Active and Proposed Tar Sand Projects

FIELD	OPERATOR	OPERATING PERIOD	PROCESS	DEPTH (FT)	NET PAY (FT)	OIL GRAVITY °API	OIL VISCOSITY CP
Cort Canyon, Santa Barbara County, California	Getty	6/76-5/81	Steam flood	2,500	80	9°	25,000
Marport, Monterey County, California	Ogle	7/77-?	Steam flood	1,406	80	9-10°	>100,000 @ 200°F
McKittrick, Kern County, California (Diatomaeous earth)	Getty	1979-84	Strip mining and surface extraction	0-1,200	NA	13-6°	
Northwest Asphalt Ridge, Uintah County, Utah	LETC	71-9/2	In-situ combustion and steam flood	300-600	10-80	14°	>1,000,000
Paris Valley, Monterey County, California	Husky	1/75-1/80	In-situ combustion with water injection	840	50	9-11°	50,000-400,000
Jones Asphalt Mine Ranch, Guadalupe County, New Mexico	American Exploration and Management Inc.	Planning 1981-?	Strip mining and solvent extraction				

A method being developed by Sandia Laboratory may improve the feasibility of injecting steam into deep reservoirs.[62] Project "Deep Steam" involves generating high pressure steam by burning hydrocarbons with air at the bottom of injection wells. This method improves the overall heat efficiency by reducing heat loss and steam condensation in the injection well. The method would be particularly effective for reservoirs deeper than 2,500 ft. At present, prototype down-hole steam generators capable of producing 5 million btu/hour are being tested in 70 feet towers, operating under simulated conditions of up to 700°F and 3,000 psi. Pilot field tests are also being carried out in the Kern River, California field.

In summary, compared to Canadian and Venezuelan deposits, U.S. tar sand resources are ill defined, much smaller and of poorer quality. Since limited U.S. reserves are amenable to surface mining methods, it is important that in-situ technology be improved to solve the problems peculiar to U.S. deposits. Moreover, better resource evaluation and pricing incentives are needed if U.S. tar sands resources are to be used as a transition fuel.

(4) Russia, Bitumen Production

Present crude oil production in Russia is estimated to be over 11 million b/d, which is the world's largest production. To maintain this production level, Russia is presently applying enhanced recovery methods to existing fields and developing new fields in remote areas of the country. For the longer term, production methods are being developed for producing very heavy oil and bitumen from lower-grade deposits. Present bitumen production is estimated to total only 20,000 to 25,000 b/d. Part of the bitumen production is used directly in road construction.

Russia is currently investigating several methods for in-situ production of bitumen, including cyclic steam injection, steam flooding, in-situ combustion, and down-hole electrical heating. Major recovery problems result from many of the bitumen deposits having relatively thin pay thickness and being at depths of 2,600-4,000 feet (800-1,200 m). Some 70 percent of the bitumen deposits are in carbonate rock formations which prevents the use of in-situ combustion methods. The pay thickness is generally from 15 to 25 feet (5 to 7 m), in the bitumen deposits in Azerbaijan, Emba, and the Tartar Republic.

Three in-situ pilot projects are being developed in the Tartar Republic, and some 100 wells have been drilled.[63] The recovery methods being tested include cyclic and continuous steam injection and in-situ combustion. Problems have been encountered in steam escaping from the formation through vertical fractures in the deposits. Effectiveness of the injected steam is also decreased by the intrusion of water into the reservoir. The formations are highly heterogeneous and have low permeability which require extensive hydrofracturing to establish a passage between wells. The in-situ combustion project has experienced difficulty in establishing continuous combustion. Initiation of combustion was established after several attempts by using a down-hole natural gas ignitor.

A fire-flood pilot project is being conducted in the Russkaze field to produce the field's super viscous crude oil. Russkaze was discovered in 1968 and is potentially a giant field covering 115 square miles. It is located above the Arctic circle, 400 miles northeast of the Middle Ob regions and is far from existing crude oil pipelines. The deposit is located at 2,625 feet (800 m).[64]

REFERENCES

IV. TAR SANDS PRODUCTION METHODS

1. Reginald D. Humphreys, F. K. Spragins and D. R. Craig, "Oil Sands - Canada's First Answer to the Energy Shortage," Proceedings of 9th World Petroleum Congress, Tokyo, Japan, pp. 17-27, Tokyo, Japan, 1975.

2. G. V. Chilingarian, T. F. Yen, Editors, "Bitumens, Asphalts and Tar Sands," Elsevier Suen pp. 107-121, Elsevier Scientific Publishing Company, Amsterdam, 1978.

3. Application of the Great Canadian Oil Sands Limited to the Energy Resources Conservation Board, Alberta, Canada, March 1979.

4. E. C. Sanford and F. A. Seyer, "Processibility of Athabasca Tar Sand Using a Batch Extraction Unit: The Role of NaOH," pp. 104-168, CIM Bulletin, March 1977.

5. George Karnofsky, "Dravo Solvent Extraction Processes," Dravo Engineers and Constructors.

6. "The Lurgi-Ruhrgas (L-R) Process," Lurgi Corporation, Belmont, California.

7. Roland W. Rammler, "One Process Retorts Three Feeds for Synthetic Crude," pp. 63-66, Mining Engineering, September 1970.

8. Roland W. Rammler, "The Production of Synthetic Crude Oil from Oil Sand by Application of the Lurgi-Ruhrgas Process," pp. 552-559, The Canadian Journal of Chemical Engineering, Vol. 48, October 1979.

9. A. W. Iyoko, "Selecting Enhanced Oil Recovery Processes," pp. 61-64, World Oil, November 1979.

10. "Enhanced Oil Recovery" - An Analysis of the Potential for Enhanced Oil Recovery from Known Fields in the United States - 1976 to 2000," National Petroleum Council, December 1976.

11. "Enhanced Oil Recovery Potential in the United States," Office of Technology Assessment, Congress of the United States, Washington, D.C., January 1978.

12. Fourth Annual Report of Alberta Oil Sands Technology and Research Authority, March 31, 1979.

13. S. T. Fisher, "Electrical Induction Heating for the Processing of Fossil Fuels," Electronics of Power, July 1978.

14. Dr. D. E. Towson, "The Electric Pre-Heat Recovery Process," presented at the First International Conference on the Future of Heavy Crude and Tar Sands, Edmonton, 1979.

15. F. R. Newbold and T. K. Perkins, "Wellbore Transmission of Electrical Power," p. 39, Technology, July-September 1978.

16. E. R. Abernethy, "Production Increase of Heavy Oils by Electromagnetic Heating," pp. 91-97, Technology, July-September 1976.

17. J. Bridges, G. Svesty, A. Taflove and R. Snow, "Radio Frequency Heating to Recover Oil from Utah Tar Sands," presented at The First International Conference on the Future of Heavy Crude and Tar Sands, Edmonton, June 1979.

18. Personal Communications with Professor T. F. Yen, Professor of Chemical and Biomedical Engineering, University of Southern California.

19. Op. cit., Reference 12.

20. E. M. Khalimov, R. K. Muslinmov and G. T. Yudin, "Bitumen Deposits of the U.S.S.R. and Methods of their Development."

21. "Mining of Viscous Crude Claimed Viable," Oil and Gas Journal pp. 46-47, January 5, 1976.

22. "Azerbaijan Underground Oil Mine to Start," Oil and Gas Journal, p. 58, May 7, 1979.

23. R. A. Dick, S. P. Wimpfern, "Oil Mining," Scientific American, pp. 182-188, October 1980.

24. Suncor Inc. 1979 Annual Report.

25. Op. cit., Reference 3.

26. Op. cit., Reference 1.

27. Laurie M. Watson, "Syncrude Beats Tar Sands, Cold with Modified Equipment," pp. 21-24, Canadian Mining Journal, June 1979

28. Application of the Syncrude Canada Limited to Energy Resources Conservation Board, Alberta, Canada, September 1979.

29. "Crude Oil Starts to Flow From the Sands of Athabasca," Process Engineering, August 1978.

30. P. R. Potkins, "Status of the Syncrude Project," presented at the Seminar on Nonconventional Oil Technology, Calgary, Alberta, May 29-30, 1980.

31. Summary Report from the Alsands Project Group, Public Affairs Department, Calgary, Alberta.

32. "Politics Mixed Canadian Oil Sands Work," pp. 36-38, Oil and Gas Journal, February 11, 1980.

33. Application of the Alsands group to the Energy Resources Conservation Board, Alberta, Canada, December 1979.

34. Application of the Esso Resources Canada Limited to the Energy Conservation Board, Alberta, Canada, October 1979.

35. Patrick Crow, "Canada Looks to Tar Sands, Heavy Oil to Fill Conventional Oil Deficit," Oil and Gas Journal, February 26, 1979.

36. "Catalogue of AOSTRA Technology," Alberta Oil Sands Technology and Research Authority, December 1978.

37. Op. cit., Reference 12.

38. "Agenda and Abstracts, Seminar on Nonconventional Oil Technology," Alberta Oil Sands Technology and Research Authority, Calgary, Alberta, May 1980.

39. T. P. Lederhouse, "Gregorie Lake Block 1 Pilot Project Description and Status, Presented at the Seminar on Nonconventional Oil Technology, Calgary, Alberta, May 1980.

40. T. G. Harding, "Marguerite Lake Project Phase A," Presented at the Seminar on Nonconventional Oil Technology, Calgary, Alberta, May 1980.

41. P. Kitzan, "Design, Construction and Operations Progress Report on the Peace River In-Sites Project," Presented at the Seminar on Nonconventional Oil Technology, Calgary, Alberta, May 1980.

42. C. Negel and J. Flipping, "Project Overview of the Kinsella Heavy Oil Pilot," presented at the Seminar on Nonconventional Oil Technology, Calgary, Alberta, May 1980.

43. J. G. Vandermeer and T. C. Presber, "Heavy Oil Recovery from the Grosmont Carbonates of Alberta," presented at the Seminar on Nonconventional Oil Technology, Calgary, Alberta, May 1980.

44. John Starr and Jose M. Prats, "Chemical Properties and Reservoir Characteristics of Bitumen and Very Heavy Oil from Canada and Venezuela," presented at the First International Conference on the Future of Heavy Crude and Tar Sands, Edmontown, June 1979.

45. T. J. Stewart - Gordon, "Prospects, Ability, Cash: Venezuela Has It All," pp. 91-96, World Oil, January 1980.

46. "The Heavy Oil Industry in Venezuela," presented at the First International Conference on the Future of Heavy Crude and Tar Sands, Edmonton, June 1979.

47. Edison Perozo, "Orinoco Oil Belt, Exploration, Evaluation and Exploration," presented at the Latin American Oil Congress, Rio de Janerio, July 1980.

48. Roger Lowenstein, "The Orinoco Belt - Venezuela's Oil Bonanza," The Energy Daily, July 30, 1979.

49. Carlos J. Borregales, "Production Characteristics and Oil Recovery in the Orinoco Oil Belt," presented at the First International Conference on the Future of Heavy Crude and Tar Sands, Edmonton, June 1979.

50. Ibid.

51. Edalfo Lanfranchi, "Result of the Application of Alternating Steam Injection in Heavy Oil in South-eastern Venezuela, Oil Sands Symposium, Canadian Institute of Mining, 1977.

52. R. F. Meldau and W. B. Lumpkin, "Producing Venezuelan Fireflood Wells," presented at the Heavy Oil Symposium, University of Zulia, Maracaibo, Venezuela, July 1974.

53. Op. cit., Reference 46.

54. Alvaro Franco, "Internal Work Humming in Venezuela," The Oil and Gas Journal, April 5, 1976.

55. Op. cit., Reference 47.

56. "The Economic Potential of Domestic Tar Sands," prepared for U.S. Department of Energy, January 1978.

57. Ibid.

58. "Work on U.S. Oil Sands Heating Up," Science, March 14, 1980.

59. C. Q. Cupps, L.C. Marchant and J. J. Stosur, "Tapping the Vast Tar Sand Reserves of the U.S.," pp. 73-78 World Oil, September 1979.

60. B. A. Kennedy, A. Edery and L. A. Readdy, "Mined Oil -A Valuable Resource Opportunity," p. 47, Mining Engineering, January 1980.

61. "Mining of Oil Could Multiply Recovery and Reserves at Conventional Fields," p. 61, E & M Journal, June 1979.

62. R. L. Cox, "Development of Downhole Steam Generation Systems and Injection Storing Modifications for Deep Steam Injection," presented at DOE 4th Symposium on Enhanced Oil & Gas Recovery and Improved Drilling Methods, August 1978.

63. R. K. Muslimov, "Methods of Development of Heavy Oils and Bitumens in the U.S.S.R.," presented at the First International Conference of Future of Heavy Crude and Tar Sands, Edmonton, June 1979.

64. "Synfuel from Tar, Oil Sands, Shale Seen as Promising," p. 86, Oil and Gas Journal, June 23, 1980.

V. Bitumen and Very Heavy Oil Upgrading

This chapter focuses on the alternatives available for upgrading heavy oils. Heavy oils for this purpose are defined as any hydrocarbons whose characteristics require extraordinary treatment in order for them to be transported or refined into conventional saleable products. Under this definition, heavy oils include tar sand bitumen, very heavy crudes, and poor quality vacuum residues.

The purpose of this part of the report is to outline the development of technology applicable to upgrading heavy oil and to assess those foreign technologies which show promise in aiding in the development of heavy oil deposits in the United States. To meet this purpose, the upgrading portion has been organized into four sections providing for: an overview, a review of the factors affecting process selection, a discussion of new and conventional upgrading processes, and an assessment of foreign technology for upgrading. An appendix is also included.

The overview section discusses the value of heavy oil with respect to its distinguishing characteristics, limitations for further use and reasons for upgrading. Minimum treatments which enable the pipeline transportation of heavy oil are briefly outlined. Upgrading objectives are reviewed and the two categories within which most upgrading processes fall are described.

The section on factors affecting bitumen and heavy oil upgrading reviews and summarizes key characteristics and relative advantages and disadvantages of each conventional and recently developed process discussed in subsequent sections.

The conventional upgrading technology section describes the limitations of traditional bottom-of-the-barrel processes and examines a sample of U.S. processes which have emerged and are applicable to heavy oil processing. This section also presents a look at the major existing and planned commercial upgrading projects for a perspective on recent trends in process selection.

The third section examines new technologies that are in various stages of development. Foreign technologies are emphasized but appropriate developments in U.S. technologies are also described.

For the sake of completeness and to prevent major process distinctions and trends in process development from being overshadowed by process details, an appendix is used. Appendix I contains detailed process descriptions and more complete process information where available. This Appendix is not meant to be a comprehensive description of all upgrading technologies. Discussions of foreign technologies cover those processes or developments on which information was readily available. Because this study focuses on foreign technologies, not all U.S. processes are described. There are many U.S. processes which show promise in heavy oil applications. Representative U.S. technologies are described to the extent that they provide a sample of current technology and exhibit certain process characteristics against which foreign technologies can be compared.

1. UPGRADING PROCESSES SELECTION OVERVIEW

(1) The Value of Heavy Oil

The value of heavy oil is restricted by the characteristics which distinguish heavy oils from conventional crudes. Bitumen and very heavy oils are mixtures of oils, resins, and asphaltenes. They have low concentrations of naphthas and relatively low proportions of middle distillates. The oils, representing the lightest component, are similar to vacuum gas oil from conventional crude oils. The resins have low hydrogen-to-carbon ratios but generally can be upgraded by hydrogenation. The asphaltenes, however, are basically solids and are very difficult to upgrade to lighter products. Heavy oil is distinguished by its low hydrogen-to-carbon (H/C) ratio and its high level of contaminants comprising various levels of solids, sulfur, nitrogen, and metals which all tend to concentrate in the heavy asphaltene fraction.

Heavy oil's low H/C ratio, as shown in Exhibit V-1, essentially places it in the same category as vacuum resid from conventional crudes. The market for vacuum resid has historically been limited to industrial and utility fuels, coke production, asphalt manufacture or use as fuel in the cement industry. The market for residual fuel over the next 10 years is expected to grow annually only 2 percent in Japan, 1 percent in Western Europe, and is expected to decline by 1 percent per year in the U.S.[1] This decline is attributed to the increasingly stringent standards on sulfur in fuels and on the increasing tendency for utilities to convert to coal.

Heavy oil is well suited to the asphalt and cement industry markets. However, it must compete with the increasing supply of resid resulting from the depressed overall market for residual fuels. The expected emergence of increasing supplies of coal and shale oil will also help dampen the demand for resid.

EXHIBIT V-1

Heavy Crude Hydrogen/Carbon Ratio

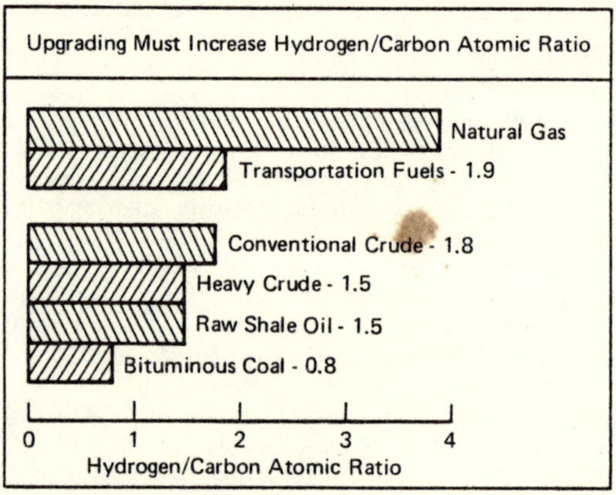

The usefulness of heavy oil as a vacuum resid is further limited by the high level of contaminants which diminish its attractiveness to most resid markets. The general characteristics of heavy oils from various sources are compared with conventional crude oils in Exhibit V-2.

In addition to the restrictions that high contaminant levels impose upon the oil's marketability, these contaminants also cause more problems for processing heavy oils in conventional refineries. The high levels of sulfur and nitrogen preclude the use of heavy oil in conventional furnaces without the use of costly stack gas scrubbers. Although sulfur removal problems are not new to the refining industry, the concentration in heavy crude can be four times that in conventional crude. In the U.S., a potential heavy oil user is more likely to switch to coal than heavy oil, providing that stack gas cleanup is necessary for either.[2] Even in countries that still use high-sulfur fuels, the long-term trend must be toward reduced sulfur emissions.

Besides the environmental problems associated with suifur emissions from fuels' combustion, sulfur in lighter naphtha products from heavy oils poisons reforming catalysts. Reforming improves the octane of naphtha to high-quality gasoline and requires both high-purity hydrogen and low-sulfur feedstocks for processing. Desulfurization is accomplished through conventional hydrotreating, which ties the sulfur up with hydrogen to form H_2S. Sulfur can be removed from the H_2S in a Claus or other conventional refinery gas cleanup process. Desulfurization is costly and requires substantial amounts of hydrogen.

Nitrogen is more difficult to remove from heavy oils than is sulfur. Reducing the nitrogen also requires costly hydrogenation processing during upgrading. Nitrogen, like sulfur, can foul catalysts; but it is a nuisance primarily for its presence in combustion products such as NOx, and its subjectivity to environmental restrictions.

Metals, primarily nickel and vanadium, poison conventional catalyst and act as dehydrogenation catalysts in cat crackers. A metals content in excess of 50 ppm will lower a cat cracker's effective capacity by increasing production of methane, hydrogen, and coke which, in turn, limits the compression capacity for byproduct gas the air-blowing capacity for burning coke.

The percentage of particulates, such as clay or sediment, is higher in very heavy oils than in most crudes. Particulate content is primarily a function of the production and recovery process used and can affect the upgrading process' ability to function properly. These contaminants affect not only the extent of upgrading processing required, but also the sequence of processing. The specific upgrading process or combination of processes used will establish the quality of the upgraded oil.

EXHIBIT V-2

Characteristics of Very Heavy Oils

		ARABIAN LT.	VENEZUELA	CALIFORNIA (OXNARD)	CANADA	UTAH UINTA
GRAVITY	°API	34.5	8.4-18	3.9	7-18	9.2-14.4
POUR POINT	°F	-15	-54-68		-40-77	
VISCOSITY RANGE	cps					
@ 100°F		5.8	200-62,000	--	--	--
@ 212°F		2.3	--	--	20-1,000	--
NITROGEN	Wt-%	0.09	0.42-0.65	0.7	0.1-0.4	0.6-1.3
SULFUR	Wt-%	1.7	1.9-5.2	8.2	3.4-5.6	0.4-0.8
METALS	ppm					
Vanadium			105-1,000	310	100-700	20-70
Nickel		16	20-150	160	40-100	70-120
Iron			7	--	--	--
CONRADSON CARBON	Wt-%	3.1	9.5-15.5	21	8-16.6	3.5-15.2
ASPHALTENES		--	4.9-17	--	12-20	11.8

Electrode-grade coke, used extensively in the manufacture of aluminum, is a valuable product of some vacuum resids. However, it requires a lower concentration of metals, sulfur, and particulates than that found in most heavy oils. In most cases, coke produced from heavy oils has no market in the aluminum industry and has limited uses as fuel. As a result, coke and other heavy residue by-products from heavy oil present significant disposition problems.

The restrictions imposed upon the use of heavy oils by their characteristically low H/C ratio and high level of contaminants present two basic options to the heavy oil producer. After primary distillation of the heavy oil to separate out the light products that can be transported and further refined by conventional means, the remaining residue having no direct uses can either be discarded or upgraded.

Discarding or otherwise disposing of the heavy oil residue from atmospheric or vacuum tower bottoms is an unacceptable choice because it is environmentally unsound, and it is usually uneconomical. Generally, there is no convenient place to store the high-carbon residue, which can amount to half the total heavy oil production. Furthermore, the high level of contaminants presents a serious potential for future problems such as groundwater contamination.

Disposal of distillation bottoms is uneconomical for two reasons. First, it would force the use of the more valuable liquid products to fuel the production step. Production fuel consumption could easily represent half the liquid product volume from heavy oil distillation. Second, the high cost of heavy oil production and alternate crude oil prices provide incentives for greater utilization of the total barrel, whether produced from tar sands or from conventional crude production.

Upgrading, on the other hand, can significantly improve the value of heavy oils. Heavy oil upgrading provides for the direct use of heavy oils or converts the heavy residue to a material which is suitable for use or further processing to conventional petroleum products.

Heavy oil upgrading can be performed at the production site or at a nearby refinery. This decision is chiefly one of economics, which involves the integration of production site demands, refinery capabilities, transportation constraints, and the flow characteristics of the oil. Pour point and viscosity are the principal characteristics affecting a fluid's ability to be transported easily.

Pour point is the temperature at which wax crystals form in the oil and inhibit its ability to flow. At its pour-point temperature, the liquid begins to gel and cannot be easily pumped or pipelined. Wax deposits also tend to form on the walls of vessels or pipelines, which leads to reduced capacities and pumping efficiencies.

The relationship between viscosity and temperature is important in pipeline system design. As temperature falls, viscosity rises, resulting in increased head loss throughout the pipeline network. Increased head losses require higher pressure pumps with greater horsepower requirements to maintain the same flowrate of oil through pipelines.

Where existing refineries have the facilities and capacity to process heavy oil, minimum treatment methods may be considered to make the oil transportable from the production site. In general, these treatments change the oil's physical properties so that it can be pipelined, but they do not impart a permanent change in the composition of the oil. This enables the oil to be moved to an off-site refinery for further upgrading to consumer products. These treatments represent the minimum effort necessary to provide a transportable product and are defined in the following section.

(2) Minimum Treatments Enabling the Pipeline Transportation of Heavy Oils

The following represent the chief methods under consideration for treating heavy oils to move them to refineries or transfer terminals:[3]

- Pour-Point Depressants - Additives modify the wax crystals in heavy oils to prevent agglomeration. By keeping the wax crystals from building up on themselves or on vessel surfaces, the fluidity of the oil is preserved. This method is generally applicable only to those oils with viscosity low enough to permit economic pipelining. There are many commercially available pour-point depressants.

- Hot Oil Pipelines - Heating is a common method to reduce viscosity or maintain a fluid above its pour point. Exhibit V-3[3] lists examples of pipelines which handle heated heavy oils. A sufficiently high temperature can be maintained by heating the oil initially and/or intermittently as needed, by flowing friction and insulation, or any combination of these methods.

Exhibit V-3
Typical Heated Oil Pipelines

Location	Length (miles)	Diameter (inches)	Fluid and Operating Temperature (°F)	First Year of Operation
Sumatra				
Cal-Tex	30	30	120 (90 pour)*	1958
California				
Alamitos-Huntington Beach Fuel Pipeline, Southern CA	16	12¾	200–210**	1959
England				
Esso Fawley Refinery to Marchwood	8	8⅝	110**	1963
Columbia				
Richmond Petroleum	300	10¾	140*	1964
Italy				
Anic Ref. at Sannazzaro to Chivasso	51	10¾	185*	1964
Italy				
Esso-CalTex-Sarpon Ref Trecate				
Chivasso	52	10¾	**	1965
Turbigo	8	6⅝	**	1965
Arluna	10	12¾	**	1965
Libya				
British Petroleum	320	30	130 (75 pour)*	1966
Australia				
Shell Refinery at Geelong to Newport	35	8⅝	175**	1966
California				
Belridge to Torrance Ref. Mobil Oil	155	8⅝	225*	1967
Holland				
BP, Botlek-Europoort	12	12¾	140**	1967
France				
Union General Des Petroles	28	12¾	185**	1967
Ref. at Nangis to Melum	15	8⅝		
Libya				
Occidental	132	40	160*	1968
Libya				
Occidental	42	24	160*	1968
North Dakota				
Rock Island Pipeline	65	4–6	175**	1969
England				
Esso Fawley Refinery to London	62	10¾	200**	1970
Belgium				
Feluy-Ruien	42	12¾	210**	1972
Italy				
Biacenza-Cazella	30	12¾	180**	1973
Alaska				
Alyeska	800	48	160 (10 pour)***	1977

*Heavy crude oil.
**Heavy fuel oil
***Crude oil.

Source: A. Sloan, R. Ingham, W.L. Mann, "Pipeline Transportation of Heavy Oils." Presented at the First UNITAR Conference on the Future of Heavy Crude and Tar Sands.

- Heavy Oil Dilution - Heavy oil can be piped at ambient temperatures if enough diluent of a sufficiently low viscosity (light hydrocarbon; i.e., condensate, natural gasoline, or naphtha) is added to reduce the overall viscosity and maintain a pour point lower than the flowing temperature. This technique, practiced commercially in Canada to move Lloydminster heavy crude, requires:

  - Extra capacity for transporting diluent
  - A diluent separation and return system
  - A diluent source

- Water Injection - Adding water to reduce viscosity has been proposed as a technique for pipeline transport of heavy crude oil. The oil/water emulsion would have to be separated at the pipeline terminus. The feasibility of this method has been questioned in several laboratory investigations, and no commercial applications have been reported.

- Oil/Solid Slurry - Marathon Oil has developed an oil/solid slurry technique which is applicable to waxy crude oils. With this method the waxy portion of the oil is allowed to crystallize into small solid particles that are kept suspended in the oil as a slurry. Besides the limitation cited for additives, the oil/solid slurry must remain below the pour-point temperature to prevent resolution of the wax in the slurry.

Venezuela is currently considering water injection, heat, and the use of diluents for treating the production from its Orinoco Heavy Oil Belt to allow transportation to downstream refineries. Venezuela has successfully transported its present heavy oil production by treating it with:

- Water addition
- Diluent mixing
- Primary and intermittent heating
- Sand covering.[4]

Husky[5,6] has found from its experience with transporting heavy California and Lloydminster crudes that:

- Blending crude with condensate before treating by raising its gravity usually aids in the treating step.

- A low temperature (130°F) atmospheric wash tank followed by settling tanks is cheaper than a heater treater for normal crudes.

. Additional benefits accrue from solids settling and emulsion breaking.

. Heater treaters (at 180°F) are necessary for tight emulsions.

(3) Upgrading Methodology

Heavy oil upgrading is the link between the resource and the market. Its purpose is to provide the necessary and economic means to bring more oil to the marketplace. Upgrading is necessary because, by definition for this study, very heavy oils and bitumens are hydrocarbons whose characteristics require extraordinary treatment in order to transport or refine them into conventional, saleable products. Upgrading must also be economic and efficient to ensure maximum utilization of the resource.

The primary objective of the upgrading process is to produce a transportable crude that can be upgraded further in existing or modified refineries. The upgrading plant may be required to provide fuel for the combined production and upgrading project. In this case, providing fuel may be considered a secondary objective of the upgrading process

Upgrading methods for improving the quality of very heavy oils range from the very simple, which may do no more than change the oil's characteristics just enough to produce a pumpable product, to the very complex, which can be as large and costly as any major refinery. Depending on their complexity, upgrading methods can add varying amounts of economic value to the heavy oils processed. Generally, value increases with hydrogen content. Exhibit V-4 shows the relationship between hydrogen content and gravity for principal petroleum fractions.

The upgrading function is performed by removing carbon and contaminants from the heavy oil, by adding substantial amounts of hydrogen to the oil, or by doing both. Carbon removal processes are generally more proven and require far less hydrogen than hydrogen addition processes. Hydrogen addition processes have higher liquid yields and fewer by-product disposition problems. Commercial applications may suggest a scheme which uses combinations of one or more processes from both categories. A simplified schematic of processing options is presented in Exhibit V-5. The options diagramed cover all the basic processing categories. However, the options are not mutually exclusive, as products from one process often become feeds for others. Exhibit V-6 lists possible dispositions for upgraded products.

EXHIBIT V-4

Increase in Hydrogen Content Required to Upgrade Bitumen to Consumer Products

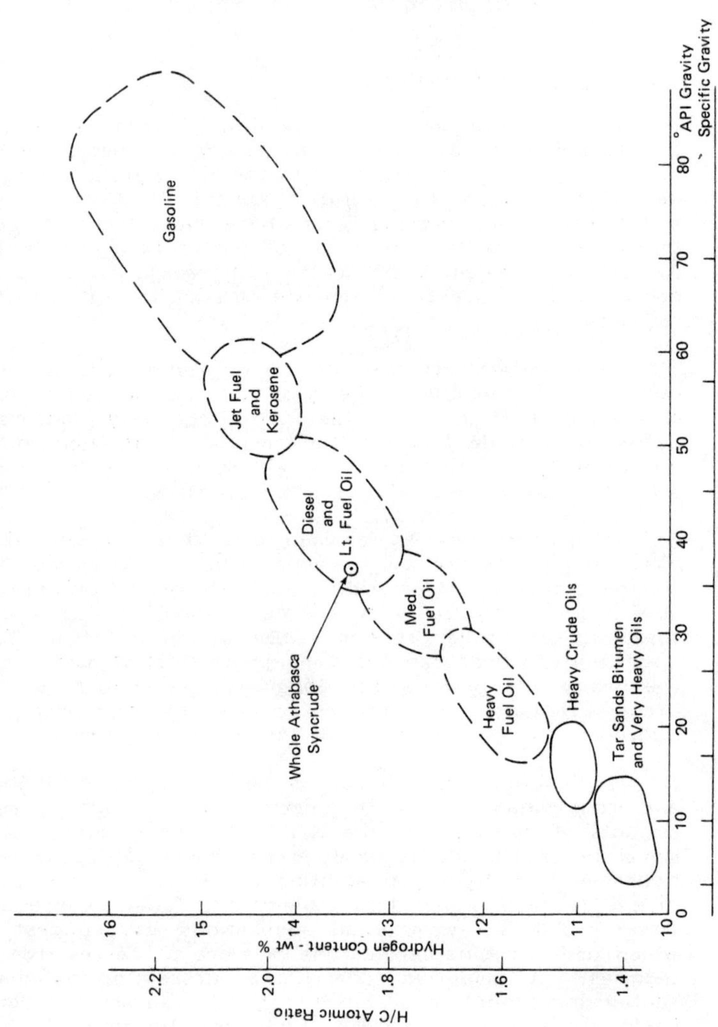

EXHIBIT V-5

Generic Processing Options for Converting Very Heavy Oils to Finished Products

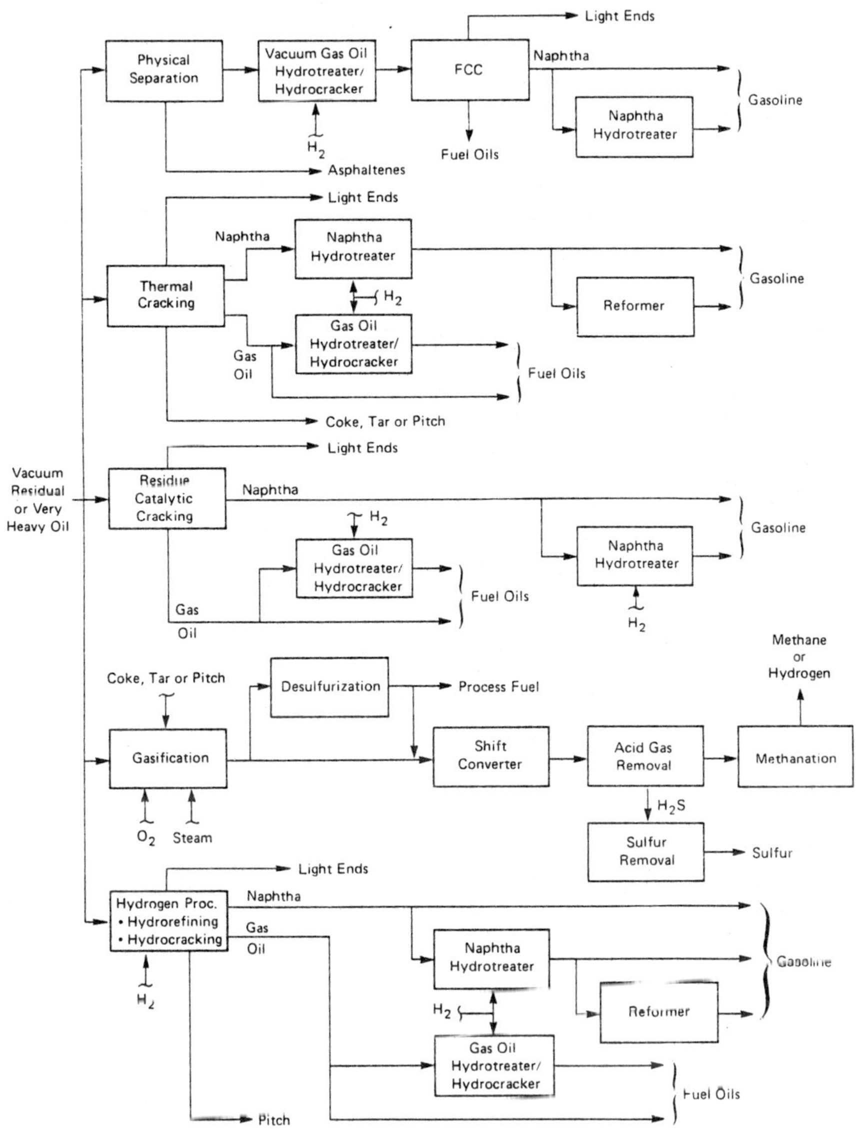

EXHIBIT V-6

Possible Disposition of Heavy Oil Upgrading Products

UPGRADING PRODUCT	DISPOSITION	
	WITHIN PLANT	MARKET
Light Ends (C_1, C_2, C_3, C_4)	• Process Fuel • Gas Plant • Hydrogen Production	Petrochemical Manufacture, LPG, Natural Gas –
Light Distillates (C_4, Naphtha, Reformate)	• Gasoline Blending • Petrochemicals Manufacture • Process Fuel	Gasoline Petrochemicals –
Middle Distillates (Kerosene, Diesel, No. 2 Oil)	• Distillate Blending	Fuel[1]
Heavy Distillates (Gas Oils, Resins)	• FCC/Hydrocracker Feed • Fuel Oil Blending • Lube Plant • Process Fuel	Fuels, Mineral Oils & Lighter Products Waxes & Lubricating Oils –
Heavy Residues (Resins & Asphaltenes)	• Fuel Blending • Lube Plant • Asphalt Blending • Gasification • Process Fuel	Fuel Oils Lubricating Oils, Petrolatum Asphalt, Road Oils Methane[2] –
Coke, Tars, Pitch	• Petrochemicals Manufacture • Gasification • Process Fuel	Fuels, Electrode Coke, Carbon Black Methane[2] –
Sulfur	–	Sales
Other (Catalysts, Drag Streams)	• Metals Recovery	Sales Disposal

(1) Fuel includes Jet Fuels, Kerosene, Diesel, No. 2 oil.

(2) Methane represents possible export product, others (i.e., hydrogen and fuel gas) are more suited to process use.

1. Carbon Removal Upgrading Processes

The inventory of carbon removal upgrading processes is listed in Exhibit V-7. Carbon removal processes are disproportionation processes which separate and/or remove the asphaltene fraction of the heavy oil, thereby increasing the hydrogen/carbon mole ratio of the liquid fraction. The liquids can then be hydrotreated to an even greater H/C ratio. The by-product, or asphaltene fraction, can be used in further processing, either by charging it directly to furnaces (with necessary stack gas scrubbers in place) as a fuel or by first gasifying it to fuel gas or to hydrogen for process use. If markets exist, the residue fraction can be sold as coke or asphalt or used in petrochemical manufacture. Generally, the carbon removal processes are less expensive and more commercially proven than hydrogen addition processes. The common disadvantages of the carbon removal processes are their relatively low liquid yield and the problem of disposition of its by-product (coke, pitch, or tar) stream.

Carbon removal processes can be catalytic or noncatalytic. Noncatalytic carbon removal processes include physical separation and thermal cracking processes as well as gasification and combustion. Physical separation and thermal cracking processes produce a high carbon residue stream that either requires further processing or can be utilized internally. Gasification and combustion processes convert the entire residue feed to either gas or energy. Fluid cat cracking processes are catalytic carbon removal processes which crack the residue feed to lighter products in the presence of a catalyst. The coke produced in the cracking reaction adheres to the catalyst and is "removed" via combustion in the regenerator.

2. Hydrogen Addition Upgrading Processes

Compared to carbon removal processes, the hydrogen addition processes are generally more complicated. A list of most processes which incorporate hydrogen addition to upgrade very heavy oils is contained in Exhibit V-8. In hydrogen addition, the entire feed is reacted in the presence of hydrogen. The reactions are generally catalytic. The H/C ratio is increased as more hydrogen atoms are attached to cracked chains of carbon hydrogen molecules. Hydrogen addition processes exhibit a wide range of operating severity (temperatures and pressures) affecting varying reductions in the molecular size of the feed. As severity increases, the process first saturates the olefins, desulfurizes then denitrifies the feedstock, and finally cracks the larger

EXHIBIT V-7 Carbon Removal Processes

PROCESS NAME	PROCESS TYPE	DEVELOPER	STAGE OF DEVELOPMENT[1] EXISTING TYPE[2]	EXISTING (NUMBER)	EXISTING START-UP DATE[3]	PLANNED TYPE	PLANNED (NUMBER)	PLANNED START-UP DATE	FEEDSTOCKS	KEY PRODUCTS
PHYSICAL SEPARATION										
Vacuum Distillation	Vacuum Distillation	Many, Non-Proprietary	C	Many					Atmospheric Residue	Vacuum Residue
Solvent Deasphalting	Solvent Deasphalting	Many	C	(41)					Vacuum Residue Whole Bitumen	Bright Stocks Cracking Feedstocks
Demex	Solvent Deasphalting	UOP Process Division	C	(3)		C	(2)		Vacuum Residue	Demetallized Oil
Residuum Oil Super-critical Extraction (Rose)	Solvent Deasphalting	Kerr McGee	C	(2)	1979	C	(1)	1981	Atmospheric & Vacuum Residue (shale oil & tar sands w/pilot plt.)	Asphalt, Lube Oils Resins, Cat Cracking Feed
Asphalt Blowing/Bitumen Manufacture	Bitumen Manufacture	Snamprogetti S.p.A., BP Trading Limited	C						Vacuum Residue	Bitumen, Asphalt
THERMAL CRACKING[4]										
Visbreaking	Mild Thermal Cracking	Non-Proprietary	C	(20+)		C	(Several)		Atmospheric & Vacuum Residue	Middle Distillates & Fuel Oil, Improved Pourpt. & Lower Viscosity
Delayed Coking	Coking	Many	C	(50+)	Pre-1960				Vacuum Residue, Tar Sands Liquids	Coke, Gas, LPG, Naphtha & Gas Oils
Fluid Coking	Coking	Exxon Research & Engineering Co.	C	(9)	Pre-1960				Vacuum Residue, Tar Sands Liquids	Coke, Gas, LPG, Naphtha & Gas Oils
Eureka Vacuum Residue Cracking	Thermal Cracking/Steam Stripping	Eureka Chemical and Chiyo Chiyoda Chemical Engineering & Construction Company	C	(1)	1975	C	(1)	1982	Vacuum Residue	Aromatic Pitch & Cracked Oils
ERCO Coker	Direct Fluid Coking	ERCO, Bechtel	BS	(1)	1979	C	(1)	1984	Tar Sand, Water	Synthetic Crude, Putty-like Coke Slurry
LR Gas	Direct Coking	Lurgi	PP	(1)	1979				Tar Sand, Heavy Oil	Med. Btu Gas, Naphtha, Middle Distillates
Taciuk	Direct Coking	UMATAC Industrial Processes Ltd.	PP	(1)		D	(1)	1982	Tar Sand, Heavy Oil	Med. Btu Gas, Naphtha, Middle Distillates
Mobil Ether Solvent	Etherification/Blending/Cracking	Mobil	BS		1976				Bitumen, Vacuum Residue, and Methyl Tertiary Butyl Ether	Butanes, Gasoline
MCC	Fluid Cracking	Mitsubishi Petrochemical and Alberta Research Council	BS, PP	(1), (1)					Bitumen, Vacuum Residue	Olefins (i.e., ethylene, propylene, benzene)
COSMOS	Thermal Cracking with Steam & Molten Salts	Mitsui Petrochemical Ind. & Mitsui Engineering & Shipbuilding Company	BS	(1)	1977	PP	(1)		Low Sulfur Crudes	Olefins

(continued)

EXHIBIT V-7 (Cont.)

PROCESS NAME	PROCESS TYPE	DEVELOPER	STAGE OF DEVELOPMENT[1] EXISTING TYPE[2]	(NUMBER)	START-UP DATE[3]	PLANNED TYPE	(NUMBER)	START-UP DATE	FEEDSTOCKS	KEY PRODUCTS
CATALYTIC CRACKING[5]										
Heavy Oil Cracking (HOC)	FCC	Pullman Kellogg and Phillips Petroleum	C	(1)	1961	C	(1)	1980	Atmospheric & Vacuum Residue	Gasoline, Fuel Blending Stock, LPG Carbon Black Feed
Resid Cat Cracking	FCC	Total Petroleum Inc.	C	(1)	1976	C	(2)	1981, 1982	Atmospheric Residue	Gas, Gasoline, Middle Distillates
Reduced Crude Conversion (RCC)	FCC	Ashland Oil Company, UOP	D	(1)	1978	C	(1)	1982	Atmospheric Residue	LPG, Gasoline, Fuel Oils
RAROP Resid Cat Cracking	FCC	Nippon Oil Company	BS	(1)	1979	PP	(1)		Atmospheric Residue	LPG, Gasoline, Fuel Oils
GASIFICATION										
Shell Gasification Process	Partial Oxidation	Shell Development	C	(5+)	1958	C			Vacuum Residue, Asphalt	Hydrogen, Med. Btu Gas
Texaco Partial Oxidation	Partial Oxidation	Texaco	C	(35+)	1956	C			Vacuum Residue, Asphalt	Hydrogen, Med. Btu Gas
M-Gas	Catalytic Gasification	Mitsui Engineering & Ship Bldg. and Mitsui Kozan Coking Ind.	BS	(1)		PP	(1)		Vacuum Residue, Asphalt	Medium Btu Fuel Gas
Flexicoking	Coking & Gasification	Exxon Research & Engineering Co.	C	(1)	1976	C	(1-3)	1982	Vacuum Residue, Asphalt	LPG, Naphtha, Gas Oils, Low BTU Gas
Dynacracking	Coking, Gasification, Cracking	Hydrocarbon Research Inc.	PP	(1)	Pre-1955	D	(1)	1982	Vacuum Residue High Sametals	Distillates Naphthas & High Quality Synfuel Gas
RAROP Resid Cracking and Gasification	Coking, Gasification, Cat Cracking	Nippon Mining Co., Kashima Oil Co., Hitachi Ltd.	BS	(1)	1979	PP	(1)	1981	Vacuum Residue	Middle Distillates & Hydrogen
DIRECT COMBUSTION										
	Steam Generation								Vacuum Residue, Heavy Oil	Steam
	Bunker Fuel Marine Transport								Bunker C	
Exxon Chemically Active	Fluidized Bed Combustion	Exxon Research & Engineering Co.				PP			Vacuum Residue	

(1) Known locations and sizes of existing and planned facilities are included in process descriptions.
(2) Plant types: BS – Bench Scale: Laboratory unit, usually not continuous, capacity $<$1 b/d.
PP – Pilot Plant: Continuous, capacity varies from 0.5 - 1,000 b/d.
D – Demonstration: Minimum commercial scale, product markets needed, operation focus is in test mode to demonstrate commercial reliability, capacity varies from 1,000 - 10,000 b/d.
C – Commercial: Economic commercial scale operation for profit, capacity $>$5,000 b/d.
(3) Date of earliest plant built to accommodate high-carbon feedstocks.
(4) Gasification and direct combustion process are categorized separately.
(5) Many companies are developing catalysts and/or process modifications to handle heavier feeds in existing FCC units, and in new process configurations.

EXHIBIT V-8 Hydrogen Addition Processes

			STAGE OF DEVELOPMENT							
			EXISTING			PLANNED				
PROCESS NAME	PROCESS TYPE	DEVELOPER	TYPE[1]	(NUMBER)	START-UP DATE[2]	TYPE	(NUMBER)	START-UP DATE	FEEDSTOCKS	KEY PRODUCTS
HYDROREFINING										
RDS & VRDS	Desulfurization	Chevron Research Company	C	(1)	1976	C	(2))	1982 1983	Atmospheric & Vacuum Resid & Whole Crude	Light & Heavy Naphthas & Gas Oils & Resid
Resid HDS	Fixed Bed Catalytic	Gulf Research & Development Co.	C	(6)	1970				Atmospheric & Vacuum Residue	Fuel Oil, Cat Cracking Feedstock/ Upgraded Resid
Residfining	Fixed Bed Catalytic	Exxon Research & Engineering Co.	C	(2)	1977				Atmospheric & Vacuum Residue	Distillates, Gas Oil, FCC Feedstock
Unicracking/HDS	Fixed Bed Catalytic	Union Oil of California	C	(2)	1976	C	(2)	1983	Atmospheric Residue	Low S Fuel Oil, Cat Cracker Feedstocks
BOC Unibon	Fixed Bed Catalytic	UOP Process Division	C	(4)	1967				Vacuum Residue, All Crude Fractions	Lt., & Hvy. Gas Oil & Pitch
RCD Unibon	Fixed Bed Catalytic	UOP Process Division	C	(1)	1973				High Metal Vacuum Resid	Hvy. Fuel Oil, Cat Cracking Feedstocks
Shell HDS	Moving Bed Catalytic	Shell Development Company	C	(2)	1973	D[3]	(1)	1981	Atmospheric & Vacuum Residue High Metals & Sulfur	Fuel Oils, Upgraded Resid
Aurabon		UOP Process Division	PP	(1)		D	(1)	1983	Heavy Oils, High Metals & Asphaltene	High Quality Syncrude
HYDROCRACKING										
LC Fining	Expanded Bed Catalytic	CE Lummus & Cities Service	C	(2)	1963				High Metals Vacuum Resid & Bitumen	Fuels & Synthetic Crude
H-Oil	Expanded Bed Catalytic	Hydrocarbon Research Inc.	C	(4)	1963	C	(1)		High Metals Vacuum Resid & Bitumen	Synthetic Crude, Liquid Fuels, Pitch
Kombie Kammer	Non-Catalytic	Veba Oil S.A., Lurgi Corp	C	(1)	Pre-1945				Coal Liquids	Liquid Hydrocarbons
Canmet	Expanded Bed Non-Catalytic	Canadian Energy Mines & Resources Dept./Petro Canada	PP	(1)	1970	D	(1)		Heavy Oil	Fuels & Syncrude
Asphaltene Cracking	Fixed Bed Catalytic	Chiyoda Chemical Eng. & Const. Co.	BS	(1)	1978				Asphaltene Residue	Pour Point Reduction/Synthetic Crude
HFC	Moving Bed	Niigata Engineering Co., Ltd.	PP	(1)	Pre-1977				Atmospheric & Vacuum Residue	Syncrude
RAROP Hydro cracker	Fixed Bed	Idemitsu Kosan Company, Ltd.	BS			PP	(1)	19f	Vacuum Residue	Fuel Gas, Naphtha, Middle Distillates, Residual Oil
OTHER HYDROGEN ADDITION										
Molten Salt Cracking	Catalytic Cracking w/Zinc Chloride Melt	Various	BS						Coal	Gasoline

(1) Plant Types: BS – Bench Scale: Laboratory unit, usually not continuous, capacity < 1 b/d.
PP – Pilot Plant: Continuous, capacity varies from 0.5 - 1,000 b/d.
D – Demonstration: Minimum commercial scale, product markets needed, operation focus is in test made to demonstrate commercial reliability, capacity - 1,000 - 10,000 b/d.
C – Commercial: Economic commercial scale operation for profit capacity > 5,000 b/d.

(2) Date of earliest plant built to accommodate high-carbon feedstocks.

(3) Demonstration as Hydrodemetallization Process in Venezuela's Carbon Refinery.

chain hydrocarbons. Exhibit V-9 shows the relative severities required for various feedstock/ processing combinations.[7] The processes produce limited amounts of by-products, but their economics are dependent on the cost of materials and auxiliary services (i.e., catalysts and hydrogen).

Hydrorefining involves a 10-50 percent reduction in feed mole size and represents the majority of hydrogen addition processes now in use. Although many hydrorefining processes were designed chiefly for desulfurization, improvements are being made to handle heavier crudes with higher metals content.

Hydrocracking is the most severe (temperatures between 400° and 800°F and pressure from 1500 to 3000 psig) hydrogen addition process whereby 50 percent or more of the feed is reduced in molecular size. Chemically, it is a combination of cracking, hydrogenation, and isomerization. Hydrocracking processes provide the greatest liquid yields for a given feed but consume the most hydrogen, which is a disadvantage because of the high cost of producing hydrogen.

2. FACTORS AFFECTING UPGRADING PROCESS SELECTION

As in the selection of a tar sand production method, upgrading process selection from amongst either carbon removal or hydrogen addition processes is dependent upon how well the characteristics of each process match up against the characteristics and requirements of the production site and the product market (Exhibit V-10). For the purpose of upgrading process selection, production site characteristics include all of the attributes of location, production method, and bitumen properties. The heavy oil product market includes the refineries, distribution systems, and intermediate processors of the upgraded oil product, as well as the end use petroleum products market.

Many of the factors which influence the selection of an upgrading process are characterized by the production site and product market, as well as the specific process. Examples include: process economics which are affected as much by feedstock characteristics and available resources as they are by market requirements; and construction lead times which are affected by the availability of materials and services, distances to market, obtaining permits and meeting regulations, and process equipment specifications. As was noted earlier in Chapter IV, the costs associated with delaying a full-scale commercial tar sand recovery plant are enormous. Any process that can minimize these delays holds a strong chance of being selected over other processes.

EXHIBIT V-9

Relationship of Hydrogen Processing Severity to Various Hydrocarbon Feedstocks

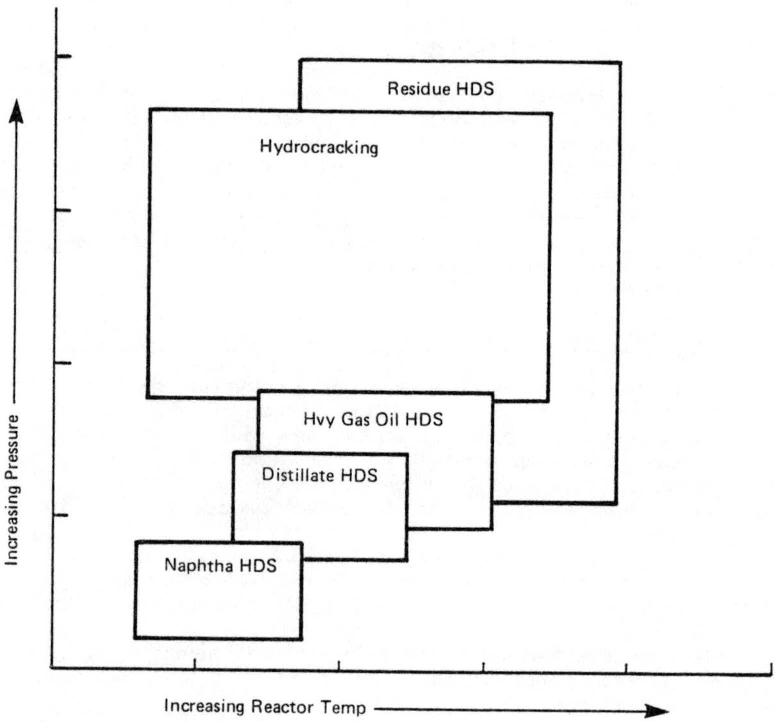

Source: 1980 Refining Process Handbook Hydrocarbon Processing September 1980, p. 98.

EXHIBIT V-10

Heavy Oil Upgrading
Process Selection

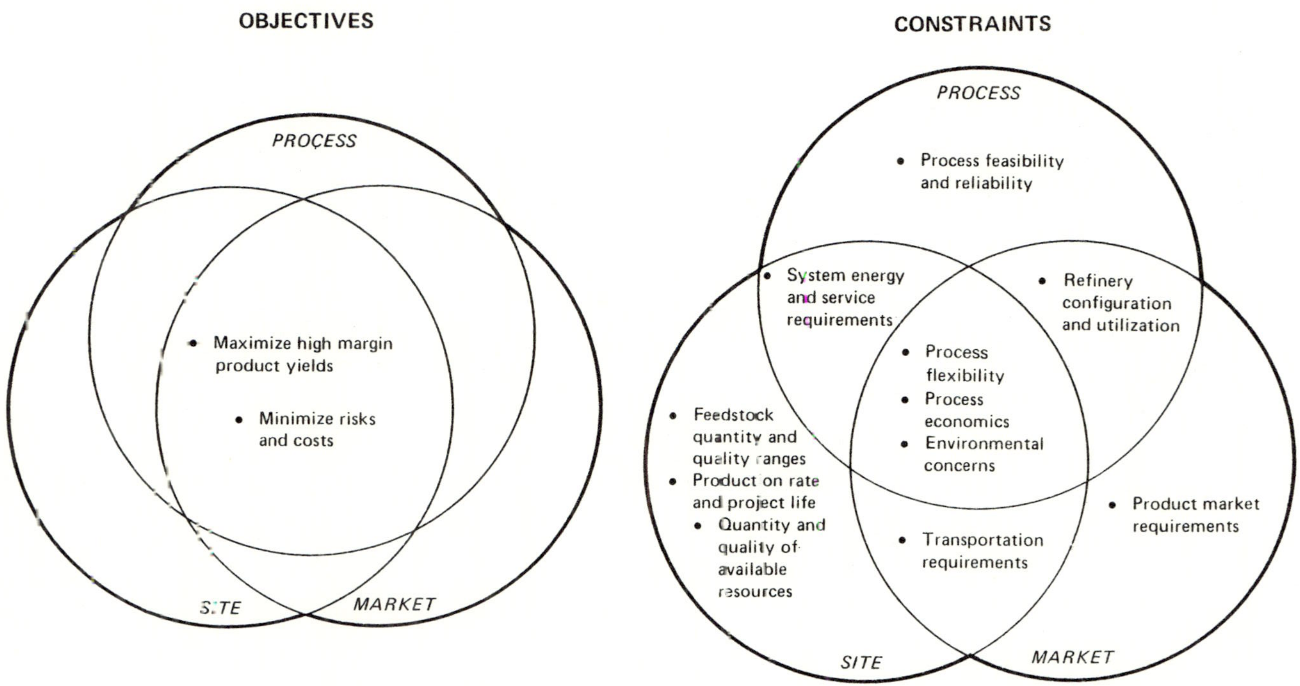

The key selection factors that are specific to each process are the technical feasibility of the process and its reliability. The best test for these criteria is the demonstrated commercial performance of a process. There is no substitute for a commercial example of the process operating at full scale.

(1) <u>Site Specific Factors</u>

Site specific factors in upgrading process selection are fixed by the production method employed. This strong influence is the reason why, in many cases, the upgrading process is selected in conjunction with the production method and why the entire tar sand-to-synthetic crude processing sequence is evaluated as a single project.

In addition to the factors specified (Chapter IV, Section 4) as important in the selection of a recovery and upgrading process, the following site-specific factors also need careful consideration:

. Feedstock quality range

- Gravity, boiling range
- Contaminants

.. Sulfur
.. Nitrogen
.. Oxygen
.. Metals

- Viscosity, pour point
- Asphaltene content, Conradson residue, H/C ratio
- Aromatic content

.. PONA (paraffin, olefin, naphthenes, aromaticity) analysis

.. Characterization factor

. Feedstock quantity ranges

- Size of the reserve
- Rate of production

. Infrastructure factors

- Availability and quality of resources (auxilliary fuels, hydrogen, oxygen, water)

- Availability of services (transportation, labor, social, etc.)

. Geographical factors including climatic conditions and extremes.

(2) <u>Market Factors</u>

The proximity of the production site to the market, the size and complexity of the market and the nature of the terrain between the site and the market contribute to the upgraded product transportation requirements. These transportation requirements influence the selection of an upgrading process. Other market related factors that influence process selection include:

. Petroleum product market, size, and growth rate
. Forecasted product prices
. Downstream refinery complexity and configuration
. Utilized percent of refinery capacity.

(3) <u>General and External Factors</u>

Many factors important to the selection process are influenced by a combination of the product site, product market, and process peculiar to one application. Examples of these factors include:

. Energy requirements for the total production, upgrading, and distribution project

. Auxiliary service requirements for the total project

. Flexibility of each project in adjusting to changes in feedstock and product quality or volume

. Environmental concerns including:

- Present and proposed regulations

- By-product production and disposition

- Potential hazards from process air and water emissions

. Process economics including:

- Fixed costs and their relation to the total project

- Operating costs

- Net energy efficiency

- Source of funding.

In addition, the same external factors cited in Chapter IV, Section 4 apply in the selection of an upgrading process.

(4) Process Selection

Once the factors for process selection have been identified, proper weights in line with project objectives must be assigned. If, for example, full utilization of the entire barrel is a key project objective, as in Cold Lake and Alsands, then this requirement may influence the selection of a process with less attractive overall economics. The ultimate decision for weighting the significance of each factor and establishing a priority of selection criteria rests with the organization responsible for the project's success.

Exhibit V-11 provides a summary of the key characteristics and relative advantages and disadvantages of each process. Exhibit V-12 summarizes each process' effectiveness in overcoming the principal barriers to the processing of heavy oil as a conventional crude. These exhibits provide a start in assigning weights to arrive at the criteria most critical to process selection.

3. CONVENTIONAL UPGRADING METHODS AND COMMERICAL FACILITIES

(1) Traditional Bottom of the Barrel Processes

Catalytic processes have typically been the most effective conventional means for converting heavier oils to lighter, more valuable products. However, these processes have historically had limited success in processing residual or very heavy oils because of the high proportion of asphaltenes and concomitant levels of contaminants usually found in these oils. Catalytic processes can usually accept feeds with minor amounts of asphaltenes and metals. However, asphaltenes tend to deactivate the catalyst through the creation of coke on fixed bed catalysts and through excessive coke make on fluid units.

In fluid catalytic crackers (FCC), excessive coke make can lead to high temperatures in the regeneration zone, where the coke is oxidized from the catalyst. High temperatures can damage the catalyst structure, resulting in a loss in activity. High sulfur levels in the coke result in intolerable levels of SO_2 in the regenerator flue gas. To accommodate high sulfur feeds, stack gas scrubbers similar to those required for many coal-fired boilers would be necessary.

EXHIBIT V-11 Tar Sands and Heavy Oil Upgrading Process Comparison

PROCESS	LIGHT ENDS YIELD (WT %)	LIQUIDS YIELD (WT %)	LIQUIDS QUALITY[1]	RESIDUE YIELD (WT %)	RESIDUE QUALITY	GAS CONVERSION (WT %)	GAS QUALITY	PRIMARY CHARACTERISTICS	RELATIVE ADVANTAGES	RELATIVE DISADVANTAGES
Solvent Deasphalting	None	55-80	L	20-45	L	None		Feed preparation step required	• High liquid yield • Separates metals from DAO • Simple	• Residue disposition required • DAO below pipeline quality; requires additional processing • DAO high in sulfur • High fuel cost • Not proven on bitumen
Visbreaking	<5	10-20	L	70-80	L	None			• Simple	• Low liquid yield • Substantial additional processing needed
Delayed Coking	8-12	60-70	L	22-32	L	None		Semi batch, thermal cracking; coke by-product	• Proven reliability • Full range feedstock flexibility	• Low liquid yield • Coke high in impurities and difficult to use • Liquids may require hydrotreating to saturate olefins • Complex coke removal
Fluid Coking	10-13	56-74	L	9-24[2]	VL	None		Most amenable to large projects; orientation to field construction	• Proven reliability • Full range feedstock flexibility • Energy self sufficient • Less coke than delayed coking	• Coke high in impurities and difficult to use • Flue gas desulfurization needed • Complex operating
Eureka	5	70-76	M	19-26	M	None		Thermal cracking plus steam stripping yields a liquid pitch	• High liquid yield • Proven reliability • Full range feedstock flexibility • Pitch may have advantages over coke • Simple	• Pitch high in impurities • Liquids may require hydrotreating
Lurgi Ruhrgas	5-15	65-80	L	–	–	–	–	Direct coking	• Replaces several separation steps in bitumen recovery	• Equipment required to cool sand • Unproven in bitumen recovery
Taciuk	5-15	65-80	L	–	–	–	–	Direct coking	• Replaces several separation steps in bitumen recovery • Lower Temperatures	• Unproven • Equipment required to cool sand • Unproven
Flexicoking	10-13	56-74	L	1-4[3]	VVL	9-26	Very low Btu	Coking plus gasification; fuel gas produced rather than coke; most amenable to very large projects; orientation to field construction	• No significant coke by-product • Demonstrated commercially • Fuel gas produced may back out valuable fuels • Wide range feedstock flexibility	• Fuel gas is low quality; requires modification of combustion equipment • Liquids may require hydrotreating • Flue gas desulfurization may be required • Unproven on bitumen • Complex operating
Dynacracking		20-75	M-H	Very Little		20-80	Med. Btu	Coking and gasification + hydrocracking, fuel gas and hydrogen produced rather than coke, liquid products more stable	• No significant coke by-product • High naphtha production • Fuel gas is usable with few equipment modifications • Wide range feedstock flexibility • Potential investment savings over flexicoking • Liquid products may not require hydrotreating	• Unproven • Oxygen plant needed • Significant compression required • High light end production, limited middle distillate production • Residue by-product?
HAROP Catalytic Cracking and Gasification	15-25	50-65	M-H	–	–	20-25	–	Fluid cracking plus gasification; produces hydrogen	• High distillate yield • No apparent coke by-product • Hydrogen production contributes to process self-sufficiency	• Unproven • Flue gas desulfurization needed • High light ends production • Significant compression required

Quality Classifications
VL – Very low L – Low M – Medium H – High

[2] Net coke make; 25-40% of gross is consumed as fuel.

[3] Net coke make; 95-99% of gross coke is gassified or consumed as fuel.

(continued)

EXHIBIT V-11 (Cont.)

PROCESS	LIGHT ENDS YIELD (WT %)	LIQUIDS YIELD (WT %)	LIQUIDS QUALITY	RESIDUE YIELD (WT %)	RESIDUE QUALITY	GAS CONVERSION (WT %)	GAS QUALITY	PRIMARY CHARACTERISTICS	RELATIVE ADVANTAGES	RELATIVE DISADVANTAGES
MCC Thermal Cracking	–	–	H[4]	>10	–	–	–	Fluid thermal cracking; produces olefins	• Produces petrochemical feedstocks • Little or no coke by-product	• Special use products; no fuels • Unproven commercially • Flue gas desulfurization needed • Special product handling needed • Limited to small projects or in conjunction with other upgrading processes
Cosmos	30-40	–	H[4]	>10	–	15-20	–	Produces olefins; coke is burned in internal combustion	• Produces petrochemical feedstocks • Little or no coke by-product	• Special use products; no fuels • Unproven commercially • Flue gas desulfurization needed • Special product handling needed • Limited to small projects or in conjunction with other upgrading processes • Strong feed sulfur limitation
HFC	1-3	60-70	H	25-35	M	None		Hydrocracking with spent desulfurization catalyst	• Good liquid yield • Low hydrogen consumption • Wide range of feedstock flexibility; high metal tolerance • May not require secondary processing • Cheap source of catalyst	• Higher conversions may require additional processing • Hydrogen required • Unproven • H_2S removal needed • High capital and operating costs
CANMET	2-6	70-94	H	4-15	M	None		Non catalytic hydrocracking employs additive	• High liquid yield • Wide range of feedstock flexibility • High conversion of heavy residual material • Pitch drag stream may have application as binder for metallurgical coke	• Drag stream must be disposed of • Hydrogen required • H_2S removal needed
Asphaltene Cracking	–	80-98	H	0-15	M	None		Catalytic hydrocracking; recycles asphaltenes to extinction with the help of solvent deasphalting	• No carbon residue[5] • High liquid yields • Wide range of feedstock flexibility • May be used as sale process in simple upgrading application • Moderate hydrogen consumption	• Hydrogen required • H_2S removal needed • Catalyst replacement and disposal needed • Not demonstrated commercially • High capital and operating cost
RAROP Hydrocracker	–	–	–	–	–	–	–	Fixed bed hydrocracker; experimental	• Assumed high liquid yield	• Hydrogen required • Catalyst required • Unproven • Assumed high capital and operating cost (from hydrogen, catalyst and compression)
Kombie Kammer	–	–	–	–	–	–	–	Hydrocracker, comparable to Canmet; developed in the 50's	• Assumed high liquid yield • Demonstrated commercially at one time	• Little data
H-Oil/LC-Fining	5-10	70-87	H	5-20	L	None		Expanded bed hydrocracking; lighter products than with fixed bed for same sulfur levels	• High liquid product yield • May not require secondary processing • Reasonable feedstock flexibility • Moderate H_2 consumption • Provision for catalyst replacement • Proven commercially	• High capital and operating costs • Liquid products unstable at high conversions • Hydrogen source requires stability • Pitch by-product with questionable stability • Catalyst susceptibility to metals poisoning • Complete operation • Some feedstock limitations • Catalyst disposal • Not commercially proven on bitumen • H_2S recovery needed

[4]Specialty product: petrochemicals

[5]With solvent deasphalting to recycle asphaltenes

(continued)

EXHIBIT V-11 (Cont.)

PROCESS	LIGHT ENDS YIELD (WT %)	LIQUIDS YIELD (WT %)	LIQUIDS QUALITY	RESIDUE YIELD (WT %)	RESIDUE QUALITY	GAS CONVERSION (WT %)	GAS QUALITY	PFIMARY CHARACTERISTICS	RELATIVE ADVANTAGES	RELATIVE DISADVANTAGES
Fixed Bed HDS	2	40-65	H	29-55	M	None		Hydroprocessing higher desulfurization and less gravity improvement than expanded bed	• Simple • Good liquid yield • High degree of desulfurization • Residue is suitable as fuel or quality coker feedstock	• Additional processing necessary for high liquid yields • Susceptible to plugging desalting and filtration are necessary pretreatment steps • High capital and operating costs • Not proven on Bitumen • H_2S recovery needed • Catalyst Disposal
Shell HDS/■DM	2	40-70	H	25-55	M	None		Similar to fixed bed system Incorporates "trickle flow" or moving bed to allow on stream catalyst replacement.	• Good liquid yield • Provision for catalyst replacement • High degree of desulfurization • Residue is suitable as fuel or coker feedstock • Commercially proven as residue desulfurization	• Additional processing necessary for high liquid yield • Hydrogen required • High capital and operating costs • Not proven on bitumen or as demetallization process • H_2S removal needed
Aurabon	2-4	40-55	H	30-50	M[6]	None		Primarily feed pretreatment step for secondary processing, feed metals provide catalyst	• Metals in feed act as catalyst • Product suitable for wide range of processing flexibility	• Hydrogen needed • May be economically limited to hgih metal feeds • Unproven • High metal and sulfur drag stream; 2-5 wt % of feed
Texaco Partial Oxidation	None	None		Little		–	Varies	Gas production; most suitable for residue by-products	• Proven technology • Full range feedstock flexibility • No residue by-product • Useful fuel or hydrogen	• No liquid yield • Oxygen plant required • Unproven on bitumen • H_2S removal needed
Shell Gasification Process	None	None		Little		–	Varies	Gas production; most suitable as secondary process for residue by-products	• Proven technology • Full range feedstock flexibility • No residue by-product • Useful fuel or hydrogen	• No liquid yield • Oxygen plant required • Unproven on bitumen • H_2S removal needed
M-Gas	None	None		Little		65-75	H	Gas production; most suitable as secondary process for residue by-products	• Potential source of quality hydrogen	• No liquid yield • Unproven • H_2S removal needed • Oxygen plant required
Heavy Oil Cracking	5-8	65-75	M	None		(Steam)	–	FCC; steam produced rather than coke	• No coke by-product • High naphtha yield • Commercial	• Feed pretreatment necessary • Flue gas desulfurization needed • Unproven on bitumen • Use for steam needed • Catalyst required • Liquids may require hydrotreating
RAROP Resid Cat Cracking	–	–	–	–	–	–	–	FCC; steam produced rather than coke	• May have novel catalyst regeneration method	• Flue gas desulfurization needed • Liquid yield limited • Unproven

[6] High metal high sulfur drag stream

EXHIBIT V-12
Effectiveness of Heavy Oil Upgrading Processes in Overcoming Barriers to Conventional Processing

BARRIERS TO CONVENTIONAL REFINING / PROCESS	FEEDSTOCK BARRIERS[1]				PRODUCT BARRIERS			ENVIRONMENTAL BARRIERS	
	SULFUR	NITROGEN	METALS	ASPHALTENES	LIQUID YIELD	DISTILLABLE PRODUCT QUALITY	RESIDUAL PRODUCT QUALITY	SO_2	PARTICULATES
Solvent Deasphalting	●	●	●	●	◐	○	○	◐	●
Visbreaking	●	●	●	◐	○	○	○	◐	●
Delayed Coking	●	●	●	●	◐	○	○	◐	●
Fluid Coking	●	●	●	●	◐	○	○	○	◐
Eureka	●	●	●	●	◐	○	◐	◐	●
Lurgi Ruhrgas	●	●	●	●	◐	○		○	○
Taciuk	●	●	●	●	◐	○		○	○
MCC	◐	◐	◐	●	●	○		○	●
COSMOS	○	◐	◐	○	●	○		◐	●
Flexicoking	●	●	●	●	◐	○	○	○	◐
Dynacracking	●	●	●	●	◐	●		◐	◐
RAROP Catalytic Cracking	?	?	◐	●	●	●		○	?
Heavy Oil Cracking	●	●	◐	◐	◐	◐		○	○
RAROP ResidCat Cracking	?	?	?	?	◐	◐		?	?
Texaco Partial Oxidation	●	●	●	●				●	●
Shell Gasification Process	●	●	●	●				●	●
M-Gas	●	●	●	●				●	●
H-OIL/LC Fining	●	●	●	●	●	●	◐	●	●
Fixed Bed HDS	●	●	◐	◐	●	●	◐	●	●
Shell HDS/HDM	●	●	●	◐	●	●	◐	●	●
Aurabon	●	●	●	◐	◐	●	◐	●	●
HFC	●	●	●	◐	●	●	◐	●	●
CANMET	●	●	◐	◐	●	●	◐	●	●
Asphaltene Cracking	●	●	◐	●	●	●		●	●
RAROP Hydrocracker	?	?	?	?	●	●	?	●	●
Kombie Kammer	?	?	?	?	●	●	?	●	●
CATEGORY KEY:	● Characteristic acceptable in feedstock in any quantities ◐ Characteristic acceptable in feedstock in limited quantities ○ Characteristic generally unacceptable in feedstock				● High ◐ Medium ○ Low			● No emmissions ◐ Minor emmissions ○ External cleanup required	

[1]Although process may accept high contaminant levels in feed, product qualities or environmental concerns may necessitate substantial additional processing.

(continued)

Exhibit V-12: (continued)

BARRIERS TO CONVENTIONAL REFINING / PROCESS	RESOURCE BARRIERS: FUEL NEEDS	RESOURCE BARRIERS: HYDROGEN	RESOURCE BARRIERS: OXYGEN	RESOURCE BARRIERS: CATALYST	TECHNOLOGICAL BARRIERS: DEMONSTRATED FEASIBILITY/ RELIABILITY	ECONOMIC BARRIERS: INVESTMENT	ECONOMIC BARRIERS: OPERATING COSTS
Solvent Deasphalting	○	◐	◐	◐	●	●	○
Visbreaking	◐	◐	◐	◐	●	●	●
Delayed Coking	◐	◐	◐	◐	●	◐	◐
Fluid Coking	●	◐	◐	◐	●	◐	◐
Eureka	◐	◐	◐	◐	●	◐	◐
Lurgi Ruhrgas	◐	◐	◐	◐	◐	●	◐
Taciuk	◐	◐	◐	◐	○	●	◐
MCC	●	◐	◐	◐	○	(?)	◐
COSMOS	◐	◐	◐	◐	○	(?)	○
Flexicoking	●	◐	◐	◐	●	◐	◐
Dynacracking	◐	◐	◐	◐	○	◐	◐
RAROP Catalytic Cracking	●	●	◐	○	○	(?)	(?)
Heavy Oil Cracking	◐	◐	◐	○	●	◐	○
RAROP ResidCat Cracking	(?)	◐	◐	(?)	○	(?)	(?)
Texaco Partial Oxidation	◐	●	○	◐	●	◐	◐
Shell Gasification Process	◐	●	○	◐	●	◐	◐
M-Gas	●	●	○	◐	○	◐	◐
H-OIL/LC Fining	○	○	◐	○	●	○	○
Fixed Bed HDS	○	○	◐	○	●	○	○
Shell HDS/HDM	○	○	◐	○	●	○	○
Aurabon	○	○	◐	◐	◐	○	○
HFC	○	○	◐	●	○	○	○
CANMET	○	○	◐	◐	◐	○	○
Asphaltene Cracking	○	○	◐	○	○	○	○
RAROP Hydrocracker	○	○	◐	○	○	(?)	(?)
Kombie Kammer	○	○	◐	○	◐	(?)	(?)
CATEGORY KEY:	● Resource Producer; ◐ Internally provides required or need relatively little resources; ○ Resource Consumer				● Commercially proven reliable; ◐ Near term Commercial; ○ Pilot plant or bench scale	● Low; ◐ Medium; ○ High	

The presence of metals accelerates the formation of coke through catalytic action and deactivates the catalyst by sintering its surface. In addition, as metals accumulate on the catalyst, they promote dehydrogenation reactions, which can reduce a cat cracker's effective capacity. The high mineral content of bitumen can cause plugging in fixed catalyst bed reactors.

For these reasons, conventional processes for upgrading bottom-of-the-barrel residues have historically been noncatalytic. Noncatalytic processes can mostly accept low-quality feeds but, they are characterized by their relatively low liquid yields and limited resolution of the by-product disposition problem.

Fixed bed reactors must cope with the added problem of plugging, which is caused by a high mineral content in the feed as is prevalent in many bitumens.

Without substantial feed pretreatment steps or the implementation of methods designed to cope with catalyst contamination and excessive coking, conventional catalytic processes are rendered ineffective in dealing with low-quality, high-carbon feeds. Some of the processes discussed later are essentially standard catalytic processing units that incorporate novel approaches to make them more compatible with heavy oils.

Because of the deleterious effects that heavy oils have on catalytic processes, most conventional refinery units that process heavy oils are noncatalytic. Physical separation processes, such as solvent deasphalting, have been used for many years for the production of lube oils and asphalts. Thermal cracking processes were the earliest methods used to increase the yields of gasoline and light distillates from crude. Shell and Texaco have both used partial oxidation extensively to produce town gas, refinery fuel gas, the hydrogen for methanation and ammonia.[8,9] Heavy residual fuel oil has traditionally been burned in commercial and marine boilers for electricity production and marine transport, respectively.

1. Solvent Extraction

Heavy oil can be extracted with solvents to separate an oil fraction from the remaining asphaltenes. This process, known as solvent deasphalting, is described in detail in Appendix I. The process was developed in connection with the manufacture of lubricating oils. More recently, deep deasphalting (refering to a deeper extraction of heavier oil from the typical crude barrel) has been used economically to extract more cracking feedstocks and leave less residual oil.[10] If lighter products are desired, the extracted or deasphalted oil (DAO) must be cracked in FCC's and/or hydrocrackers.

The primary advantages and disadvantages of solvent deasphalting for various product dispositions are listed in Exhibit V-13. The main advantage is that it reduces vacuum residue to approximately half its original volume to yield a higher quality oil for further upgrading. DAO has typically lower levels of sulfur, nitrogen, metals, and carbon residue than FCC feed coming straight from the distillation units. The paraffin content of DAO is also higher than straight-run vacuum gas oil.[11] The key disadvantage with the process is the large remaining volume of very low-quality 975+°F residue, suitable only for combustion, gasification, or road paving, and the concomitant low yield of distillable products. The conventional process is also energy intensive, requiring large amounts of steam to extract the solvent from the DAO. Kerr McGee's ROSE process reduces this steam requirement considerably but does not exhibit any higher liquid yield than conventional solvent deasphalting.[12,13]

Unless there is a market for asphalt, the ROSE process in a heavy oil upgrading application would necessitate the inclusion of additional processing to make use of the asphalt. The asphaltenes could be:

- Made into asphalt and sold
- Burned as fuel
- Processed further in a smaller coker than would otherwise be needed
- Gasified.

All of these options have their requisite advantages and disadvantages. Selection would be based on the economic savings that would accrue over other (probably fewer, but larger) processing combinations.

2. Thermal Cracking

Thermal cracking was the earliest method used to improve liquid yields from a barrel of crude.[14] The primary thermal cracking processes visbreaking, coking, and thermal steam cracking, all of which are designed to process are untreated vacuum resid. Visbreaking is the mildest form of thermal cracking. Resid is cracked in a furnace designed to prevent coking in the tubes. The furnace effluent is allowed to soak for a short time at elevated temperatures. Visbreaking products include some cracked distillate (10-20 wt percent of vacuum charge) and a very heavy vacuum residue.[15]

EXHIBIT V-13

Solvent Deasphalting Relative Advantages and Disadvantages of Primary Product Disposition Options

PRODUCT	PRODUCT DISPOSITION	RELATIVE ADVANTAGE	RELATIVE DISADVANTAGE
Deasphalted Oil (DAO)	• FCC/Hydrocracking Feed	• High Quality	• Low Liquid Yield
			• Conversion of 975°+F material
Asphaltenes	• Asphalt Blending	• Simple	• Low Liquid Yield
		• Cost Effective	• Limited Asphalt Market
		• High Quality	
	• Fuel	• Simple	• Low Liquid Yield
			• Requires Stack Gas Cleaning
			• Low viscosity precludes transportation and use any appreciable distance beyond unit site
	• Coker Feed	• Higher Liquid Yield	• Coke By-Product
		• Smaller Coker Required Than Without SDA	• Incremental increase in cost can only be justified by associated savings in cost from coking without SDA
			• Low Liquid Yield
	• Gasification	• Can Provide Needed Hydrogen	• H_2 or fuel must be needed.
			• High Cost (Inv & Op)

The degree of cracking depends on the extent of thermal decomposition of the high molecular weight hydrocarbons into lighter components. Conversion is limited by the residue's tendency to decompose too far. Operation must be maintained at severities below the level where the fuel oil products would become unstable in storage. A high asphaltene content reduces the processes' ability to produce a stable fuel.[16] Coking is the major concern at even higher conversion levels.

Although visbreaking is a relatively low cost process, its very low conversions would necessitate extensive further processing. However, it should be considered as an addition to other upgrading processes where economics favor such a scheme. It may also be considered as an alternative to the minimum treatments for pipeline transportation described earlier. Since visbreaking is best suited for low asphaltene feedstocks, it is not considered a viable heavy oil upgrading process for this study.[17]

Coking has continued to be the process of choice for many vacuum resid and heavy oil upgrading projects. The reasons for this are fundamental. Coking has a record of demonstrated commercial process reliability for a wide range of feedstocks, including heavy vacuum residues, tar sand bitumen, and FCC cycle oils; it has a relatively high liquid yield; and the process exhibits little dependence on costly ancillary operations. The primary drawback of coking, as in other thermal cracking processes, is the low-quality coke by-product.

In the past, delayed coking was often chosen to process the abundant supplies of high-quality vacuum residues to yield a low sulfur, marketable coke, and greater quantities of light products. Delayed coking, a semibatch process, requires intensive coke-handling facilities to remove the coke from the drum and to handle it in preparation for further use.

The coke produced from sour heavy feedstocks, however, has a high sulfur and metal content which severely limits its marketability. Coke's usefulness as anodes in the aluminum industry is established but limited if it contains high levels of contaminants. Although sulfur does not lower the cell's efficiency, it is released as SO_2 during electrode manufacture. This exacerbates the emissions problem for the aluminum producer. Metals contribute to alloying and increased power consumption and catalyze harmful oxidation reactions of the anodes above the cell. A coke suitable for use in the manufacture of high-grade electrodes must be low in sulfur, low in metals, highly crystalline, and as dense as possible. Coker feeds should be aromatic and low in solids and asphaltenes. Most of the feed components should boil at temperatures over 800°F.

Other uses of coke are as fuel in coal-fired boilers equipped with stack gas scrubbers and as feed to gasification units. In both applications, the coke must be valued competitively with coal, which is an alternate source for these users. This relationship provides a significant incentive to produce greater quantities of higher valued liquid products and lesser quantities of low-value coke.

Typical aluminum anode coke specifications are listed for green and calcined (volatiles removed) coke in Exhibit V-14.[18]

Fluid coking is a recent variation of the delayed coking process which has been developed by Exxon Research and Engineering Company. Fluid coking uses higher temperatures and is a more severe process than delayed coking. The process incorporates a continuous fluidized bed of coke particles which become coated by the fresh oil feed. The oil cracks and deposits its coke residue on the particles in an onionskinlike manner until the particles are withdrawn from the bed. As with delayed coking, the disposition of low-quality coke is a serious problem. Fluid coke even from sweet crudes is much less acceptable than delayed coke for anode production. However, fluid coking yields 30-50 percent less coke (with a higher sulfur content) and more gas and distillates than delayed coking. The liquid products from fluid coking are more olefinic than their delayed coking counterparts. Because heavy oils are not suited to the production of high (electrode-grade) quality coke, and the liquid yields are better for fluid coking, leaving a smaller coke-disposition problem, fluid coking is considered a more viable process than delayed coking for heavy oil upgrading purposes. A detailed process description for fluid coking as well as a yield comparison with delayed coking is contained in Appendix I.

The primary advantages and disadvantages of the products produced from coking over other processes are outlined in Exhibit V-15. The critical disadvantage is the disposition of the coke by-product. The process is well demonstrated and reliable. Although coker distillates are generally unstable (resulting from the unsaturated hydrocarbons), with sufficient hydrotreating they can be further

EXHIBIT V-14

Typical Aluminum Anode Coke Specifications

	GREEN	CALCINED
Volatiles, % wt (maximum	12-14	0.5-0.6
Real density, g/cc		2.0-2.05
Metals, ppm maximum*		
Silicon	200-300	200-300
Iron	300	300
Vanadium†	200-250	200-250
Other metals	300	300
Ash, % wt (maximum	0.4-0.6	0.4-0.6
Sulfur, % wt (maximum)	2.5-4.0	2.5-4.0
Moisture, % wt (maximum)	1.3-1.7	0.5
+4 Mesh, %		35
+48 Mesh, %		95
−200 Mesh, %		2
Bulk density, lb/cu ft		45-55

*Some specifications limit total (Si + Fe) at 400-500 ppm.
†This limit is being stretched up to 400 ppm.

EXHIBIT V-15

Coking: Relative Advantages and Disadvantages of Primary Product Disposition Options

COKER PRODUCT	PRODUCT DISPOSITION	RELATIVE ADVANTAGES	RELATIVE DISADVANTAGES
Light Ends	• Process Fuel or LPG Recovery	– May back out more valuable fuels	– Less valuable than liquid products
		– May have attractive market	
Distillates	• Processed to finished products	– Valuable liquid product	– Unstable, require substantial hydrotreating
Coke	• Fuel	– Inexpensive	– Combustion equipment requires stack gas cleaning
		– May back out more valuable fuels	
	• Gasification	– Provide hydrogen for processing	– Costly
		– May back out more valuable fuels	– Equipment modification may be necessary to use product as fuel
			– Other fuels may be more cost effective

refined to finished products. Nevertheless, an attractive means for dealing with the coke by-product holds the future for fluid coking's application in a heavy oil upgrading facility.

3. Partial Oxidation

Partial oxidation (POX) is a time-tested, noncatalytic process that not only removes carbon but converts it to gas. Although it can be considered a carbon removal method it is not usually thought of as a primary conversion step. Partial oxidation has been developed chiefly by Texaco and Shell to process a variety of feedstocks for the ultimate production of high-purity hydrogen, ammonia, methanol, urea and oxo chemicals. Since the entire feedstock is converted to various percentages of H_2, CH_4, CO, CO_2, H_2S, and N_4, depending on the feedstock, oxidant, and downstream processing incorporated, partial oxidation is particularly well suited for processing very heavy oils. Its primary drawback is that it yields no liquid hydrocarbon products. Also a need must exist for the gas product.

The most promising application for partial oxidation in heavy oil upgrading is as a final processing step for the most difficult low-quality carbon residues that remain from other upgrading processes. Although liquid feeds are typically processed by this method, POX has shown promise in pilot operations for gasifying coals.[19] Texaco has related technology for gasifying coke, residues, solid wastes, and biomass materials in water slurries.[20] Descriptions of both the Texaco and Shell processes are contained in Appendix I. Exhibit V-16 lists the relative advantages and disadvantages of selected feedstock and product options for partial oxidation applied to heavy oil upgrading.

Partial oxidation of whole bitumen would only be attractive in instances where it is the most economic source of fuel or a large economic market for SNG existed which would favor the production of methane over liquid products. The former conditions would exist where regulation or availability precludes the use of natural gas and no carbon residue is available for gasification. The latter condition is unlikely, in light of the price differentials between gas and liquid products.

Partial oxidation used in conjunction with other processes, such as a hydrocracker, is a likely combination. It must be economically superior to other complete upgrading systems (i.e., flexicoking, dynacracking, hydrocracking) and have advantages over other gasification processes. These other processes are discussed in more detail in later sections of this report.

EXHIBIT V-16

Partial Oxidation
Relative Advantages and Disadvantages of Primary Feedstock/Product Options

FEED STOCK	DISPOSITION	RELATIVE ADVANTAGES	RELATIVE DISADVANTAGES
Bitumen, Very Heavy Oil	N.A.	– Suitable for large scale gas product needs – Simple – No feed pretreatment necessary – No other upgrading needed – Provides clean fuel – Potential solution to by-product disposition – Sulfur cleanup advantages over direct burning	– No liquids recovered – Excessive energy imbalance may force uneconomic export of gas product – Oxygen plant needed
Carbon Residue (i.e., Tar, Pitch, Coke)	N.A.	– Allows liquid recovery – No feed pretreatment necessary – Provides clean fuel – Sulfur cleanup advantages over direct burning	– Lower liquid yield than H_2-addition processes – Product may not be competitive to other fuel/gas sources (i.e., direct burning, natural gas) – Oxygen plant needed
PRODUCT			
Fuel Gas	• In plant use	– May back-out more valuable fuels – Least costly	– May exceed fuel balanced needs
Hydrogen	• In plant use	– Can be used in additional upgrading units – May eliminate reformer needs	– Reformer may still be needed
Methane	• Substitute natural gas	– May have market	– Most costly

4. Direct Combustion

The direct combustion of very heavy oils as an attractive alternative to heavy oil upgrading must be subjected to the same selection requirements as partial oxidation. Heavy oil combustion poses no serious technical problems. Rather, the issues are more environmental and economic than technical. In a study on boiler fuel markets that Booz, Allen completed for the Department of Energy, it was determined that boilers could be made to operate effectively (from an energy and reliability standpoint) on much more difficult fuels than heavy oils (i.e., pitch, woodchips, coal liquids, tar, etc.).[21] The chief concern of the 30-odd boiler manufacturers contacted was the availability and price of the fuel, not the burner modifications that would be necessary.

The direct combustion of bitumen would require stack gas cleanup very much like that required for coal. Rather than charging whole bitumen to a furnace, it would, in most large-scale cases, be more practical to burn only the heavy carbon residue that is left from some other upgrading process. This scheme would allow the recovery of valuable liquids if their quantity is enough to justify such recovery. The combustion reactions could provide the energy necessary for the total production and upgrading project. The closer the balance between the total energy available in the form of carbon residue and the total energy requirements, the more attractive this alternative becomes. Generally its use is limited by the total project energy requirements and is dependent on flue gas desulfurization and scrubbing.

As with partial oxidation, direct combustion of whole bitumen is attractive in instances where it represents a cost-effective method of supplying energy over other fuel sources. The incremental cost of stack gas scrubbing and burner modifications must be compared with the entire gasification, gas cleanup, compression, and firing sequence required for partial oxidation. Other fuel sources should be evaluated on the same basis.

Also, as in the case of partial oxidation, direct combustion may be a better choice for disposing of the carbon residue by-product from other carbon removal processes. In this sense, for example, coking plus coke combustion can be compared with fluid cat cracking and flexicoking. Final selection depends on economics and each variable's sensitivity to uncertain changes.

(2) Developments in the U.S.

The efforts of most U.S. industries attempting to process heavier feedstocks have been focused primarily on catalytic processes. Two exceptions to this are Flexicoking and Dynacracking. Both of these processes combine coking with gasification to solve the coke disposition problems. Each of these processes is described in more detail in Appendix I and later in this report.

The impetus for U.S. developments in heavy oil upgrading technology stems from the increasing quantities of sour crude supplies in relation to sweet crudes, as well as fears that an imbalance may exist between short-term and long-term supply and demand, which make refiners hesitant about adding overall refining capacity. As a result, the major developments have been aimed at increasing gasoline production without investments in major expansions which require long payout periods.

Since the introduction of zeolite cracking catalysts in the early 1960s, significant strides have been made to adapt fluid catalytic cracking to heavier feedstocks. Zeolite catalysts have proven to be less sensitive to metals, more stable, and stronger, with greater structural integrity than early catalysts.[22]

Although these catalysts have increased the range of gas oils that are now acceptable cat cracker feed, they still generally fall short of accepting the metal and sulfur levels that exist in most heavy oils. More recent research that shows promise for extending catalytic cracking technology to heavy oils includes:

- A catalyst that will retain sulfur during the regeneration step and expel the sulfur with the cracked products, thus avoiding costly SOx scrubbers[23]
- Passivators that reduce the deleterious effects metals have on catalysts[24,25]
- Continuous catalyst treatments that extend catalyst life while maintaining activity levels.[26]

In addition, Kellogg and Phillips, Ashland, UOP, and Total Petroleum, Inc. have independently announced that they have developed processing innovations that allow catalytic cracking of reduced crudes.[26-29]

Little has been published to define the feedstock limitations for these variations in cat cracking. Kellogg and Phillips' Heavy Oil Cracking Process is described in more detail in Appendix I. This process is essentially an FCC with enough steam-generating capacity to remove the heat of combustion from the regenerator.

Feed pretreatment is necessary for feeds much in excess of 50 ppm wt percent metals. UOP claims its process can handle Gach Saran reduced crude with a metals content of 300 ppm wt percent.[30] In both cases, though, flue gas desulfurization is needed. Ashland claims to have a Reduced Crude Conversion (RCC) process that requires no feed pretreatment. A 200 b/d demonstration plant has been in operation at Ashland's Catlettsburg Refinery since 1978. The company is planning a 40,000 b/d commercial unit for a 1982 start-up date. Although Ashland has not elaborated on its method of sulfur removal, it says some "pretty tough crudes," including Arabian medium, Mexican Mayan, and Alaskan North Slope have been processed with the help of a proprietary metals passivating technique.[31]

Total claims that its modified cat crackers (Alma, Michigan; Arkansas City, Kansas; and Ardmore, Oklahoma) will be able to run 10-15° API crudes and is planning to test its Kansas unit on Lloydminster early next year. Total says its units will operate on catalysts that are currently available from several companies.[32,33]

Although great strides are being made to handle lower quality feedstocks, the maximum liquid yield from cat crackers is set by the carbon content of the feed. High carbon feeds result in high-coke make on the catalyst which finally ends up as steam that must have a use within the upgrading complex. Exhibit V-17 itemizes the key advantages and disadvantages associated with the application of FCC technology to heavy oil upgrading.

The range of catalytic hydrogen processing technologies available in the U.S. is covered by Shell's moving bed or "trickle flow" Residue Hydrodesulfurization Process and HRI's H-Oil Process (similar to CE Lummus' LC-Fining). Both processes are described in Appendix I.

Shell's process is currently being developed to serve as a combined hydrodesulfurization (HDS), hydrodemetallization (HDM) process. A demonstration-scale unit is planned to test Tia Juana Pesado vacuum residue in Venezuela by 1981. The process' key feature is its moving bed design, which allows the changing of catalyst during unit operation. The process is less severe than the H-Oil Process and consequently is not classified as a hydrocracking process.

The H-Oil (HRI) or LC-Fining (CE Lummus and Cities Service) is presently the only proven, commercial resid hydrocracking process in operation. The process charges atmospheric residue feed and hydrogen in an upward direction through the catalyst bed. The upward flow expands the bed and distributes the feed, hydrogen, and catalyst uniformly throughout the bed. This expanded bed concept allows the continuous addition and withdrawal of catalyst, thus permitting the processing of low-quality feeds.

EXHIBIT V-17

Residue Catalytic Cracking
Relative Advantages and Disadvantages for Use in Heavy Oil Upgrading

RELATIVE ADVANTAGES	RELATIVE DISADVANTAGES
• Quality liquid products	• Feedstock limitations still exist; pretreatment (i.e., desulfurization, demetallization, metal passivators) may be needed
• No high carbon by-product	• By-product steam must have useful disposition
• May eliminate need for vacuum unit	• Liquid yield not as great as with hydrogen addition processes
	• Stack gas scrubbing may be required

A third hydrogen addition process meriting consideration is Exxon's Residfining process. Originally developed as a standard desulfurization process, Residfining has incorporated Exxon's high-activity RT-621 catalyst to extend its feedstock flexibility. Up to 50 percent conversion of 1050+°F material can be achieved through a number of reactions at a relatively high partial pressure. The process employs a fixed bed of catalyst and has demonstrated an ability to desulfurize and convert heavy oils to lighter products.[34] The catalyst is reported to have withstood high contaminant levels over reasonable durations in several pilot plant runs.[35] The process utilizes a reactor plugging protection system to allow processing of these low-quality feeds.[36]

All three hydrogen addition processes require substantial amounts of hydrogen and rely on at least periodic changes of catalyst. Hydrogen addition is likely to play a role in almost all heavy oil upgrading schemes. Its natural application would exist in downstream operations to saturate and desulfurize and further correct the H/C imbalance associated with liquid products from carbon removal processes. It can, however, replace or precede the carbon removal step in any upgrading scheme. The processing sequence will depend chiefly on incremental costs and savings associated with unit sizes, catalyst costs, hydrogen and fuel availability and consumption, and the need for ancillary services. Exhibit V-18 suggests the merits of using hydrogen processing before and/or after carbon removal processes.

Once the need for and position of a hydrogen processing unit in an upgrading application have been established, selection of the type of hydrogen addition process can be made. Conventional hydrodesulfurization, an appropriate process for downstream hydrotreating of gas oil products, is a rather expensive primary pretreatment step with limited resulting gravity improvement. Residual hydrodesulfurization processes are offered by a number of licensors, including Chevron, Exxon, Gulf, UOP, Union, and others. Although a number of installations charge atmospheric resid, commercial experience is limited for processing vacuum resid. As a primary heavy oil upgrading facility, a conventional HDS process would achieve a probable maximum of 70-75 percent sulfur removal. Nevertheless, the very high metals in the feed or the concomitant economic savings associated with the secondary processing step may warrant such a selection. The Shell HDS/HDM demonstration process will test just such a processing arrangement in Venezuela.

Because of its ability to process atmospheric bottoms, thereby eliminating the need for a vacuum tower, hydrocracking is more suited than hydrotreating as a primary upgrading process. H-Oil/LC-Fining conversion is limited by the stability of the liquid products. At conversion levels, much greater than 80 percent of the 1050+°F vacuum feed, the remaining asphaltenes

EXHIBIT V-18

Relative Advantages/Disadvantages of Utilizing Hydrogen Process Options as Primary or Secondary Upgrading Process in Combination with Carbon Removal Processes

UPGRADING PROCESS APPLICATION SELECTION CRITERIA	PRIMARY	SECONDARY	COMMENTS
Liquid yield	?	–	Almost total conversion of 1050°F material can be achieved in primary application
Investment	?	?	Secondary application may require more but smaller units; primary application may not offset the need for downstream hydrotreating, ancillary processing may overshadow any investment savings from unit size.
Hydrogen Consumption	–	+	Secondary application is more selective.
Catalyst Consumption	–	+	Secondary application is more selective.
Feedstock Pretreatment	–	+	Fixed bed operation most affected
Fuel	+	?	Dependent on many factors
By-Product Disposition	?	–	Heavy resid volume is reduced but not eliminated in the primary application's. May require Pox or direct combustion and flue gas desulfurization
Complexity	–	+	High degree of process integration is required to balance fuel, sulfur removal, and hydrogen requirements.

\+ Advantage
– Disadvantage

should begin to precipitate out. Its feedstock flexibility keeps many processing options open for downstream production of lighter products as product markets change over time.

Each developer claims superior advantages of its process over others. Selection will depend on the overall ability to integrate the process with the specific production and refining environment.

Although no definitive judgment which recommends the moving bed reactor over the fixed or expanded bed reactors is available, the following differences can be noted:

- The expanded bed reactors are more efficient than the fixed bed reactors; they can achieve greater conversion (70 percent vs 50 percent) of the 1050+°F material.

- Sulfur reduction is greater for fired and moving bed reactors.

- While the Shell HDS/HDM process offers no advantage over the expanded bed hydrocrackers (H-Oil and LC-Fining) in metals tolerance, it would achieve the same degree of desulfurization with a smaller reactor volume. This is possible because the trickle flow reactor has a larger catalyst holdup than an expanded-bed reactor which allows a higher level of activity per unit of reactor volume.

- An expanded bed system may also require a higher rate of catalyst replacement.

- Operation of a hydrogen addition process can be seriously affected by temperature runaways (loss of temperature control) in the reactors. The exothermic hydrocracking reaction presents the potential for a temperature runaway. Theoretically, an expanded or fluidized bed reactor should have a fairly uniform temperature throughout, as is characteristic of the near -perfect mixing achieved with fluidization. In practice, fluid beds have exhibited channeling and stagnant zones which could contribute to temperature runaways. Difficulties in scale-up present uncertainties which make it impossible to predict the behavior or a large fluid bed which has not been tested at that scale. The behavior of fixed or moving beds, on the other hand, is easier to predict, as scale-up parameters are better understood. However, fixed and slow-moving beds are more prone to develop hot spots (isolated areas of very high temperatures) than fluid beds would be.

- Both expanded bed and moving bed systems offer methods for replacing catalysts during operation. Fixed bed reactors do not. The moving bed allows a more selective withdrawal of inactivated catalyst than does the expanded bed system. Both the fixed bed and moving bed reactors, however, are more susceptible to plugging than are the expanded bed reactors.

- Fixed bed units generally provide a higher quality liquid product, and a second fixed bed hydrotreater may be needed to upgrade light products from the expanded bed unit.

- Gravity improvement at the same level of desulfurization is generally better for the expanded bed concept.

(3) Commercial Upgrading Projects

Before assessing foreign developments in heavy oil upgrading technology, a review of the processes that have been selected for commercial upgrading projects will provide a useful perspective on selection trends. Canada and Venezuela are the most active nations pursuing the development of their respective heavy oil resources. To date almost all commercial processes existing or planned for upgrading tar sand bitumen or very heavy oils have been developed in the U.S. Two exceptions to this rule are the Shell HDS/HDM demonstration project planned for Venezuela, and Petro Canada's CANMET process, which is being considered in an application to demonstrate its ability to upgrade Venezuela's Boscan Crude.[37] The CANMET process, though, is still four years away from licensing.[38]

Canada took the first steps toward commercially demonstrating the technical feasibility of upgrading bitumen when Great Canadian Oil Sands Ltd. (GCOS) was formed in 1953. This led to the start up of the commercial CGOS (now Suncor) oil sands project in 1967. Exhibit V-19 lists the commercial heavy oil upgrading projects that are operating or have been announced in Canada. Exhibit V-20 schematically compares the key upgrading processes employed in each project. Coking of one form or another has been the process of choice in each Canadian commercial application thus far.

1. Suncor

In the Suncor facility the bitumen extracted from the mined tar sands is thermally cracked in a delayed coker. In 1967 delayed coking was the standard method used by refineries to squeeze more liquids out of the bottom of the crude barrel. Its application as a process to produce refinable products from bitumen is well suited to the high

Exhibit V-19
Operating or Announced Commercial Very Heavy Oil Upgrading Projects

Location	Affiliation	Upgrading Process	Licensor	Capacity b/d (t/d)	Type	Start-Up Date	Feed	Products	Disposition
Athabasca Region, Canada	Suncor Project (formerly Great Canadian Oil Sands) Sun Oil Co. Ltd.	Delayed coking	—	(240) 45,000	PP C	(1963–65)* 1967	Mined Athabasca Athabasca	Syncrude (hydrogenated distillates)	Pipelined to refineries
				58,000	C	1981	Bitumen	Coke	Discarded
Athabasca Region, Canada	Syncrude-Esso et al	Fluid coking	Exxon Research & Engineering Co.	 50,000	PP C	(1959–62)* 1978	Mined Athabasca Bitumen	Syncrude (hydrogenated distillates)	Pipelined to refineries
				110,000 129,000	C C	1979 1982	—	Coke	Stockpiled
Athabasca Region, Canada	Alsands-Shell et al	Fluid coking	Exxon Research & Engineering Co.	125,000 to 140,000	C	1986 (construction suspended)	Mined Athabasca Bitumen	Syncrude (hydrogenated distillates)	Pipelined to refineries
		Gasification			PP	1964	—	Coke	Stockpiled**
Cold Lake Region, Canada	Cold Lake-Imperial Oil, Ltd. (Esso Resources Canada, Ltd.)	Flexicoking	Exxon Research & Engineering Co.	160,000	C	Planning suspended	In-situ Athabasca Bitumen	Syncrude (hydrogenated distillates)	Pipelined to refinery
Saskatchewan, Canada	Husky Oil Ltd.	Undecided (may not upgrade)	—	20,000 production	—	—	Lloydminster	—	—
Alberta, Canada	Pacific Petroleum et al	Undecided	—	100,000	C	Feasibility stage	Lloydminster	—	—

*Period of operation.
**Coke will be stockpiled until gasification unit comes on stream.

EXHIBIT V-20

Schematic Comparisons of Operating or Announced Canadian Commercial Very Heavy Oil Upgrading Projects

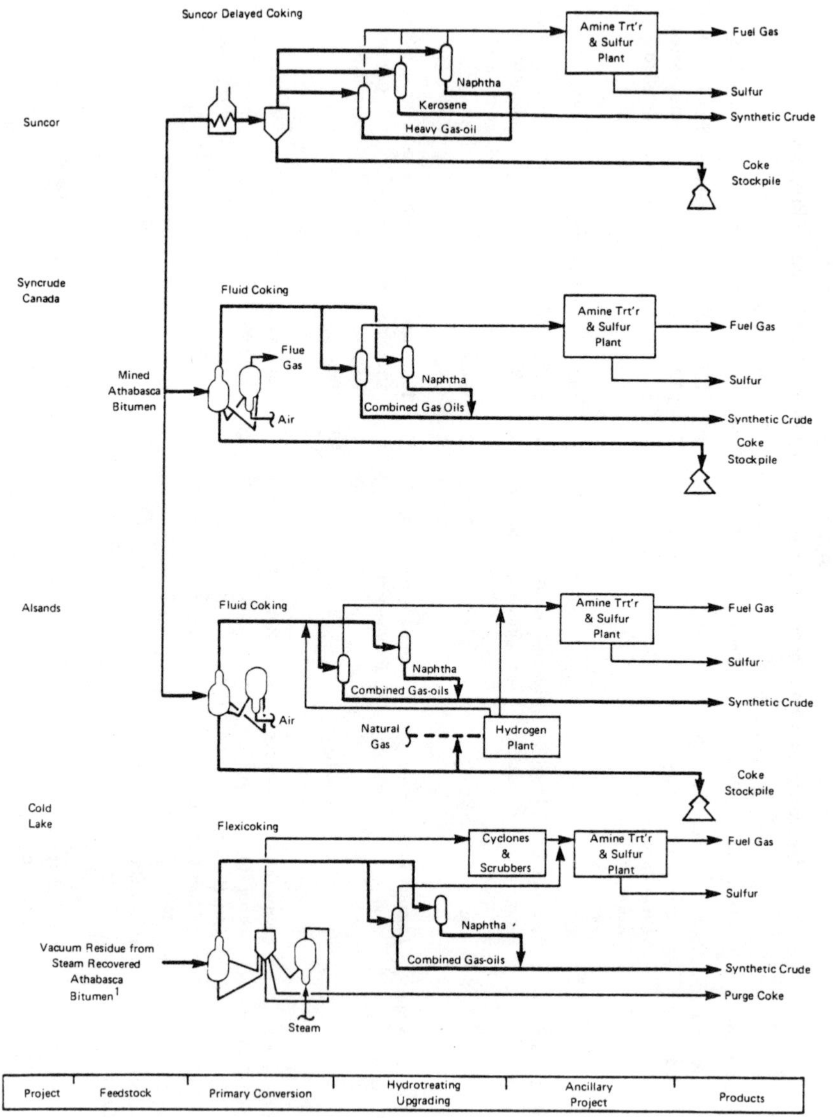

1. After atmospheric and vacuum distillation

——— Primary Heavy Oil Disposition

– – – Undecided or temporary streams

sulfur, metals, and particulate content of mined Athabasca tar sands. At the time the process was selected, few alternatives to delayed coking were available commercially.

Suncor's coker distillates are hydrotreated as separate naphtha, kerosene, and gas oil fractions, blended to a 30° API syncrude, and pipelined to Edmonton, where much of it enters the Interprovincial Pipe Line and is transported east. The low quality of the coke by-product precludes its sale to commercial markets. The Alberta environmental authorities have permitted the plant to use the produced coke as boiler fuel.[39] However, coke production at Suncor exceeds the amount required for system fuel. This fuel imbalance necessitates the stockpiling of about one third of the coke produced. Current plans call for the burying of any unsold coke and sulfur at the end of the life of the project lease. Approval from the Canadian Energy Resources Conservation Board (ERCB) would be required before this plan could be carried out.

2. Syncrude

Fluid coking, commercially available since the mid-fifties, offered a second alternative to heavy oil upgrading. Although the Suncor project has successfully demonstrated the commercial feasibility of delayed coking as a heavy oil upgrading process, fluid coking is capable of yielding 6 percent more distillate products. A yield comparison between delayed coking and fluid coking is contained in Appendix I.

Fluid coking was chosen as the primary upgrading step used in the Syncrude project. The total 146 M b/d of design processing capacity at Syncrude represents one half of the world's fluid coking capacity.[40] The naphtha and gas oil streams leaving the fluid coker are each hydrotreated separately and reblended to a 33° API syncrude. The steam reforming of purchased natural gas provides the needed hydrogen. The lower coke make attributed to fluid coking over delayed coking is an advantage not only for the associated higher liquid yields, but also because the coke has no useful market. About 40 percent of the coke produced is consumed in supplying reactor heat. The remaining Syncrude coke is slurried with recirculated process water and piped to diked storage ponds. Since the 5 wt percent sulfur coke is not suitable as project fuel, the plant uses natural gas and desulfurized fuel gas from the cracked products as the system energy source. The Canadian Energy Resources Conservation Board requires a continuous study of coke utilization to review developing technologies for possible application to the Syncrude project.[41]

Current operations at Syncrude require the addition of approximately one new storage pond each year. Syncrude is studying, also at the request of the ERCB, the possibility of increasing the depth of the coke deposited in each pond, thereby increasing the per pond volume of coke that could be stored and reducing the frequency of new pond additions.[42]

As was already stated in the production section of this study, Syncrude did not startup and line out without experiencing its share of problems. A list of the major upgrading problems and solutions is reviewed in Exhibit V-21.[43]

Although fluid coking can be considered a well demonstrated, technically reliable process, the problems experienced at Syncrude illustrate the difficulty of applying even a well-proven technology to a new application as severe as bitumen upgrading. Most of Syncrude's problems have been solved. Production for the first six months of 1980 was 12.1 million barrels, or twice last year's production for the same period. Year-end output is expected to be 30 million barrels.[44]

3. Alsands

The proposed Alsands project closely resembles the existing Syncrude facility. As in Syncrude, primary upgrading will be carried out in a fluid coker. The principal difference lies in the disposition of the coke by-product. Alsands plans to gasify the coke in a partial oxidation unit.

Coker distillates will be hydrotreated with hydrogen generated in the partial oxidation step. Initially, natural gas or coker gas oil will be fed to the gasification process. Eventually, coke may be the chief gasifier feed. Synthetic crude oil with a gravity of 34-36° API will be pipelined to markets. Initial coke production will be stockpiled. If planned test demonstrations in coke gasification prove successful, all the coke will be gasified. Exhibit V-22 provides a representative picture of the disposition of one cubic yard of oil sand through the proposed Alsand facility.[45]

The proposed Alsands upgrading scheme is representative of the trend towards utilization of the entire petroleum source. The Alsands project group claims that coke gasification was selected as an efficient use of the coke by-product to contribute to a higher yield of liquid products. The group is evaluating both the Texaco and the Shell-Koppers processes for high-pressure coal gasification. Process selection was scheduled for this year. Alsands claims it would be ready to generate hydrogen from coke after

EXHIBIT V-21

Syncrude's Fluid Coking Process: Selected Major Problems and Solutions

PROBLEM	SOLUTION
• Coke carryover	– Modifications to cyclones – Improvements to piping runs
• Coke formation & deposition	– Modifications to feed nozzles – Higher reactor temperature
• Coke utilization	– Increased capacity of storage ponds – Study utilization alternatives & economics
• Electrostatic Precipitator plugging	– Modifications to transport system – Modifications to hopper heat tracing

Source: R.A. Potkins, "Status of the Syncrude Project" Presented at the First Unitar Conference on the Future of Heavy Crude and Tar Sands.

EXHIBIT V-22

Alsands Project Final Disposition of Oil Sand

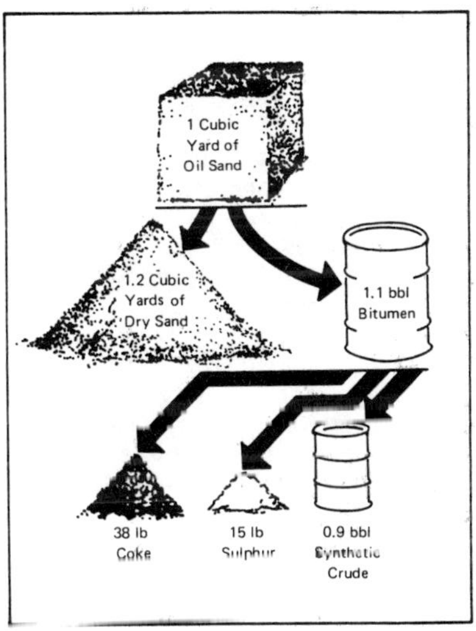

8 years (from plant start-up date) of testing in a prototype gasifier. A process flow diagram of the planned prototype highpressure gasifier is shown in Exhibit V-23.[46]

Delayed coking, flexicoking, hydroconversion, and hydrovisbreaking were all considered in the primary upgrading process. As was stated in the Production Section of this report, the Taciuk direct coking process was originally considered part of the extraction as well as upgrading sequence to produce distillates. However, uncertainties associated with scale-up, heat loss, and sulfur emissions were too great to justify further consideration of this processing scheme. The Taciuk process is discussed further later. Final decision was made on the ground of liquid product yield, energy efficiency, cost, reliability, commercial status, and environmental acceptability.[47]

Alsands emphasized the demonstrated proven commercial reliability of each process as a critical factor in the overall selection process. Since the viability of the project hinges on the operation of the primary conversion process, it was felt that fluid coking was the best choice. Alsands claims that more stringent environmental standards, which may require flue gas desulfurization of the coker off gas, would not shift their choice to another process.[48] It is also interesting to note that the Alsands project group felt that the mineral content of the mined bitumen was great enough to cause problems in other processing alternatives. For example, the group was not comfortable with the effect this presence of solids could have on the smooth operation of a large-scale Flexicoker.[49]

Final go-ahead for the Alsands project awaits an oil pricing and revenue-sharing agreement with the federal and provincial governments of Canada. Recent delays have pushed the project cost to $8 billion, with 1988 as the earliest start date.[50] A project of this magnitude is too large for Shell to undertake without reasonable assurances of stable product prices. Since government intervention is a necessary requisite for project viability, it is reasonable to assume that the project factors which concern the appropriate government agencies will also affect the process selection. In this case, full utilization of by-product coke and environmental emissions are two stated concerns held by the Canadian Energy Resources Conservation Board.[51] Processes that do not address these concerns, cost effective or not, would stand little chance of being selected.

EXHIBIT V-23

Alsands Heavy Oil Upgrading Hydrogen Plant -- Proposed Prototype

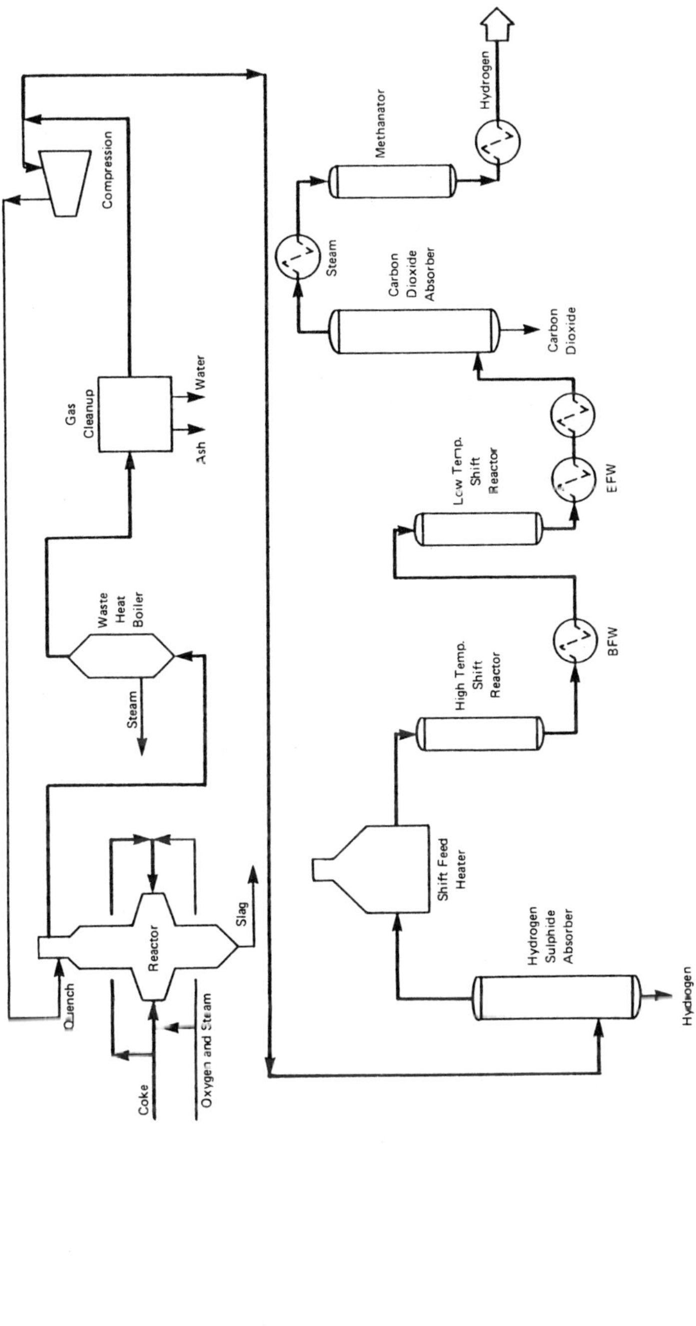

4. Cold Lake

The Cold Lake project, proposed by Esso Resources Canada Ltd., has selected Flexicoking, an upgrading process that is different from either of the two existing commercial projects. Flexicoking combines the fluid coking and gasification steps in one proprietary process developed by Exxon. A detailed description of the process is presented in Appendix I. The process would gasify approximately 98 percent of the coke produced. The remaining purge coke would be stockpiled, as in Syncrude, awaiting a future application. Coker distillates would be hydrotreated and pipelined to downstream refineries.

The objective in upgrading at Cold Lake is to produce a transportable crude that is similar to other crudes in the existing pipeline network and will not necessitate major modifications in downstream refineries. Esso Resources Canada's interpretation of this objective is tabulated in Exhibit V-24.[52]

EXHIBIT V-24

Cold Lake Project Objectives Bitumen Upgraded Product Yield Comparisons

	VOLUME PERCENT		
	COLD LAKE BITUMEN	TYPICAL ALBERTA LIGHT CRUDE	UPGRADED CRUDE OBJECTIVE
Butane (C_4)	–	3	0 – 4
Naphtha (C_5 - 180°C) (C_5 - 350°F)	2	30	15 – 30
Distillate (180 - 345°C) (350 - 650°F)	16	30	30 – 50
Gas Oil (345 - 565°C) (650 - 1050°F)	40	30	25 – 30
Residuum (565°C+) (1050+°F)	42	7	0

To satisfy this objective, Esso considered the following processes:

- Delayed coking
- Fluid coking
- Flexicoking
- Solvent deasphalting
- Partial oxidation
- Direct combustion
- Fixed bed hydrodesulfurization (Residfining/Gulf HDS)
- Expanded bed hydroconversion (H-Oil/ LC-Fining).[53]

A comparative analysis of liquid yield, commercial status cost, energy efficiency, and environmental factors led to the recommendation of Flexicoking over the others. Imperial Oil contended, "Flexicoking was commercially proven, was the most economic alternative, and would result in the highest liquid product yield, highest overall efficiency, lowest sulfur dioxide emissions, and minimum coke production."[54] Coker yield comparisons of Flexicoking with delayed coking and fluid coking are presented in Exhibit V-25.[55]

EXHIBIT V-25

Cold Lake Coker Yield Comparison

	FLEXICOKING	FLUID COKING	DELAYED COKING
C_4 & lighter, wt%	12	12	11
$C_5/180^oC$ ($C_5/350^oF$), LV%	15	15	18
$180/345^oC$ ($350/650^oF$), LV5	22	22	33
345^oC+ (650^oF+), LV%	33	33	15
Gross Coke, wt%	21	21	32
Coke Gas, wt%	26	–	
Net Coke, wt%	0.5	15	32
Coke Sulphur Content, wt%	2.0	6.6	5.7

Exhibits V-26 and V-27 are Esso Resources Canada's comparisons of the Flexicoking alternative versus alternative carbon removal and hydrogen addition processes, respectively. The principal reasons Esso gave for rejecting each alternative are summarized in Exhibit V-28.[56]

A few observations on the Cold Lake upgrading process selection procedure can be made. The comparisons with the hydrogen addition processes were based on the use of liquid products as the injection steam balancing fuel. Also, hydrogen for the Flexicoking alternative was assumed to be provided by the steam reforming of natural gas, while it was felt that expensive partial oxidation was necessary for the hydrogen addition cases in order to eliminate the unwanted residue. At the same time, flue gas sulfur removal was deemed necessary to cope with excess residue that would be burned as fuel.

If, instead of costly (twice that of steam reforming) partial oxidation, steam reforming of natural gas was considered as the hydrogen source, this would free up more carbon residue to be available for burning. Investment would not go up significantly, as the case already provides for flue gas scrubbers. More available fuel would reduce the amount of valuable liquid product that would otherwise be needed for fuel, thereby increasing the liquid yield and possibly eliminating the costly investment for partial oxidation. Furthermore, if natural gas were allowed as the make-up fuel, as was the case in the final Cold Lake selection, even more liquid products would be available for sale. It is clear that an energy-intensive project would be a more suitable setting for a carbon removal process that generates a large amount of low-quality gas with little or no other use. In addition, a hydrogen addition process with less low-quality energy available for use in the overall project is penalized if it is expected to satisfy process fuel requirements with its high-cost liquid product.

Less stringent liquid product standards may have allowed the selection of a less costly alternative. Additional optimization may have further eliminated costly additional processes and led to additional investment and operating savings to be passed on to the refineries in the form of reduced syncrude price. A lower syncrude price would warrant investments to modify refinery equipment to handle the slightly poorer quality crude, if indeed any modification is necessary. Also, a high overall sulfur plant recovery is more easily achieved by processes that concentrate the sulfur in fewer places. This is one reason why the hydrogen

EXHIBIT V-26

Cold Lake Project: Comparison of Carbon Removal Processes

	FLEXICOKING	FLUID COKING
	Base	(+FGDS)[2]
Balancing Fuel for Injection Steam	Coal	Coke
Bitumen Rate, $10^3m^3/d$	25.4	25.4
MB/D	160	160
Upgraded Crude Yield, LV%	88.4	82.4
Overall Energy Efficiency, %	68.5	68.0
SO_2 Emissions, t $SO_2/10^3m^3$ feed	2.88	3.69
LT SO_2/MB feed	0.45	0.58
Relative Investment, %	Base	104
Relative Operating Cost, %	Base	102

[1]All upgrading process comparison studies are based on constant product quality (target 30 LV% max. gas oil, 0 LV% bottoms).
Data represent total plant (upgrading and production facilities).
[2]Flue gas desulfurization

EXHIBIT V-27

Cold Lake Project: Comparison of Flexicoking with Hydrogen Addition Processes

	FLEXICOKING		HYDROGEN ADDITION PROCESSES	
			Fixed Bed	Expanded Bed
	Base	Alternate	(RESIDFINING)	(H-OIL)
Balancing Fuel for Injection Steam	Coal	Liquid	Liquid	Liquid
Bitumen Rate, $10^3 m^3/d$	------	------	25.4 ------	------
MB/D	------	------	160 ------	------
Upgraded Crude Yield, LV%	88.4	78.0	75.0	76.3
Overall Energy Efficiency, %	68.5	65.5	65.5	66.9
SO_2 Emissions, t $SO_2/10^3 m^3$ feed	2.88	2.48	3.20	3.45
LT SO_2/MB feed	0.45	0.39	0.50	0.54
DCF, %	Base	-0.3	-1.5	-1.0
Relative Investment, %	Base	96	95	95
Relative Operating Cost, %	Base	77	91	87

*All upgrading process comparison studies are based on constant product quality (target 30 LV% max. gas oil, 0 LV% bottoms). Data represents total plant (upgrading and production facilities).

EXHIBIT V-28

Cold Lake Project: Esso's Reasons for Rejecting Various Upgrading Process Alternatives to Flexicoking[1]

PROCESS CONSIDERED	REASONS FOR REJECTION
CARBON REMOVAL	
• Solvent Deasphalting	– No conversion of vacuum residue
• Delayed Coking	– Lower liquid yield – Large amount of sulfur coke produced
• Fluid Coking	– Lower yield
• Partial Oxidation	– Greater investment & operating costs – Lower return
• Direct Combustion	– Lower energy efficiency
HYDROGEN ADDITION	Lower return
Fixed Bed HDS and Expanded Bed	– Lower liquid yield – Higher sulfur emissions – Potential handling and desulfurization problems of gas oil and bottoms fractions attributed to the high viscosities and carbon forming tendencies of the high conversion expanded bed processes. (which may lead to precipitation of asphaltenes).
	– Considered only as secondary conversion step
	– Considered only as secondary conversion step
	.. Limited by project energy requirements .. Limited experience: uncertainties associated with applications of required flue gas desulfurization processes to Cold Lake environment

[1]Flexicoking with Stretford gas clean up compared to Fluid Coking and Coke burning with flue gas desulfurization.

addition combination of partial oxidation and stack gas scrubbing was more costly and less efficient than was the Flexicoker. Elimination of partial oxidation or stack gas scrubbing would reduce the Flexicoking advantage in this area also.

One other observation which may suggest an order of preference in upgrading process selection is in the processes that were selected at Cold Lake for secondary treating. They are: Hydrofining, Gofining, and Hydrocracking, all licensed by Exxon Research and Engineering Company.[57]

Turning to Venezuela, we find that Petroleos de Venezuela S.A. (PDVSA), Venezuela's national oil company, is undertaking a massive program to process more of the country's heavy crude and produce more light products. Although PDVSA and its four integrated subsidiary companies are totally state-owned, the technology selection procedure for each one is still strongly influenced by its former ownership. As seen in Exhibit V-29, which tabulates the country's commercial upgrading plans, Lagoven and Maraven have relied heavily on Exxon or Shell, their respective former owners, for technology selections.

Upgrading activity in Venezuela is centered on balancing refinery production with reserves and demand. Venezuela's refinery crude slate is approximately only 35 percent heavy crudes, while 44 percent of its reserves are heavy or extraheavy crudes. Also as shown in Exhibit V-30, local consumption grossly favors light products, while over half of the refined product in 1979 was residual fuel oil. At times, to make enough of the needed light products, production is increased while the resid, representing 50-60 percent of the crude barrel, is stored in huge earthen reservoirs. Since 1978 Venezuela has already spent or committed almost $2 billion on refinery upgrading.[58] Schematic comparisons of the key upgrading processes are shown in Exhibit V-31.

EXHIBIT V-30

Venezuelan Refining Output Short of Local Consumption

	YIELD 1,000 b/d	%	CONSUMPTION 1,000 b/d	%	% OF PRODUCTION CONSUMED LOCALLY
Gasoline and naphthas	193.4	19.6	150.3	51.6	77.7
Kero/jet fuel	40.3	4.1	21.7	7.5	53.8
Gasoil/No. 2 F.O.	150.7	15.3	66.1	22.7	43.9
Residual F.O.	541.2	54.8	53.0	18.2	9.8

EXHIBIT V-29

Operating or Announced Commercial Heavy Oil Upgrading Projects

LOCATION	AFFILLIATION	UPGRADING PROCESS	LICENSOR	TOTAL CAPACITY b/d	Type	Start-Up Date	FEEDSTOCK	KEY PRODUCTS	DISPOSITION
El Palito Refinery, Carabobo Venezuela	Corpoven S.A.	Vacuum Unit	Foster Wheeler	65,000	C	1981	Atmospheric Resid	Gas Oil, Resid	FCC, Utility
		FCC	UOP	42,000	C	1981	Vacuum Gas Oil	Gasoline	Sales
		Electricity & Steam Generator		(40 MW & 900 Mlb/h)	C	–	Vacuum Resid	Electricity & Steam	Plant Use
Amuay Refinery Paraguanan, Venezuela	Lagoven S.A. formerly Exxon	Flexicoker	Exxon Research & Development Co.	52,000	C	1982	Orinoco, Cerro Negro-crude	Coke Gas, Coker Gas Oils, Gasoline	Plant Fuel, HDS Feed Sales
Cardon Refinery Paraguanan, Venezuela	Maraven, S.A. formerly Shell	Hydrodemetallization	Shell Development Co.	2,500	D	1981	Tiajuana Pesado Vacuum Resid	Med Sulfur/Metal Residue	Hydrocracker Feed
		Atm. & Vac. Dist.		100,000	C	1985	Tiajuana Pesado Crude	Diesel	Sales
		Deasphalting		–					
		Hydrotreating		–					
		Hydrocracking		–					
		Partial Oxidation		–					
Bajo Grande Refinery Lake Maracaibo, Venezuela	Corpoven S.A.	Aurabon	UOP	2,500	D	1983 pending	Boscan Vacuum Resid	Conventional Processing Feedstock	Conventional Processing
Puerto La Cruz Refinery Venezuela	Meneven, S.A., formerly Gulf	Heavy Oil Cracker	Pullman Kellogg	52,500	C	1985	Orinoco	Gasoline	Sales

EXHIBIT V-31

Schematic Comparison of Venezuelan Commercial Heavy Oil Upgrading Projects

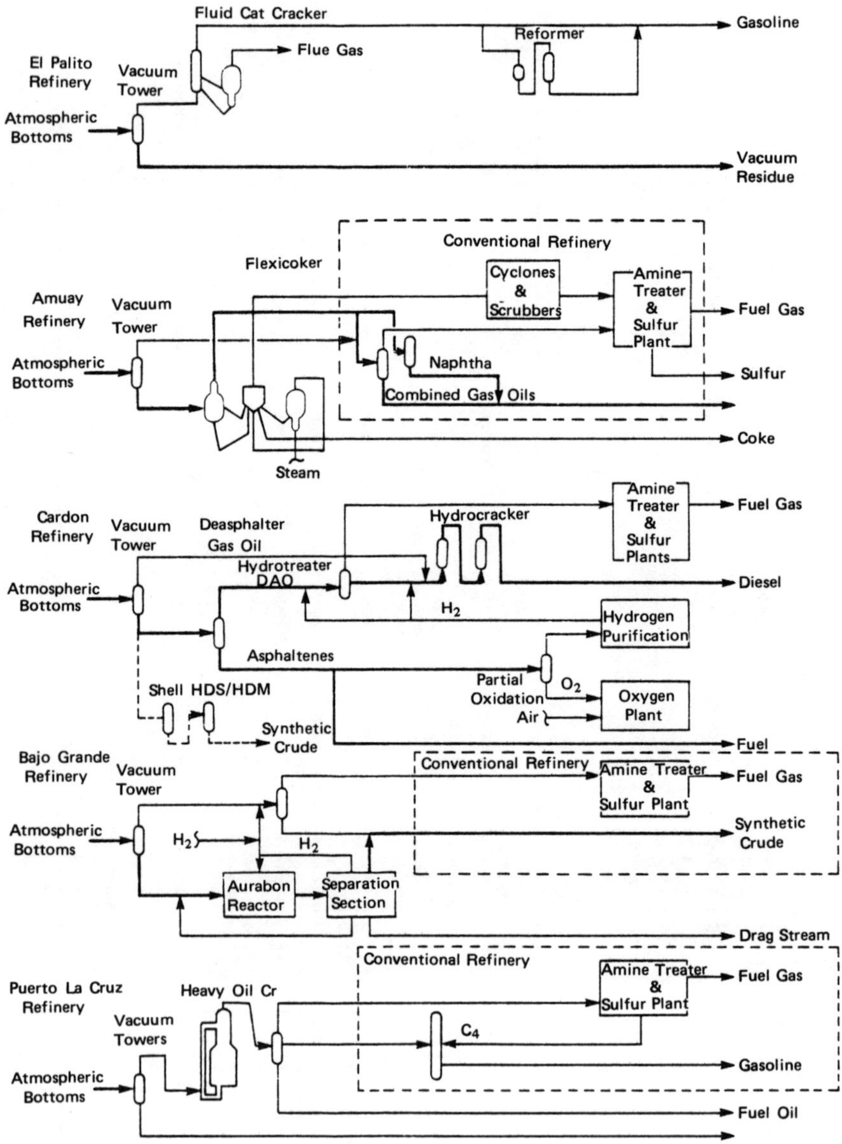

— Primary Heavy Oil Disposition
--- Temporary Streams

5. El Palito

Corpoven's El Palito Refinery is scheduled to start up a 65 Mb/d vacuum unit as well as a 42 Mb/d FCC and a 22 Mb/d HF alkylation unit. The additions will not increase capacity but will drop resid output from 60 Mb/d to 29 Mb/d.[62] More details are not available, but it is assumed that the FCC feed will be the vacuum gas oil. Vacuum resid will be pipelined to a nearby electric utility. Eventually, a 40 megawatt generator and a 900 M lb/hr steam plant will be added to accept this reduced charge.

6. Amuay

The Amuay (formerly Creole or Exxon) Refinery is currently constructing or planning to construct:

- A Flexicoker (52 Mb/sd)
- A Flexicracker (FCC) (74.3 Mb/sd)
- An Alkylation Unit (14.5 Mb/sd)
- A Butane Isomerization Unit (7.5 Mb/sd).[59]

Construction of this project was 20-percent complete at the end of the 1980 summer.[60]

The Flexicoker will process vacuum resid (@1.042 specific gravity, 3.5 wt percent sulfur, 871 wppm metals, 20.5 wt percent Conradson Carbon) that is presently used as fuel oil and asphalt. Purge coke will contain 7 wt percent vanadium, which represents about 10 percent of world vanadium demand. The economics of recovering this vanadium are being assessed.[61] The Flexicoker is expected to be on line during the first quarter of 1982.

7. Cardon

Maraven's (formerly Shell) Cardon Refinery at Punto Fijo is planning two upgrading projects designed to handle Tiajuana Pesado (heavy) resid (12° API, 800 wppm metals, 1.2 wt percent sulfur). The two projects are:

- A demonstration of a hydrodemetallization version of the Shell HDS process

- A 100 Mb/d processing scheme consisting of:

  - Atmospheric and vacuum distillation
  - Solvent deasphalting
  - Hydrotreating (metals removal)
  - Hydrocracking
  - Partial oxidation.

The 2500 b/d Shell HDS/HDM demonstration unit will come onstream in 1981. Shell has developed a regenerable catalyst made in Germany for this test. The process will first demetalize the resid to 150 wppm and then desulfurize it somewhat through a series of Shell's "bunker" reactors.

The commercial-scale project is in its first phase of design, with start-up planned for 1985.[62] The deasphalter will yield 43 wt percent DAO from the vacuum resid. The DAO will contain 70 to 75 wppm metals.

8. Bajo Grande

Corpoven may be the first to test a demonstration version of UOP's Aurabon process. The unit would significantly reduce a high vanadium 10° API Boscan vacuum resid from 1480 to 54 wppm vanadium. The process is under consideration along with Petro Canada's CANMET process and would not be in operation until the mid-1980s in Corpoven's Bajo Grande Refinery.[63]

9. Puerto La Cruz

A 52.5 Mb/d Pullman Kellogg Heavy Oil Cracker will replace the old FCC at Meneven's (formerly Gulf) Puerto La Cruz Refinery by 1985. The unit will receive its charge from two new vacuum units designed to increase heavy oil refining capacity by 25,000 b/d.

In addition to refinery upgrading projects, Venezuela is planning a full-scale upgrading project to process 150 M b/d of heavy oil from its Orinoco Belt. Production is expected to begin in the late eighties. As of this writing, an upgrading process had not been selected for this application.

4. RECENT DEVELOPMENTS IN FOREIGN HEAVY OIL UPGRADING TECHNOLOGIES

(1) Overview of Major Development Efforts Outside The United States

Research and development of heavy oil upgrading technology outside the United States have taken place largely in Canada and Japan. Motivated by the large deposits of tar sands and heavy oil and by the progress already made in developing these resources, Canada organized the Alberta Oil Sands Technology and Research Authority (AOSTRA) in 1975. In Canada, essentially all of the significant R&D activity that is directed toward finding methods for upgrading heavy oil is in some way associated with AOSTRA. Although Japan does not have the heavy oil reserves that Canada has, it is developing the processing flexibility to cope with the increasing range of foreign crude quantities on which it must depend. The Japanese counterpart to AOSTRA is RAROP (Research Association for Residual Oil Processing).

Much of the information and project results collected through AOSTRA & RAROP are proprietary or available only on a cost-sharing basis (prices range from $3000-75,000).[64] Processes are described to the extent information is available later in this section.

1. Canada

Among the principal objectives of AOSTRA is a goal to find "more effective, efficient, and environmentally acceptable bitumen and heavy oil upgrading technology."[68] In its effort to achieve this goal, AOSTRA has been engaged in the following activites:

- Development of the Taciuk "dry" retorting Process.
- Preliminary evaluation of the Lurgi-Ruhrgas "dry" distillation process.
- Assessment of the merits of hydrocracking by awarding contracts to:
  - Initiate a study to quantify the potential advantage of hydrocracking over competitive technology.
  - Determine the size, design and site selection for a large pilot unit.
- Development of operating data for Flexicoking Athabasca bitumen via a test run in Exxon's Baytown demonstration unit.
- Construction of a bench scale version of Mitsubishi's ethylene cracking process to test the viability of producing ethylene and other petrochemical feedstocks from bitumen.

In its continuing effort to find uses for bitumen, AOSTRA began a project in 1979 to determine the feasibility of using bitumen and heavy oil as a solvent in coal liquefaction. The association found that part of the coal will dissolve in these hydrocarbon mixtures. AOSTRA claims that yields are similar to those obtained with the reference solvent, tetralin.[66]

AOSTRA is also spearheading a detailed technical and economic evaluation of 14 upgrading processes each in various stages of development. The association is not considering processes which have already been commercially applied to oil sands or heavy oil in Alberta. The project is a joint venture with at least nine major U.S. oil companies and one Canadian company, however, the membership list is not final yet. The study is expected to take over a year to complete and cost will exceed $1 million. Additional funding will most likely be necessary for bench scale tests that will be required for agreement on individual process yield data and to provide a consistent evaluation basis.

The study is in its early stages of scope and staffing determinations and the members are still identifying processes for consideration. It is hoped that the study will lead to the identification of one, two or three processes each with distinct advantages over coking that may be applied to small upgrading plants. If successful, approval may be granted for Alberta government funding of one or more large scale (1000-5000 b/d) demonstration plants costing in the order of $200 million each.[67]

2. Japan

While the demand for middle distillates for domestic heating and transportation is increasing steadily in Japan, the ratio of heavy crude oil to total imports is also increasing on a yearly basis. This relative increase in heavy oil supplies has been brought about by the OPEC nations' policy to sell heavier crudes and by the purchase of heavy Chinese crude through long-term Japan-China trade agreements.[68] These events have encouraged Japan to anticipate a continuing surplus of residual oil and a worsening shortage of middle distillates throughout the foreseeable future. To cope with this petroleum product supply and demand imbalance, Japan has formed an organization to develop processing technology that will yield more middle distillates from vacuum residue. Most Japanese research directed toward this end has been organized by this Research Association.

RAROP was formed in May 1979 with 26 members comprising representatives from the petroleum refining, iron and steel, and engineering and construction industries as well as a juridicial foundation which represents the electric utilities. Exhibit V-32 lists the Research Association members.[69] RAROP has two objectives that it hopes to achieve within a 4-year period. These objectives are to develop both:

EXHIBIT V-32

Research Association for Residual Oil Processing

MEMBERS LIST (26 MEMBERS)

1. PETROLEUM REFINING COMPANIES (14)
   Asia Oil Co., Ltd.
   Daikyo Oil Co., Ltd.
   General Sekiyu Seisei K.K.
   Idemitsu Kosan Co., Ltd.
   Kashima Oil Co., Ltd.
   Koa Oil Co., Ltd.
   Kyushu Oil Co., Ltd.
   Maruzen Oil Co., Ltd.
   Mitsubishi Oil Co., Ltd.
   Nippon Mining Co., Ltd.
   Nippon Oil Co., Ltd.
   Shell Sekiyu K.K.
   Showa Oil Co., Ltd.
   Toa Nenryo Kogyo K.K.

2. IRON AND STEEL COMPANIES (7)
   Kawasaki Chemical Industry Co., Ltd.
   Kawasaki Steel Corp.
   Kobe Steel, Ltd.
   Nippon Kokan K.K.
   Nippon Steel Corp.
   Nisshin Steel Co., Ltd.
   Sumitomo Metal Industries, Ltd.

3. CENTRAL RESEARCH INSTITUTE OF ELECTRIC POWER INDUSTRY
   (Representative of 9 Electric Power Companies)

4 PLANT MANUFACTURING COMPANIES (4)
   Hitachi, Ltd.
   Ishikawajima Harima Heavy Industries, Co., Ltd.
   Kawasaki Heavy Industries, Ltd.
   Mitsubishi Heavy Industries, Ltd.

Source: Reference 6

- A residual oil cracking process to produce middle distillates, and

- A process to utilize super-heavy residual oil in the iron and steel and the electric power industries.[70]

The primary thrust of RAROP's work in developing residual oil upgrading technology has been in the area of catalytic cracking. The association is actively developing a method to catalytically crack residual oil to produce middle distillates and hydrogen, and a method to hydrocrack reduced crude. A few RAROP members are also engaged in developing catalysts that will withstand the high levels of impurities contained in most reduced crudes. Details that have been released on these activities and others in Japan are presented later in this section of this report.

In addition to its process development efforts, RAROP is also actively involved in investigating catalyst improvements. In a parallel effort to RAROP's cat cracking process development, a catalyst research program has been established to find a catalyst which will be more resistant to asphaltene poisoning and will render the accumulated metals inert. A related effort has focused on the development of a process to regenerate or reactivate a poisoned catalyst.

On the hydrocracking side, Maruzen Oil is working to develop a catalyst that can resist poisoning and will not be prone to fouling in a fixed bed reactor. Another objective of Maruzen's work is to maintain catalyst activity and keep hydrogen consumption low. Maruzen hopes to screen the catalyst in 1981 and have it ready for commercial evaluation by 1982.[71] The catalyst may be applicable to the fixed bed catalytic hydrocracking process that Idemitsu Kosan is developing.

RAROP also plans to design an inexpensive catalyst and a related process for pretreating residue. It is hopeful that this processing step will remove most harmful contaminants from the residual feed before it encounters the hydroprocessing step.[72]

Outside of RAROP, hydrovisbreaking is being explored by Kashima Oil Company Ltd., and the University of Tokyo as a prehydrodesulfurization step to permit conversion of bitumen into distillates for catalytic processing with reduced coke formation. The research is being carried out in a conventional bench scale reactor in Kashima's Research and Testing Laboratory.[73]

Chiyoda is developing a hydrodenitrogenation (HDN) catalyst which selectively removes nitrogen from residual feeds. Chiyoda's interest in nitrogen removal is based on a need to reduce NOx emisssions from fuel oils that are high in nitrogen content. Much of the Chinese oil imported by Japan is high in nitrogen. Chiyoda claims that its HDN-11M and HDN-11S catalysts consume less hydrogen and maintain higher activities with less pore mouth plugging than conventional catalysts. Additional work is being performed to extend the lives of these catalysts.[74]

3. France

France, which currently relies heavily on residual oil and middle distillates for heating and power has adopted a policy which calls for greater use of nuclear power and coal and a shift in oil consumption from heating and electric power generation to transportation fuels and petrochemicals. A French Industry Ministry white paper on the subject outlines a plan to:

- Increase nuclear's contribution of the nation's electricity from today's 21 percent to 55 percent by 1985.
- Convert enough industries from oil to coal to reduce net demand from today's consumption, for heavy fuel oils by 8 percent in 1985 and 10 percent in 1990.
- Reduce overall dependency on oil as a fuel source from today's 59 percent to 30 percent in 1990.
- Restrict oil use to "specialized, high-value-added sectors."[75]

The plan, if carried out, would necessitate the addition of substantial cracking capacity to convert the fuel oil that would be freed up to transportation fuels. The investments made and planned by firms such as Elf Aquitane and Total are for such modifications. The cracking additions planned to date are for conventional catalytic, thermal cracking, and alkylation units. Although "these projects are only the first in a long term overhauling of their refineries," no selection has so far been made for any unconventional very heavy oil processing units such as the ones described in this study.[76]

Although France has not developed technology that is currently available for processing heavy crudes, the Institut Francais du Pétrole (IFP) has been working on a commercial application of using pentane to deasphalt vacuum residue. A pilot solvent deasphalting plant has been in operation for the last three years. IFP is also in the engineering stage of developing a metals resistant catalyst. Elf Acquitaine is working on a catalyst that requires less hydrogen consumption. The company is also experimenting with thermal processing but will not disclose details on its work.[77]

In a joint effort, ELF, CFR, IFP, and Total are planning a small, 2-3 ton per hour facility to test many kinds of heavy crude and vacuum resid. The pilot plant will include: desalting, deasphalting, visbreaking, and hydrotreating. The project is now in its engineering phase and is scheduled to begin operation in the late part of 1982.[78] Additional details on these developments have not been released.

4. Germany

BASF A.G. and the Institut Francais du Pétrole are co-licensors of a hydrocracking process to convert deasphalted vacuum residues and lighter feeds to gasoline, jet fuel, diesel, and lubes. The process has not been designed for heavy oils with high asphaltenes, sulfur, or metals and could only be used in a secondary upgrading configuration after substantial pretreatment of the feed. For this reason, this study does not consider the process any further.

BASF is also engaged in developing catalyst for applications in heavy oil upgrading. For example, BASF catalyst K8-11 was developed to promote the water gas shift reaction which is necessary in most high-carbon gasification reactions.[79] BASF also manufactures catalysts for several hydrotreating processes (i.e. BASF catalysts M8-10, M8-21, and M8-30). Exhibit V-33 lists several characteristics of some new BASF catalysts.[80,81] More details on these catalysts were not available at this writing.

Lurgi Corporation is developing its Ruhrgas Process for application to tar sand processing. This process is discussed later with direct coking. Lurgi is also in the very early stages of developing a hydrovisbreaking process that may be suited to heavy oils. Specifics on this development have not been released.[82]

EXHIBIT V-33 Characteristics of BASF Catalysts Applicable to Heavy Oil Upgrading

CATALYST	APPLICATION	COMPOSITION	OPERATING TEMPERATURE	THERMAL STABILITY
K 8-11	A catalyst for CO conversion at elevated pressure in the presence of sulfur compounds. At pressures above 70 atm, the catalyst may be used only in conjunction with the Texaco partial oxidation process.	Oxides of cobalt and molybdenum on a special carrier.	530-930°F	Up to about 1000°F
M 8-10	Hydrotreating of petroleum fractions for removal of sulfur and nitrogen compounds and metallic impurities, hydrogenation of olefines, and stabilizing heating oils.	MoO_3 ca. 13.5 wt% CoO ca. 5 wt% SiO_2 ca. 2 wt% SO_4 3 wt% Na_2O 0.7 wt% Al_2O_3 balance (carrier) Loss on heating (600°C) ca. 3 wt%	200-400°C (500-750°F)	Up to 500°C (950°F) Notice catalysts can be supplied in the sulfided form.
M 8-21	For hydrotreating of petroleum fractions with a high activity to remove sulfur and oxygen compounds as well as metallic impurities. It has been proven especially suitable for the hydrogenation of nitrogen compounds.	MoO_3 ca. 15 wt% NiO ca. 3 wt% Al_2O_3 balance (carrier)		Up to 500°C (950°F) Catalyst can be supplied in the sulfided form.
M 8-30	A catalyst for pressure refining of crude benzene by the BASF-Scholven process, generally used in conjunction with BASF Catalyst M 8-10. This process will yield a product suitable for the extraction of high purity aromatics, e.g., benzene with a sulfur content of less than 0.5 ppm.	MoO_3 ca. 15 wt% SO_4 ca. 3 wt% Na_2O 0.7 wt% Loss on heating ca 2 wt%	300-400°C (550-750°F)	Up to 500°C (950°F)

Also in Germany, Shell has announced that it is making a new regenerable catalyst that it will use in the Shell HDM demonstration project scheduled for Venezuela.[83]

5. The Netherlands

The Ketjen Catalyst Group in Akzo N.V. introduced new cobalt/molybdenum and nickel/molybdenum hydrodesulfurization catalysts which exhibit 20-30 percent better activity and allow operation at reduced temperatures.[84]

Shell's hydrodesulfurization and demetalization process and its partial oxidation process are both described in this study.

6. Russia

The Moscow Institute of Petrochemical and Gas Industry has conducted studies to determine the influence of asphaltenes present in hydraulic tar on coke yield and structure.[85]

(2) Carbon Removal Processes

The development of carbon removal processes has been directed toward increasing liquid yields and producing more acceptable by-products or eliminating unacceptable by-products completely. In noncatalytic processes, the primary focus of developers has been on product quality and disposition while, in catalytic processes, research has centered on increasing feedstock flexibility. Some of these processes are relatively new, others are old but have received renewed attention in heavy oil applications.

The thermal cracking processes studied include: direct coking (LR-Gas, Taciuk), Eureka, and two crude cracking processes (MCC, COSMOS) which produce olefins for petrochemical manufacture.

1. Direct Coking

Both Lurgi and UMATAC (Underwood, McLellan and Associates, Ltd. (UMA) of Calgary, Alberta) are developing direct coking processes which thermally crack tar sands directly without requiring prior separation. In a tar sand mining operation, direct coking would replace the conventional hot water extraction, froth treatment, diluent recovery, diluent preparation, and bitumen coking steps. Since direct coking is a combination of bitumen separation and upgrading, its selection in an upgrading scheme would typically be most suited to situations where tar sands are mined. By its use of hot circulating sand as a carrier, direct coking can also be applied to upgrading alone.

The Lurgi-Ruhrgas process was described as an alternative to hot water extraction in the production section of this report. The process was first used on a pilot scale as a coal devolatization process to recover gas and oil from coal before firing it in the utility boiler. The first pilot plant (LR-I) was built in Herten, West Germany, in the late forties. The unit was designed for 0.2 tons per hour of coal feed. Experiments in cracking crude and naphtha to ethylene led to a large pilot plant (LR II) which was operated by Bayer. A semicommercial plant (LR III) was coupled to a utility boiler and it processed 10 t/hr. of coal until 1960 when the availability of the North Sea gas and Mid East oil made the devolatization of coal uneconomical. Interest in the recovery of oil from tar sand and shale has led to the start-up this year of LR IV, a one t/hr. pilot plant in Lurgi's Frankfort laboratory.[86]

AOSTRA has contracted with Lurgi to test the commercial viability of producing Athabasca tar sand on an Essen, Germany, pilot unit. The process economics will be compared with those of the state of the art in hot water processing.[87]

In a related study, AOSTRA is studying the UMATAC or Tacuik process in a pilot unit that was completed in the spring of 1978.[88] The process has been under development by UMA since 1975. The program involves the testing of various hydrocarbon feedstocks, including a sample of the Sunnyside Tar Sands in Utah which AOSTRA is processing for Great National Corporation of Dallas.[89] Data from these tests will be made public in January 1981 when AOSTRA will be ready to look for partners for possible commercial ventures.[90]

Syncrude Canada Ltd. has built and operated a bench scale direct coking unit called Direco. The unit was built to study the key variables affecting coking and product quality and reconcile differences in previously reported yields.[91]

A process schematic for the Lurgi-Ruhrgas gas process is shown in Exhibit V-34.[92] The main difference between the LR-Gas process and the Taciuk process is in equipment. In the Taciuk process, the bitumen is retorted in a proprietary rotating kiln rather than the mixer and surge hopper used in the LR process. The rotating kiln allows for the recovery of some process heat by the heating of the bitumen feed with recycled spent sand.

EXHIBIT V-34

Lurgi-Ruhrgas Process Schematic

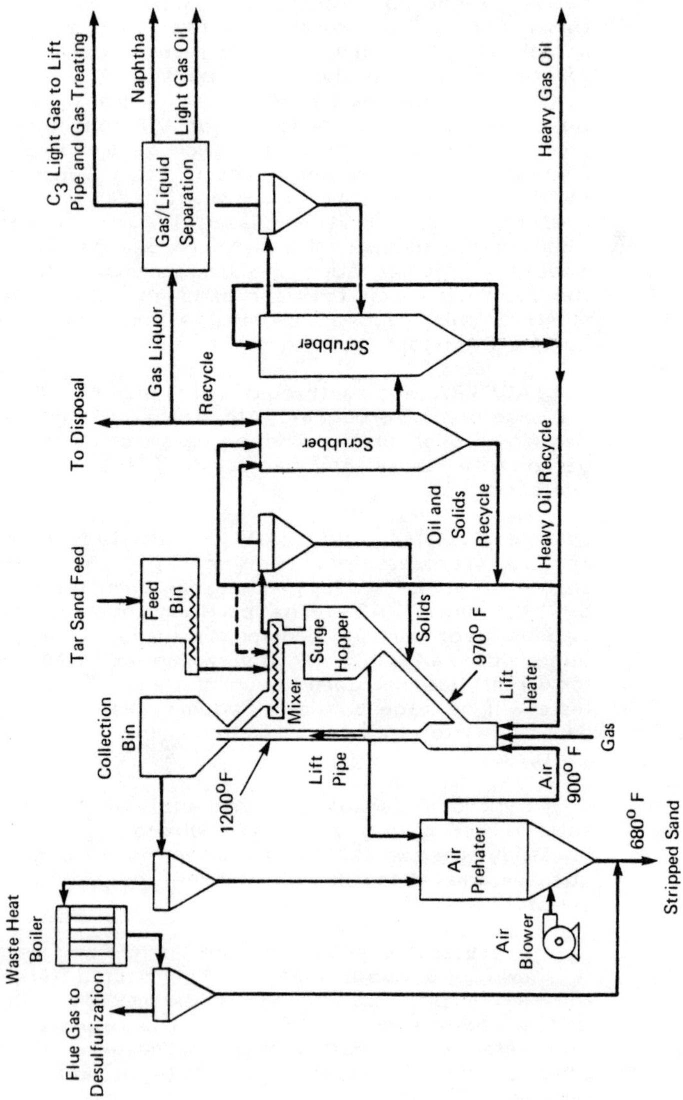

Tar sand (or heavy oil and a heat carrier) is fed to the mechanical screw mixer where it is mixed with recirculating hot carrier (usually sand and reaction residues at 850-1,200°F). Temperature equilibrium between the heat carrier and the feed is reached within seconds so that a large part of the thermal reaction takes place in the mixer. Depending on the unit operating conditions, the retort may consist of low temperature distillation or severe thermal cracking.

As in fluid coking, the asphaltene portion of the feed remains on the carrier as a carbon residue. The mixture of distilled sand and heat carrier passes from the screw mixer to the surge hopper where retorting is completed. The cracked products leave the mixer, pass through a series of cyclones for particle removal, and are scrubbed, condensed, and separated. The coked carrier continues to the bottom of the life pipe where it is contacted with air. The air (@ 750°F) pneumatically lifts the mixture to the collecting bin while it simultaneously burns off the carbon residue. The combustion gas is separated from the carrier in the collection bin which operates at 1,100-1,300°F. The vapors are withdrawn from the top of the bin while the heat carrier and tar products are returned to the screw mixer.

Combustion gases surrender their heat in a waste heat boiler and a feed water preheater. Then they are cleaned in a secondary cyclone, humidifier, and an electrostatic precipitator before being discharged to the atmosphere.

Liquid yields in direct coking range from 70 to 80 percent depending on the supplemental heat requirements and heavy oil and light gas burned in the process. This yield compares with 65 percent bitumen recovery for hot water extraction. Direct coking gives similar yields of gas, liquid, and coke similar to the yields from commercial bitumen coking processes.[93] Product quality and yield are affected most by vapor residue, time and cracking temperature.[96] AOSTRA claims that the lower temperatures attributable to the Taciuk dry retorting process allow a higher percentage Fisher Assay recovery, and a reduced gas yield than is possible from the LR-Gas process.[94]

A common objection to any direct coking process is the severe heat loss associated with heating up the large quantity of sand. Heat recovery from coke combustion is not sufficient to keep the unit in heat balance, so use of a supplemental fuel is required. However, the net energy efficiency is favorable when compared to current extraction technology. Heat losses from direct coking are similar to the combined losses for conventional hot water extraction with froth treatment.[95] Nevertheless, the ability to recover more heat from the expanded sand would make direct coking more attractive. At present, added investment is necessary to cool the sand for ultimate disposal.

Air preheat accounts for very little of the heat load contained in the expanded sand and a considerable cost in investment and operation would be involved in preheating the feedstock. An increase in the oil-to-sand ratio in the feedstock would reduce the overall heat required and increase the process' thermal efficiency and upgrading capacity.

Rammler has reported on the use of a feed pretreatment system to reduce the volume of sand while not materially affecting the flow rate of bitumen.[96] Exhibit V-35A shows the effect of sand removal as a pretreatment to direct coking.[97,98] The sand reduction step is similar to the hot water extraction step in conventional bitumen separation with the exception that only a fraction of the sand needs to be eliminated.

Exhibit V-35B is a diagrammatic comparison of direct coking with conventional bitumen separation plus coking.[99] Since the cost of sand reduction increases with the degree of separation required, the benefit of alternative A over B is apparent.

The Taciuk Process has a different approach to utilizing the heat from the spent sand. The rotating horizontal kiln used in the process is similar to that used in iron ore sintering. The kiln is divided into a preheat zone which is surrounded by the cooling spent sand and a reaction zone in the center of the kiln, as well as a combustion zone surrounding the other two. Fresh feed moves through the center of the kiln, first through the preheat zone which gets its heat from the cooling spent sand in the combustion zone; and next, through the reaction zone where it is cracked and flashed. Cracked products leave the reaction zone through a disengaging section. Heavy ends and coked carrier enter the combustion zone in the outer periphery of the kiln. In the combustion zone, the carbon is burned off the carrier, while combustion gas and hot sand release their sensible heat to the fresh feed.

AOSTRA is hopeful that the countercurrent recycle of spent sand will improve net efficiency enough to significantly reduce the cost of Taciuk operation. Little can be drawn from the similar yield claims from the developers of each process. The major process differences are in operability. While the Taciuk offers some capacity advantage in

EXHIBIT V-35

Processing Steps Required for Direct Coking vs. Conventional Bitumen Extraction and Upgrading

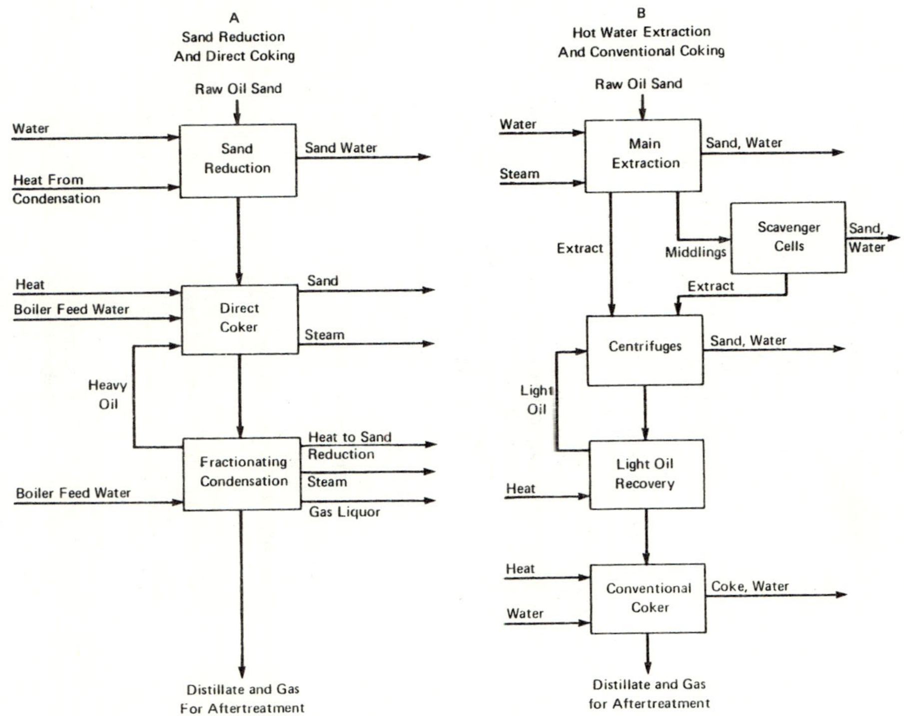

size over the Lurgi, it presents additional potential mechanical problems. Feed preheat, however, would be much more complex and expensive for the L-R system. Without a detailed comparison of each unit, no selection of preference can be made for one process over the other.

Other advantages attributed to direct coking in general are:

- Direct coking is not sensitive to feed quality as is hot water extraction.

- Sand leaving the unit is clean, thereby reducing the tailings disposal problem.

Nevertheless, the problems associated with direct coking are significant. The challenges of solids handling and mechanical operation are continually imposing obstacles to process commercialization. The relative advantages and disadvantages of direct coking in relation to product disposition are shown in Exhibit V-36. Direct coking products are similar to those from other thermal cracking processes with the exception that no net coke is produced.

2. Eureka

The Eureka process is a liquid phase thermal cracking process with simultaneous steam stripping under atmospheric pressure. The process is the joint development of Kureha Chemical Industry Co., Ltd., and Chiyoda Chemical Engineering & Construction Co., Ltd. It is currently licensed by Mitsui & Co., Ltd. The first commercial plant (20,000 b/sd) built in 1975 in Chiba, Japan, has been operated for 5 years to upgrade high sulfur refinery bottoms. A second 20,000 b/sd plant is scheduled for startup in 1982 in the People's Republic of China to upgrade Shanghi heavy crude.

The Eureka Process is an extension of delayed coking except that a thermally stable liquid pitch is produced instead of coke. This difference results from the injection of superheated steam into the bottom of the reactor to strip out cracked oils and to supply reaction heat. The striping also suppresses coke forming polymerization reactions and overcracking reactions which form light gases. As a result, Eureka offers a slightly greater liquid yield than either delayed coking or fluid coking.

EXHIBIT V-36

Direct Coking[1]
Relative Advantages and Disadvantages of Primary Product Disposition Options

DIRECT COKER PRODUCT	PRODUCT DISPOSITION	RELATIVE ADVANTAGES	RELATIVE DISADVANTAGES
Light Ends	• Process Fuel or LPG Recovery	– May back out more valuable fuels – May have attractive market	– Less valuable than liquid products
Distillates	• Processed to finished products	– Valuable liquid product	– Unstable, require substantial hydrotreating
(Coke)[2]	• Consumed Entirely as fuel within process	– Simple	– Establishes ceiling for maximum liquid yield – Requires flue gas desulfurization

[1]Main advantage in direct coking is the combined production and upgrading step.

[2]Coke is produced and consumed internally; there is no net coke production.

A simplified process flow diagram of the Eureka Process is shown in Exhibit V-37.[100] Vacuum residue is heated to 650°F (340°C) in the preheater and fed to the bottom of the product oil fractionator where it mixes with heavy recycle oil. This feedstock is pumped from the bottom of the tower to the charge heater, heated up to about 910°F (490°C) and fed to one of the two paired reactors which alternate in operation (similar to the continuous flow, semibatch switching of coke drums in delayed coking).

As in delayed coking, charge furnace temperature and feed rate through the tubes are controlled to prevent coking in the furnace or transfer lines. Both cracking and poly-condensation reactions take place in the reactor where most of the heat required for cracking is supplied by the hot feed. Additional heat is supplied by superheated steam (@1,200°F) which is sprayed into the bottom of the reactor. The steam continuously strips cracked product oil from the residual liquid, leaving a viscous pitch.

The cracked products, oil and gas, are drawn off the top of the reactor with the steam and sent to the fractionator. Entrained pitch is washed to the fractionator bottom to join the heavy oil feed. The cracked oil is separated into light oil and heavy oil fractions, drawn off the tower and sent to desulfurization after heat removal. Cracked gas is sent to gas cleanup and used as fuel gas.

When the liquid pitch reaches a predetermined softening point, the reaction is stopped by quench water which is sprayed directly into the reactor. The pitch is drawn off into the stabilizer. From the stabilizer, the highly viscous pitch is continuously pumped to the pitch flaker, solidified, flaked, and then conveyed to storage.

Process waste water is treated first in a sour water stripper, followed by a coagulation and filtration unit and finally a bed of activated carbon.

Feedstock and yield data for the Eureka process are listed in Exhibit V-38.[101]

The liquid state of the pitch residue in the reactor allows continuous operation of the Eureka process. Process developers claim that the pitch is an excellent additive for manufacturing high-grade metallurgical coke from low grade coking coal. The relative advantages and disadvantages of Eureka's product disposition are listed in Exhibit V-39.[102]

EXHIBIT V-37

Eureka Process: Simplified Process Flow Design

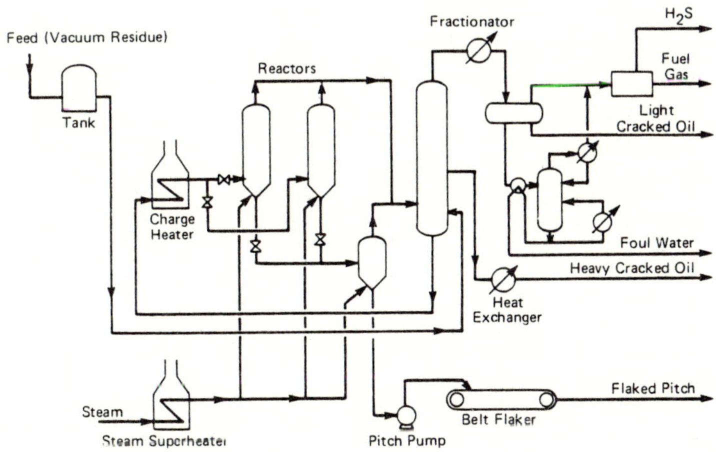

EXHIBIT V-38

Eureka Process: Feedstock Properties and Product Yields

FEEDSTOCK		Iranian Heavy (Gach Saran)
Type		Vacuum Residue
Gravity, °API		6.73
Kinematic Viscosity, cST @ 100°F		550
H/C Ratio Atomic		1.5
H/C Ratio wt		.125
Sulfur, wt %		3.43
Nitrogen, wt %		0.58
Metals, ppm		
V		345
Ni		115
Asphaltenes, wt %		11.8
Conradson Carbon Residue		20.2
PRODUCTS, WT%	**YIELD**	**SULFUR**
Gas	5.4	10.7
Naphtha	11.3	1.7
Gas Oil	52.9	2.5
Pitch	30.4	4.4

EXHIBIT V-39

Eureka Process:
Relative Advantages and Disadvantages of Primary Product Disposition Options

EUREKA PRODUCT	PRODUCT DISPOSITION	RELATIVE ADVANTAGES	RELATIVE DISADVANTAGES
Light Ends	• Process Fuel or LPG Recovery	– May back out more valuable fuels	– Less valuable than liquid products
		– May have attractive market	
Distillates	• Processed to finished products	– Valuable liquid product	– Unstable, require substantial hydrotreating
Pitch	• Binder for metallurgical coking coals	– Allows use of weaker/noncoking coals in metallurgical coke production	– Market may be small
	• Fuel	– Inexpensive – Thermally stable – May back out more valuable fuels – More volatiles, easier to burn than coke	– Requires flue gas desulfurication
	• Gasification	– Can provide hydrogen for downstream processing	– Costly
		– May back out more valuable fuels	– Equipment modifications may be necessary to use produce as fuel
		– Less oxygen consumption than with coke gas gasification	– Other fuels may be more cost effective

The process has been tested in bench or pilot plant scale on a variety of feedstocks, including:

- Oil sand bitumen
- Other Canadian heavy oils
- Orinoco heavy oil.

3. MCC Process

Mitsubishi Petrochemical is working with the Japanese Ministry of International Trade and Industry (MITI) and the Research Council of Alberta in the development of a fluid bed thermal heavy oil cracking process. The process, MCC, is similar to fluid coking in that it accepts raw bitumen or vacuum residue and cracks it in a fluid bed coking process. It differs from fluid coking in that it produces a high yield of olefins (i.e., ethylene, propylene, benzene) and other aromatics rather than gasoline and middle distillates. A small (kg size) bench scale unit in Canada and a 100 t/d pilot plant in Tokyo have been operated to collect performance data on Alberta feedstocks. The project was completed in April 1980 with no continuation expected. The results are being assessed by AOSTRA and Mitsubishi while AOSTRA will continue In-house research on the bench scale.

The process operates at higher temperatures and lower residence times than fluid coking. As in fluid coking, the coked carbon residue provides the heat of reaction. Gross coke make is less than in fluid coking as a result of the reduced residence time. Net coke production is minimal (little or no coke by-product) because the higher temperatures require greater heat requirements from the already reduced quantities of coke.

Process viability would depend upon the products' worth as petrochemical feedstocks competing with ethane or naphtha. Although the ethylene yield would not be as high as the yields from an ethane cracker, the process becomes more attractive as chemical plants rely on heavier feeds such as naphtha and even whole crude for production of its chemical building blocks.

A possible application for the process would be at small sites or in parallel with a large fuels dedicated upgrading plant, since it would not be practical to produce petrochemicals in the same quantities (i.e., 140 Mb/d) as transportation fuels currently produced at Syncrude or Suncor. A disadvantage is the relatively limited use of petrochemical products compared to fuels. Production of olefins would require in-place distribution, processing, and marketing systems typically found in nearby petrochemical complexes

4. COSMOS

The COSMOS (Cracking Oil by Steam and Molten Salts) Process, developed jointly by Mitsui Petrochemical Industries and Mitsui Engineering & Shipbuilding Company, is a crude oil cracking process that produces ethylene. Its developers claim considerable savings over conventional naphtha cracking in spite of higher capital costs.[103] A 5 kg/hr bench-scale unit has been used in tests since 1977. Mitsui has designed a 400 kg/hr pilot plant, which it plans to build and operate to study the feasibility of commercial scale-up.

The COSMOS system is a conventional naphtha cracking furnace with an added melt circulation system. Melt is introduced to a high velocity stream of 1400°F dilution steam. The steam shears and atomizes the melt causing some of the melt to coat the inside wall of the reactor tube with a flowing liquid film. As in naphtha cracking, a mixture of preheated heavy oil feed and atomized steam is injected into the reactor tube. The feed undergoes high temperature thermal cracking which produces carbonized pitch as well as ethylene product. The molten film lining the reactor wall covers any surfaces on which the pitch would ordinarily deposit, thus preventing carbon agglomeration and reactor plugging.

The pitch is gasified via the water-gas shift reaction in the presence of the melt which acts as a catalyst. The mixture of reaction products, steam, and melt are first quenched, then separated in a cyclone which recycles the melt back to the cracking furnace and the gas to products.

Suitable feedstock quality is limited by the process metallurgy. With existing alloys, a 0.5 wt percent sulfur concentration in the melt can be tolerated. This concentration can be estimated roughly as the sulfur concentration in the feed. Paraffin-rich feeds are most suited to the process, and their yields are comparable to those from naphtha. A list of possible feeds and yields are listed in Exhibit V-40.[104] Mitsui is working with Kurimoto Iron Works Ltd. to develop a high-temperature, corrosion-proof, and heat-resistant metal to withstand higher sulfur levels.

A higher steam ratio, a CO_2 removal system, reduced quencher efficiency, and the melt system requirements are all characteristics of a crude cracking process that contribute to a higher capital and operating cost than that required for a conventional naphtha cracker. As the naphtha-crude price differential widens, this investment becomes more attractive. It may be some time though before COSMOS can accommodate high-sulfur, high-metal heavy oils such as those from tar sands.

EXHIBIT V-40

COSMOS Crude Cracking Process: Feedstock Properties and Product Yields

FEEDSTOCK	NAPHTHA	CRUDE	CRUDE	CRUDE
Type	—	MINAS	TACHING	ARABIAN LIGHT
Specific Gravity 60°F	0.718	.86	.866	.854
Distillation, °F				
IBP	117	217	158	127
5%			320	181
10%	153	340	397	232
30%	190	576	635	379
50%	241			572
70%	273			676
H/C Ratio, wt	.173	.155	.157	.145
Sulfur wt%	0.04	0.12	0.17	2.04
Conradson Carbon wt%	0.01	2.5	2.7	3.1
Pona analysis				
Aromatics	7.2	8.2	9.5	17.2
Olefins	0.3	0.3	1.5	0.4
Parafinns	73.7	91.5	89.0	82.4
Naphthenes	18.8			
PRODUCT YIELD WT %				
H	1.12	2.01	2.09	2.33
CH	11.55	13.70	12.48	13.28
C_2H_2	0.54	0.31	0.30	0.20
C_2H_4	27.79	25.74	25.19	19.57
C_2H_6	2.29	2.23	2.21	2.51
C_3H_6	16.63	11.71	11.19	9.67
C_3H_8	0.31	0.17	0.16	0
C_4s	11.37	6.11	6.65	4.85
C_5+	13.55	9.31	8.09	5.82
CO	0.20	1.10	1.30	1.67
CO_2	1.68	11.75	15.30	22.01
Fuel Oil	14.50	25.86	28.30	37.13

Most of the research activity in applying fluid cat cracking technology to heavy oil upgrading has been directed toward catalyst development. Many catalyst and process companies (UOP, Total, Davidson, Ashland, Phillips, etc.) have claimed recent increases in FCC feedstock flexibility. As mentioned earlier, Total claims to have discovered a processing methodology to handle high-metal feeds. The company plans to demonstrate its processing knowhow on Lloydminster heavy oil in 1981. Ashland has made similar claims. An assessment of these technologies cannot be made until tests have been performed and data is released.

Foreign research in FCC technology is concentrated in Japan. Nippon Oil's work in RAROP is described briefly below. AOSTRA has indicated that few, if any, of the major oil companies with whom the Association is evaluating upgrading processes, have expressed an interest in fluid cat cracking.[105] Beyond the catalyst poisoning nuisance, the FCC is further limited by the ceiling on liquid yield that all carbon removal processes possess. In addition, sulfur imposes costly removal problems for the regenerator off gas.

5. Resid CAT Cracking--Nippon Oil

Nippon Oil, a member of RAROP, has built a 1 b/d bench scale test FCC unit for cracking high-metal resids. A simplified process flow for the unit is shown in Exhibit V-41.[106] The process will strip, regenerate, and recycle a new catalyst which was developed for the direct cracking of atmospheric residue. The process, if successful, will eliminate the need for a vacuum unit. Nippon Oil plans to design, construct, and operate a larger pilot plant scale unit soon. Additional details on the process or type of catalyst have not been released as of this writing.

Most of the newer heavy oil gasification processes being developed are, like Flexicoking, a combination of cracking and gasification. The one exception to this is Mitsui's M-Gas process which is being designed to catalytically gasify vacuum residue into a high hydrogen content fuel gas. M-Gas is similar to the Texaco and Shell partial oxidation processes, while HRI's Dynacracking and RAROP's catalytic residue cracking process more closely resemble Flexicoking.

Both Dynacracking and the Japanese process employ gasification, not only as a means of disposing of the carbon, but also to further beneficiate the liquid products. Dynacracking produces hydrogen and hydrocracks the feed with it, all in the same vessel. The RAROP process produces hydrogen to be used in downstream hydrotreating processes.

EXHIBIT V-41

Nippon Oil's Resid Cat Cracker
Simplified Schematic Flow Diagram

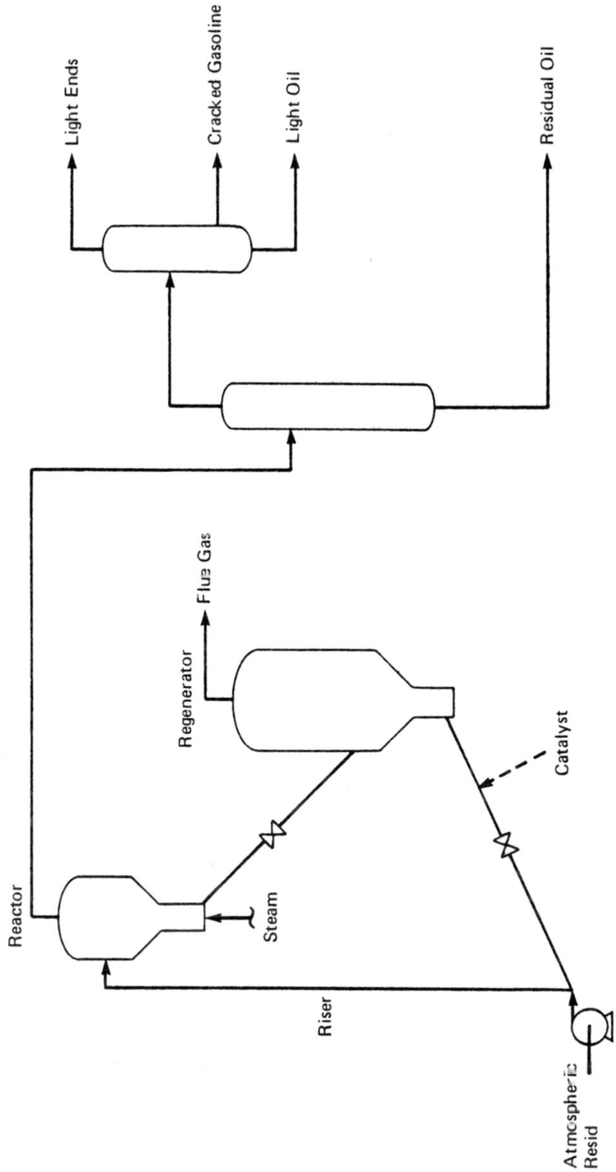

6. Shell Gasification Process

The Shell Gasification Process (SGP) is a continuous noncatalytic partial oxidation process for converting any hydrocarbon feedstock, from natural gas to bitumen into a water gas mixture of hydrogen and carbon monoxide. As in Texaco's partial oxidation process, SGP products can be further converted to ammonia, methanol hydrogen, town gas, reducing gas for the reduction of iron ores or process fuel gas. The process has been available commercially since the early 1950s. Exhibit V-42 lists the location, startup date, and process application for each of the commercial units which currently process very heavy oil feedstocks.[107]

A simplified process flow diagram of the Shell Gasification Process is shown in Exhibit V-43.[108] Hydrocarbon feed and oxidant (air or oxygen) are preheated separately and mixed in the combustion chamber where part of the feed is burned, and the remainder is thermally cracked by the heat of combustion. The cracked product reacts with the combustion product in the reactor to form the desired gas mixture. Reactor temperature ranges from 1800 to 2900°F and is moderated by either nitrogen (when air is the oxidant) or steam (when oxygen is used) which is introduced to the reactor. The moderator controls the temperature to ensure simultaneous combustion and cracking of the feed in the desired ratio.

The gas mixture leaving the reactor enters a waste heat boiler where it gives up much of its sensible heat. Gas purification consists of a water wash system which yields a pumpable carbon slurry and a virtually carbon-free gas. Most of the carbon entrained in the gas mixture is removed in the carbon catcher, slurried to a carbon separator where it is pelleted for fuel or other carbon-based products. The water is recycled from the carbon separator back to the carbon catcher. Exhibit V-44 gives process conversion data for some reference feedstocks processed in the SGP including residue from propane solvent deasphalting.[109]

Shell claims that unit operation is simple, with a minimum need for labor and supervision. The process has demonstrated continuous operation with a 95 percent on-streat factor. Shell also asserts that the plant can operate at design conditions with 70 percent of design capacity throughput and that feedstocks can be changed within 30 minutes.

EXHIBIT V-42

Processing Very Heavy Oil

SHELL GASIFICATION								
COMPANY	LOCATION	FEEDSTOCK	REACTOR TYPE	DESIGN CAP. IN Nm^3 $CO + H_2/sd$	UTILIZED FOR	START-UP DATE	OPERATING PRESSURE $kg/cm^2(g)$	CARBON RECOVERY
Union Rheinische Braunkohlen Kraftstoff A.G. Wesseling	Wesseling (Germany)	Heavy fuel oil + Naphtha + Tar oil Heavy Residue	2 x 200	400,000	Methanol	Dec. 1958	29.0	Pellets Remixing Recycle
Union Kheinischu Braunkohlen Kraftstoff A.G. (III)	Wesseling (Germany)	Heavy cracked residue	1 x 300	300,000	Methanol	Jul 1962	29.0	Pellets Remixing Recycle
Union Rheinische Braunkohlen Kraftstoff A.G. (IV)	Wesseling (Germany)	Heavy cracked residue	1 x 300	300,000	Methanol	Nov 1964	29.0	Pellets Remixing
Union Rheinische Braunkohlen Kraftstoff A.G. (V)	Wesseling (Germany)	Heavy cracked residue	1 x 300	300,000	Methanol	Nov 1964	29.0	Pellets Remixing
Zavody Pro Zpracorani Ropy a Uhio	Most Czechoslovakia	Vacuum residue (Asphalt)	6 x 700	4,200,000	Oxo-CH_3OH-NH_3-Synthesis and Town gas and H_2	Dec 1970	30	Pellets Remixing Recycle
Union Rheinische Braunkohlen Kraftstoff A.G. (VI)	Wesseling (Germany)	Heavy cracked residue	2 x 700	1,400,000	Methanol	Feb 1969	36	Pellets Remixing Recycle
VEEA Chemie A. G.	Gelsenkirchen (Germany)	Vacuum Residue	4 x 1000	4,000,000	Synthesis gas	1973	60	Pellets Remixing + Recycle
SIPM	Amsterdam (Netherlands)	Full range	1 x 50	50,000	Demonstration plant	Aug 1971	100	Pelletizing Remixing + Recycle

EXHIBIT V-43

Shell Gasification Process: Simplified Process Flow Diagram

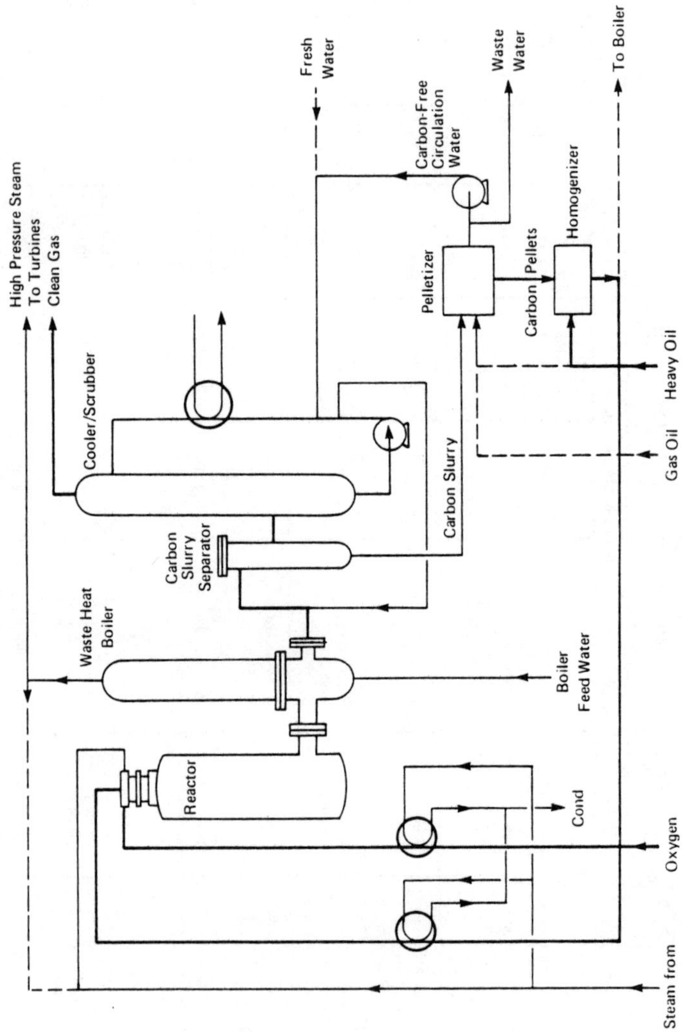

EXHIBIT V-44

Shell Gasification Process: Feedstock Properties and Product Yields

FEEDSTOCK				
Type	PDA ASPHALT	VACUUM RESIDUE	CRACKED RESIDUE	BUNKER C GRADE FUEL
Gravity °API				
Specific Gravity 60/60°F	1.07	0.99	0.95	0.97
Kinematic Viscosity c^S				
@ 212°F	–	862	37	40
@ 300°F	700	73	10	11
Composition, wt %				
Carbon	83.60	85.00	86.50	84.60
Hydrogen	9.40	10.60	11.20	11.30
Sulfur	6.56	3.30	1.67	3.50
Nitrogen	0.29	0.50	0.35	0.40
Oxygen	0.05	0.45	0.25	0.13
Ash	0.10	0.15	0.03	0.07
Metals, wppm				
Vanadium (V)	169	235	78	112
Nickel (Ni)	51	17	22	35
Sodium	21	30	33	50
PRODUCTS				
Composition, vol %				
CO	47.74	47.21	46.83	46.14
CO_2	4.51	4.20	4.15	4.32
H_2	44.22	45.81	46.67	46.73
CH_4	0.60	0.60	0.60	0.60
N_2	0.58	0.64	0.60	0.62
A	0.74	0.76	0.76	0.77
H_2S	1.55	0.75	0.37	0.79
SO_2	0.06	0.03	0.02	0.03

[1]Propane deasphalter

The process is comparable to the Texaco partial oxidation process in terms of commercial reliability, feedstock flexibility, and product application. Selection of one process over the other is more a matter of fit or integration of process facilities with existing equipment or needed services than a question of feedstock or product specification.

7. M-Gas

M-Gas is a process undergoing joint development by Mitsui Engineering and Shipbuilding Company, Ltd., and Mitsui Kozan Coking Industry Company, Ltd., for the catalytic gasification of vacuum residual oil into medium BTU fuel gas in a fluidized bed. Process development has progressed from bench scale to a small 250 kg per day pilot plant that is not fully integrated. The commercial process is in the conceptual design phase. Mitsui is continuing work to extend process application to very heavy thermally cracked pitch and coke and to establish technical and economic limitations of the process.

A schematic diagram of the M-Gas pilot unit is shown with a schematic of a typical process application in Exhibit V-45.[110] Vacuum residue feed is introduced to a bed of fluidized catalyst in the gasifier. Catalyst is continuously circulated from the gasifier to the regenerator and back. Regenerator heat is supplied by a gas oil stream fed to the vessel. Heated catalyst transferred to the gasifier provides the heat of reaction. The regenerator temperature is maintained at about 2000°F while the gasifier temperature is kept at 1750°-1830°F. Steam which serves as the gasifier fluidization medium and the feed/solids transport medium is introduced with the catalyst at the bottom of the gasifier and with the heavy oil feed at the side of the vessel.

Gas leaving the gasifier is quenched to 750-850°F in a waste heat boiler which produced medium pressure steam. The quenched gas is scrubbed and cooled further in an oil-scrubbing system which slurries the entrained particles. After additional cooling to about 100°F, the gas is fed to an acid gas removal system for H_2S removal. A portion of the slurried effluent is recycled as feed to the gasifier and fuel to the regenerator.

Hot regenerator off gas (@ 2100-2200°F) is quenched to 1100-1300°F in a second waste heat boiler. After filtering, the gas is expanded to recover power for compression. The cooled gas must then be desulfurized.

EXHIBIT V-45

M-Gas Process: Schematic Diagram of Small Pilot Plant

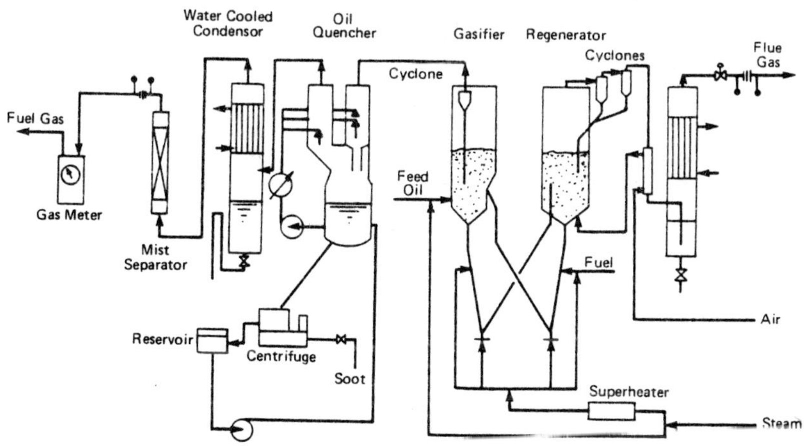

M-Gas Process: Schematic Diagram of Small Pilot Plan with Application

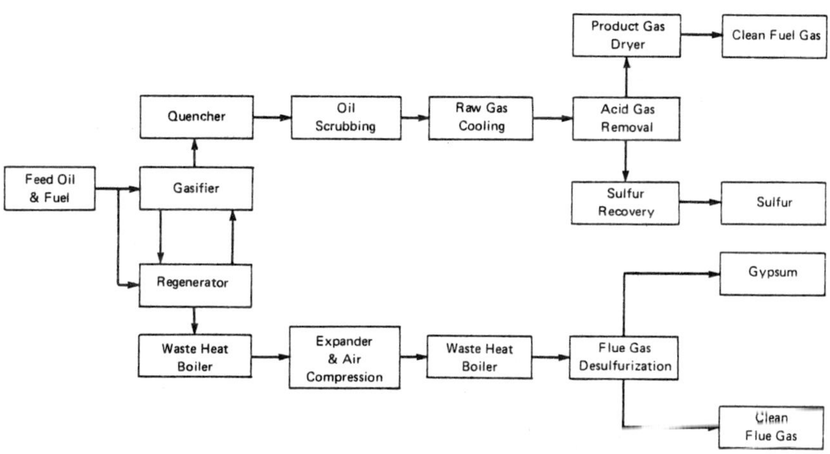

Exhibit V-46 presents feedstock and yield data for the bench scale gasification of Gach Saran vacuum residual oil.[111] From the yield data presented, M-Gas appears to have the potential to produce a higher quality syngas than either the Texaco or Shell processes. However, the latter processes have demonstrated their commercial reliability over a long period of time, while the Mitsui process is still in the early stages of development.

EXHIBIT V-46

M-Gas: Feedstock Properties and Product Yields

FEEDSTOCK		Iranian Heavy (Gach Saran) Vacuum Residue
Type		
Gravity °API		6.0
Specific Gravity 60/60°F		1.029
Viscosity cp		
@ 300°F		210
Pour Point, °F		48
Composition, wt %		
Carbon		84.6
Hydrogen		11.1
Sulfur		3.4
Nitrogen		0.8
Ash		0.1
Metals, wppm		
Vanadium (V)		492
Nickel (ni)		143
Iron (Fe)		25
Sodium		0.8
Chlorine		45
Conradson Carbon Residue, wt %		21.4
Asphaltenes, wt %		8.3
CATALYST / PRODUCTS	KM-18	M_gO Clinker
Composition, vol %		
CO	8.3	6.8
CO_2	18.4	11.0
H_2	55.2	55.6
CH_4	10.5	16.6
C_2^+	7.6	10.0
Gas Yield, wt %	131.5	86.1
Gross Heating Value (HHV), Kcul/nms	4080	4962

In addition to direct coking, thermal crude cracking, catalytic cracking, and gasification, a number of "hybrid" carbon removal processes were also reviewed, including Dynacracking catalytic residue cracking.

8. Dynacracking

The Dynacracking Process is a noncatalytic heavy oil conversion process that upgrades vacuum residues and heavy oils with high metals and sulfur content. The process combines coking, gasification and hydrocracking to produce distillate, naphtha, and synthetic fuel gas. The process was developed by Hydrocarbon Research Incorporated (HRI) in the 1950s and was tested then in a 15 b/d pilot plant unit. The process was shelved in 1955 because industry had no need for heavy oil processing at the time. A small 5,100 BPD commercial unit is planned for the upgrading of vacuum residue on the West Coast. Scale-up is being made directly from the data produced in the 50s as no other data exists. The project is in its final stages of detailed engineering design. Start-up is scheduled for 1982.

The Dynacracking Process centers around a unique fluid-bed reactor which is divided into three process zones separated by internal baffling. A schematic of the reactor and process flow is shown in Exhibit V-47.[112] Preheated feed enters the lower part of the hydrocracking zone in the top of the reactor. Operating conditions in the hydrocracking zone include temperatures from 900-1400°F and pressures from 400-600 psig. According to HRI, this pressure allows sufficient hydrogen partial pressure (less than 175 psig) to provide desulfurization and hydrocracking to form gases, liquid products, and coke.

The coke, together with any metals from the feedstock and some very heavy liquids, is deposited on an inert carrier which makes up the fluid bed. The coke laden carrier moves downward through the stripping zone in the central part of the reactor. The stripping zone is packed with inert stationary packing that separates the hydrocracking section from the gasification section of the reactor. In this zone, the heavy liquids are stripped from the falling coke particles.

As the stripped coke particles leave the stripping zone, they are contacted with steam and oxygen at temperatures up to 1800°F in the gasification zone. The gasification reactions yield a hydrogen-rich synthesis gas that serves as the stripping medium in the stripping zone and part of the fluidizing gas for all sections of the reactor. The hydrogen provides the environment for the hydroconversion reactions in the hydrocracking zone. The heat requirement for all the reactions in the vessel are supplied by the combustion of coke with oxygen in the gasification zone.

Reactor products leave the reactor from the knockout space above the hydrocracking zone and proceed to gas-liquid separation and further processing. The decoked carrier from the gasification section of the reactor is recirculated to the hydrocracking zone via a center (or external) transport tube which uses steam or product fuel gas as the propellant

EXHIBIT V-47 Dynacracking Process Flow Scheme

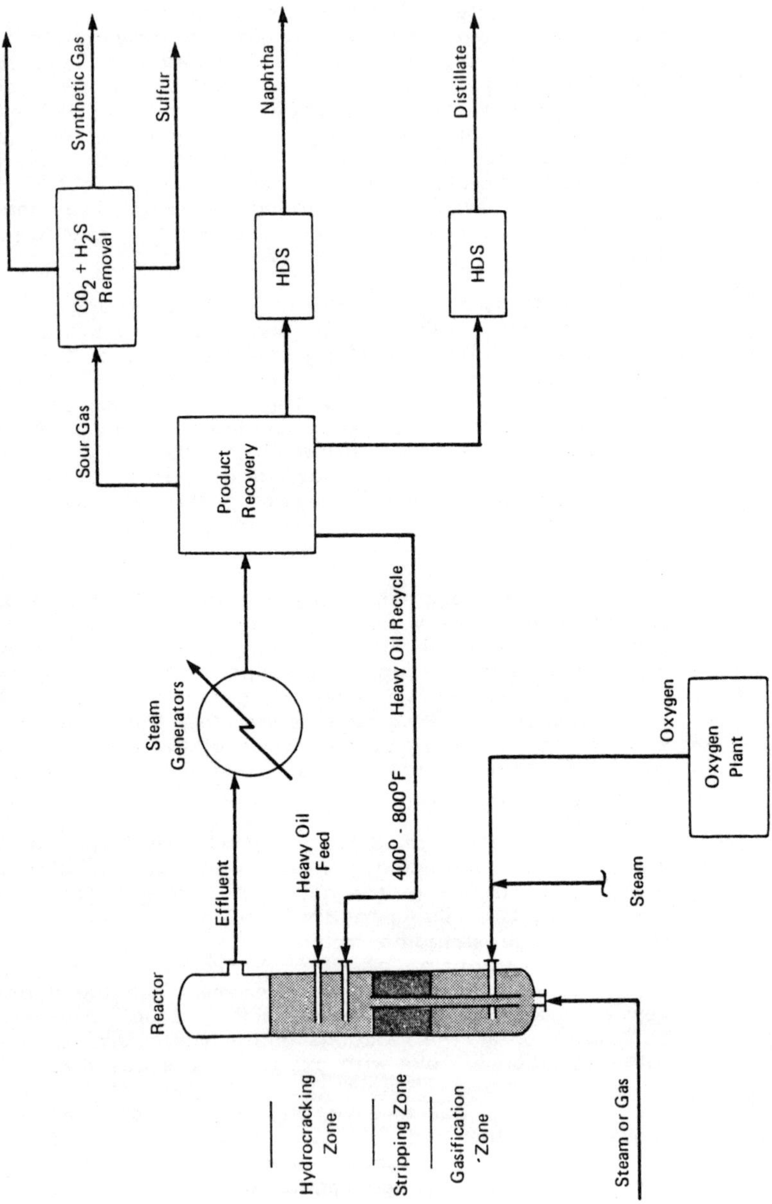

Feedstock properties and yield data for the Dynacracking process are listed in Exhibit V-48. The process does not appear to have any serious feedstock limitations that would require pretreatment steps. The liquid products are similar to those from other thermal cracking processes.

Dynacracking is a combination of coking, gasification and hydrogen addition processing. Approximately 8 percent of the carbon content is converted to coke compared to 23 to 30 percent for delayed coking. This carbon is combusted to produce hydrogen in the gasification section of the reactor. The cracking reaction is thermal and noncatalytic as in other coking processes. However, the reaction takes place at a higher severity and results in higher naphtha yields. The ability to operate at increased severity is purported to be due to the hydrogen atmosphere in the hydrocracking section. As the reaction is noncatalytic, desulfurization of products is not comparable to the "hydrogen in" processes. In the yield data that is available only 59 percent of the sulfur is removed and a relatively high sulfur naphtha is produced.

Exhibit V-49 lists the per barrel investment and operating costs for a small Dynacracking unit. The economics of the process are strongly related to the degree of integration possible with the location site. Indeed, costs can vary up to 75 percent or more depending on the size, location, and complexity of the plant and the downstream processing that may be required.[112] A nearby demand for the product fuel gas will enhance the attractiveness of the process, while a limited need for fuel gas would tend to favor a process with higher liquid yields. The combination of coking and gasification in one vessel offers attractive economies of scale over Flexicoking and partial oxidation. However, the combination also provides an internal heat balance which reduces the degrees of freedom available for controlling the unit. This may reduce the flexibility in feedstock selection. For example, a resid with a high asphaltene content would result in a high coke production. This would increase the volume of gas and concomitant heat leaving the gasification zone. A higher distillate recycle rate would be required in the hydrocracking section to maintain equilibrium temperatures. This could result in lower liquid yields and increased fuel gas production which may negatively affect process economics.

EXHIBIT V-48

Dynacracking Yields from Vacuum Resids

FEEDSTOCK	
Type	La Basin Vacuum Residue
Gravity °API	7.0
Carbon Residue wt %	16.0
Sulfur wt %	2.26
C/H mol ratio	1.43
Visc., SFS @ 210°F	910
PRODUCTS (Percent)	
Naphtha (C_4-400)	44
Distillate (400-650)	5
Gas Oil (650-900)	8
Resid (900+)	16
Fuel Gas C_4, FOEB	34

Operation @ 425 psig, 1000°F hydrocracking zone
1800°F gasification, H_2 partial pressure 100° psi

EXHIBIT V-49

Dynacracking Investment and Operating Costs

INVESTMENT COST ($/bbl Daily capacity)	1976	1980
Plant & Equipment	$1,813	$5,280
Interest During Const		500
Working Capital		820
Total Investment		$6,600

OPERATING COST ($/bbl)	1st Quarter 1980
Oxygen	1.12
Utilities	.43
Maintenance (4%)	.62
Taxes & Insurance (3%)	.46
Administrative	.59
Operating Labor & Supervision	.48
Contingency	.18
Total Operating Costs	3.90

The major yield advantages of Dynacracking are:

- Feedstock flexibility with a high tolerance for metals.
- An 83 percent reduction in residue volume.
- A high naphtha production.
- A product fuel gas (after cleanup) that requires few combustion equipment modifications.
- The absence of coke production.
- Potentially low capital and operating costs.

The major disadvantages of the process are:

- Need for an oxygen plant or oxygen supply.
- High fuel gas production and low middle distillate production.
- Need for gas compression.

Dynacracking, like Flexicoking, is a combination of coking and gasification. However, the gasification and cracking reactions take place in the same reactor. The hydrogen produced in the gasification section contributes to the saturation of olefins in the "hydrocracking" section. However, it is not clear how much conversion takes place as a result of hydrocracking. Dynacracking does not appear to have any serious feedstock limitations that would require specific pretreatment steps. Since it is noncatalytic, it is not restricted by metals or sulfur content in the feed.

9. Catalytic Residue Cracking Process--RAROP

In a joint venture within RAROP, Nippon Mining Company, Kashima Oil Company, and Hitachi Ltd. are constructing a 250 b/sd catalytic cracking pilot plant that will produce hydrogen as well as middle distillates (kerosene and diesel) from heavy sour crudes. A March 1981 construction completion is scheduled. Start-up will follow shortly after that.[111]

A simplified process flow diagram for the cracking process is shown in Exhibit V-50. Exhibit V-51 presents process feedstock and yield data obtained from the bench scale operation.[115] The process flow is very similar to the Flexicoking flow scheme. It is not clear that the catalyst enhances either the gasification reaction or the cracking reaction. It may only act as a carrier to move the carbon from the cracking reactor to the gasifier. The preliminary data indicates a very low liquid yield and substantial production of light ends; however, additional testing and data collection are needed before conclusions can be drawn.

EXHIBIT V-51

RAROP Cracking/Gasification Process: Feedstock Properties and Product Yields

FEEDSTOCK		
Type	Vacuum Residue	
Gravity	7.2	
Carbon Residue wt %	21.1	
C, wt %	84.1	
H, wt %	10.8	
S, wt %	5.1	
V, wppm	110	
N, wppm	33	
PRODUCTS WT %	ATMOSPHERIC PRESSURE	HIGH PRESSURE
$\leq C_4$	25	20
C_5	7	5
350-1050°F	42	54
Gross Coke	26	21
Total	100	100
Conversion of 1050 + °F, wt %	74	79

EXHIBIT V-50

RAROP Cracking/Gasification Process
Simplified Process Flow Diagram

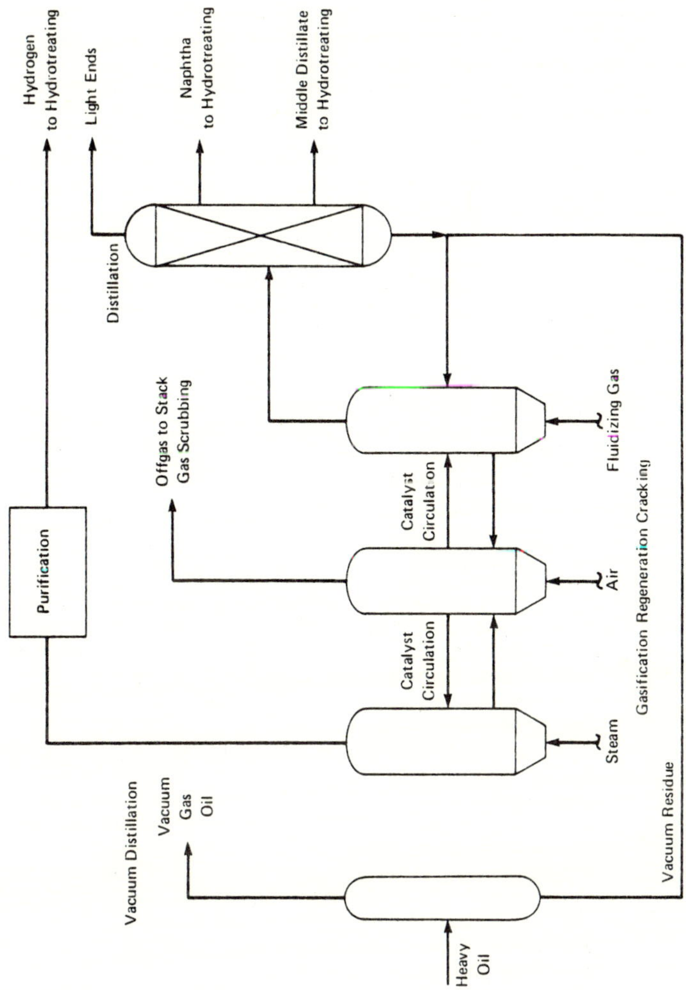

(3) Hydrogen Addition Processes

The emphasis in hydrogenation process developments is directed toward increasing feedstock flexibility. Most developing hydrogen addition processes are being designed to accept reduced feed from either the vacuum tower or, more commonly, from the atmospheric tower. Hydroprocessing atmospheric reduced crude can eliminate the need for a vacuum unit.

Research activity has focused primarily on catalysis. Developers are striving to make cheaper metals-tolerant catalysts and improve the methods for replacing and regenerating catalysts. A few process developers are using proprietary passivators which reduce the adverse effects impurities in the feed have on catalyst activity and life. Others are using additives in place of catalysts entirely. This section looks at the following processes:

- Aurabon (UOP)
- HFC (Niigata)
- Canmet (Petro Canada, EMR)
- Asphaltene Cracking (Chiyoda)
- Fixed Bed Hydrocracker (RAROP)
- Kombie Kammer (VEBA).

1. Aurabon

The Aurabon Process is a catalytic hydrogenation process which accepts heavy oil feedstocks with high metals and asphaltene content. The process produces light ends and a low asphaltene, low metal product which can be hydrocracked to transportation fuels. A 5 percent by-product drag stream is also produced. Aurabon was developed by UOP and is currently being tested in the newest of three small (.25-.5 b/d) pilot plants. This pilot plant is fully integrated and has been operating for 4 years.[116] All the available data on this process has been based on results from this pilot plant. UOP in cooperation with Corpoven in Venezuela is designing a 2,500 b/d demonstration unit for Corpoven's Bajo Grande Refinery. The unit is being designed to process vacuum resid from 10° API Boscan Crude. Although other processes are also being considered, engineering design for the Aurabon Process will be completed in 1980 and UOP is optimistic that the unit will be built and ready for a 1983 start-up.[117]

As of this writing, specifics on process flow and operating conditions have not been released by UOP. A schematic representation of the Aurabon Process is shown in Exhibit V-52.[118] The process schematic resembles Chiyoda's Asphaltene Cracking process, which links a reactor

EXHIBIT V-52

Aurabon Process: Schematic Presentation

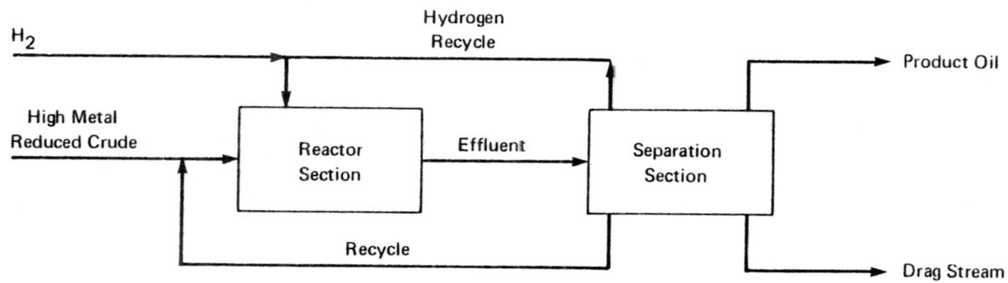

and separation section with recycle gas and liquid streams. Details of the reactor and separation sections are not known for either process. The major difference in the two processes is the selection of catalyst. Aurabon relies on the high vanadium content of certain feeds to catalyze the hydrocracking reaction, while Asphaltene Cracking employs a heterogeneous catalyst.

After conventional desalting and topping to remove brine, inerts, and lighter hydrocarbons, fresh feed at 600°F is combined with unconverted recycle liquid, recycle gas, and make-up hydrogen and charged to the reactor. UOP claims that the operating conditions under which the heavy oil conversion reactions take place in this reactor "are in the range of those presently used in the desulfurization and hydrocracking of petroleum atmospheric column bottoms.[119]

Recycled gas is separated from the total reactor effluent in a series of separators. These separators also yield both a gaseous and light, liquid product. Product specification and recycle composition are the parameters used to establish the operating conditions of this separation step.

A by-product, "drag stream," is withdrawn from the process to remove contaminants and to keep the product slate in balance with the feedstock. The drag stream is composed chiefly of metals and unconverted asphaltenes that would otherwise accumulate in the process. UOP claims that the process will remove over 95 percent of the asphaltene and organometallic compounds from the feed.

An analysis of desalted, topped Boscan and Lloydminster heavy oils that were processed in the Aurabon pilot plant are given in Exhibit V-53.[120] The conversion of 1050+°F material in the feed is only about 30 percent. As a result, the total product mix (Aurabonate) has a gravity that is still below pipeline specifications (less than 15° API).

One advantage of the Aurabon Process is that the heavy residue product has a low metals content which is adequate pretreatment for further processing. Exhibit V-54 shows the downstream processing options that are available to Aurabon residue products which are characterized in the exhibit under Aurabonate.[121] The Aurabon process is not practical if the residue is required for gasification or combustion for on-site energy.

Exhibit V-53
Aurabon Process: Feedstock Properties and Product Yields

Feedstock	. Atmospheric Reduced Crude . .	
	Boscan	Lloydminster
Properties		
API gravity	6.2	7.3
Specific gravity at 60°F/60°F	1.028	1.020
Distillation (D-1160)		
IBP°F	567	541
5%	739	757
10%	790	801
20%	883	847
30%	963	919
EP	986	1,002
Percent over volume	35.0	40.0
Microcarbon, wt %	81.0	84.5
Microhydrogen, wt %	10.1	10.3
Sulfur, wt %	5.8	5.4
Conradson carbon, wt %	17.7	15.93
Total nitrogen, wt %	0.7	0.45
Heptane insolubles, wt %	17.9	14.2
Pentane insolubles, wt %	23.4	19.1
Vanadium, wt ppm	1,484	220
Nickel, wt ppm	130	84
Viscosity at 210°F, cs	1,620	1,020
Viscosity at 250°F	390	270
Drag stream		
Yield, wt % of feed	2.4	2.9
Specific gravity at 60°/60°F	1.329	–
Sulfur, wt %	9.0	–
Vanadium, wt %	6.0	
Nickel, wt %	0.5	–
Hydrogen consumption		
Weight percent of feed	1.1	1.0
Standard cubic feet per barrel	750	675
Products		
NH_3, wt %	0.2	–
H_2S, wt %	1.7	–
C_{1-5}, wt %	2.5	–
C_6-450°F		
Yield, wt % of feed	1.0	2.8
API gravity	53	45.7
Specific gravity at 60°F/60°F	0.7669	0.9077
Sulfur, wt %	1.3	3.98
Nitrogen, wt %	0.36	0.22
450°-650°F		
Yield, wt % of feed	8.1	8.5
API gravity	25.2	24.2
Specific gravity at 60°F/60°F	0.9030	0.9077
Sulfur, wt %	3.85	3.98
Nitrogen, wt %	0.39	0.22
650°-1050°F		
Yield, wt % of feed	41.8	43.8
API gravity	17.4	16.4
Specific gravity at 60°F/60°F	0.9503	0.9567
Sulfur, wt %	3.95	3.09
Nitrogen, wt %	0.32	0.26
Conradson carbon, wt %	0.43	0.64
Viscosity at 98.9°C, cs	10.81	14.32
1050+°F		
Yield, wt % of feed	43.4	39.2
API gravity	7.7	7.6
Specific gravity at 60°F/60°F	1.0165	1.017
Sulfur, wt %	4.64	4.13
Nitrogen, wt %	0.70	0.61
Conradson carbon, wt %	16.53	16.7
Heptane insolubles, wt %	0.72	0.48
Viscosity at 121.1°C, cs	335	315
Vanadium, wt ppm	54	16
Nickel, wt ppm	19	9

EXHIBIT V-54

Aurabon Refinery Processing Options

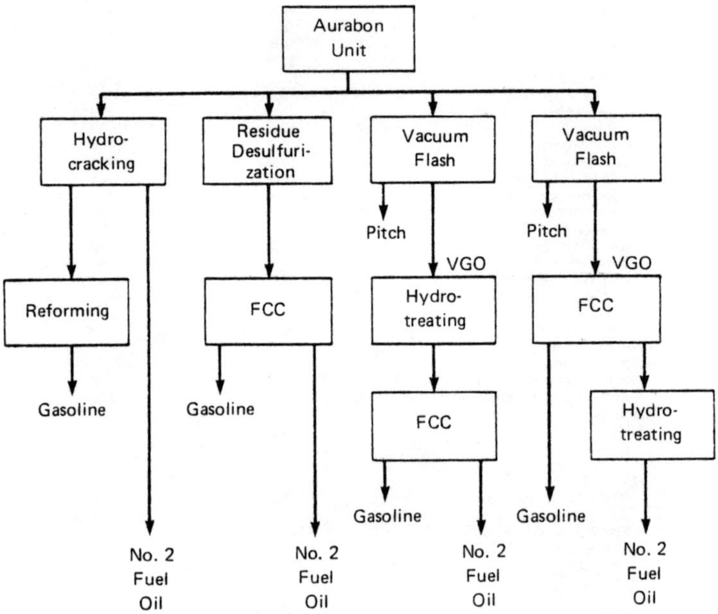

Source	Total Aurabonate
Yield, Wt-% of Feed	94.3
API Gravity	13.7
Specific Gravity at 60°F	0.9748
Sulfur, Wt-%	4.2
Nitrogen, Wt-%	0.54
Conradson Carbon, Wt-%	7.8
Heptane Insolubles, Wt-%	0.3
Vanadium, Wt. ppm	24.8
Nickel, Wt. ppm	8.7

Another advantage of the process is the potential savings on catalyst over other catalytic hydrogenation processes. A comparison of the Aurabon process over other hydrogenation processes must balance the higher quality products available through more severe hydrogenation against the additional cost in catalyst that is required.

2. HFC

Niigata Engineering Company, Ltd., and the National Research Institute for Pollution and Resources, Kawaguchi, Japan have experimented with hydrocracking atmospheric and vacuum residues with spent catalyst from residue desulfurization. A 1.5 b/sd "pilot" plant has been operated for more than 1,000 hours. The biggest merit claimed for the process is a cheap, available catalyst with high metal acceptance. Any plans for commercialization are unknown at this time.[122]

Exhibit V-55 shows the HFC pilot plant process flow diagram. A pressurized (30-200 kg/sq cm) mixture of heavy oil feed, 3 wt percent catalyst and make-up and recycle hydrogen is fed to the two reactors in series. Both reactors are tubular vessels. Flow is in an upward direction to ensure thorough mixing. Reaction temperatures range from 800-850°F. Feed is cracked to lighter products which leave the top of the second reactor and enter a series of hot and cold separators which separate the light ends, liquid products, and catalyst. Hot separator bottoms contain the catalyst which is centrifuged and recycled.

Exhibit V-56 lists the feedstock and product properties for hydrocracking Gach-Saran vacuum residue.[123] The objective of the process is to convert vacuum residue to a synthetic crude of high enough quality to be processed in an atmospheric tower with conventional crudes. Interest in the process was generated by the observations that vanadium sulfide catalyst, resulting from the thermal decomposition of organic vanadium with heavy crude, has shown a good activity in hydrogenation reactions.[124]

The process claims the following operating characteristics:

- Lower hydrogen consumption (300 to 1050 scf/bbl) than for conventional hydrocracking processes.

EXHIBIT V-55

AFC Pilot Plant Process Flow

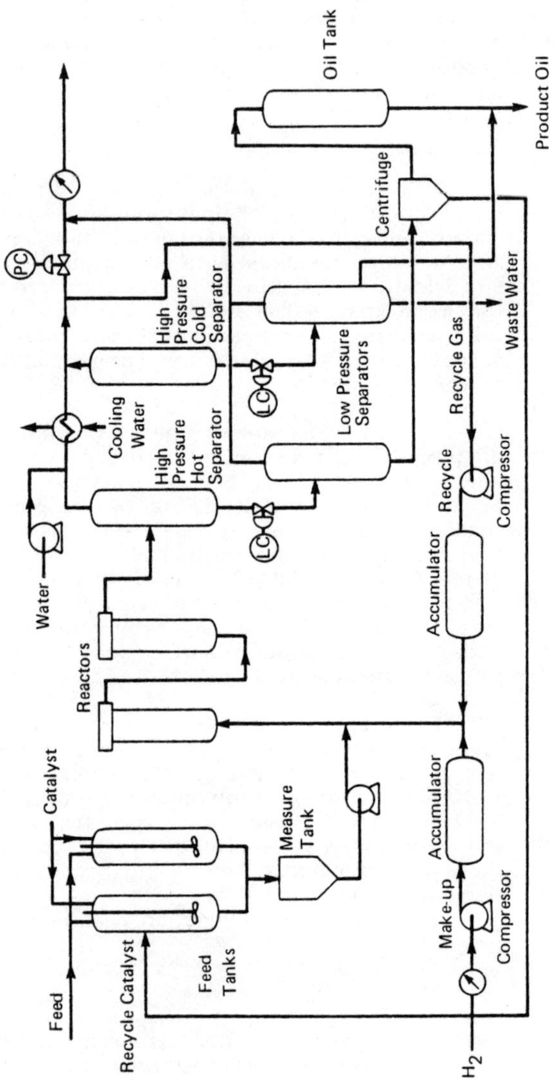

EXHIBIT V-56

HFC Process: Feedstock Properties and Product Yields

FEEDSTOCK	GACH-SARAN
Type	Vacuum Resid
Specific Gravity 60/60°F	1.0203
Distillation, °F	
100%	750
H/C Ratio, atomic	1.43
Sulfur, wt %	3.61
Nitrogen, wt %	0.67
Metals, wppm	
V	316
Ni	101
Conradson Carbon Residue, wt %	21.7

PRODUCTS	YIELD WT %	SP. GRAV. 60/60°F	VISC CS + @122°F	S WT %
$H_2S + NH_3$	2.5	–	–	–
C_1 - C_4	1.4	–	–	–
C_5 - 200°F	2.0	0.6940	–	0.07
200 - 340°F	6.1	0.7466	–	0.13
340 - 450°F	7.7	0.8075	1.10	0.22
450 - 550°F	10.2	0.8501	2.07	0.79
550 - 650°F	10.9	0.8729	4.01	0.92
650 - 750°F	10.0	0.9065	9.10	1.22
750 - 930°F	19.9	0.9663	65.4	1.58
930° + °F	30.0	1.073	–	2.85
Total		–	–	–

Operating Parameters	
Desulfurization	61 wt %
Hydrogen Consumption	6600 scf/t feed (177 Nm^3/t)

- Lower operating pressures which enable
  - Higher degree of cracking
  - Reduced desulfurization
- Feedstock flexibility
- High metal acceptance capacity of the catalyst
- Easy withdrawal of spent catalyst
- High temperature operation
- An increasing availability of potentially inexpensive catalyst.[125]

Although the process boasts of low pressures and high temperatures which permit operation at high severities, pressure reduction is limited by the coking tendency of individual feedstocks. Exhibit V-57 demonstrates the process' relative insensitivity to high vanadium levels in the feed.[126] Desulfurization and conversion levels remain relatively constant and are significantly greater than that observed at much lower vanadium levels. Consequently, however, hydrogen consumption would also be much higher at these conditions.

EXHIBIT V-57

HFC Process Insensitivity to Vanadium

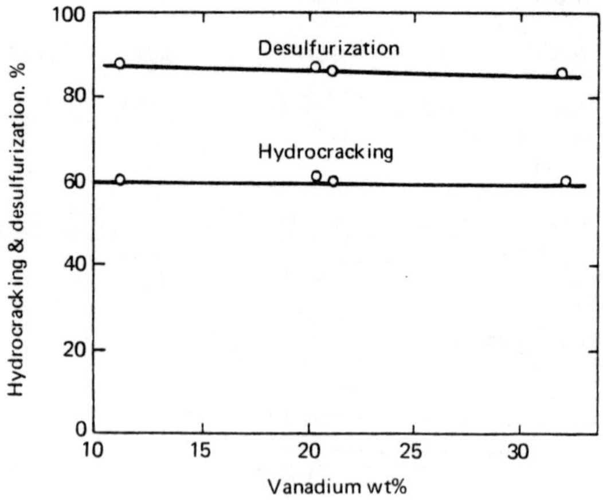

3. CANMET

CANMET, a process being developed for commercialization by Petro-Canada and a major engineering company, is an expanded bed thermal hydrocracking process in which heated heavy oil feedstock is mixed with a propriety additive to allow cracking at reduced pressures in an upflow tubular reactor. Overhead product is separated into light ends, distillates, hydrogen for recycle, and a residual pitch fraction. The pitch contains solid residues from the additive and the bulk of the feed metals. The process employs a non-precious metal additive which serves as a catalyst substitute. Claimed process features include high tolerance to metals in the feed and better conversion of .975+°F material at lower operating pressures than straight thermal hydrocracking. Two 1 barrel per day "pilot" units have logged over 20,000 hours of operation in Ottawa. A comprehensive commercialization program includes a 2,000 to 10,000 b/d demonstration plant targeted for 1981 start-up and availability for licensing by 1984. All available data has originated from pilot plant runs.

A simplified schematic of the CANMET process is shown in Exhibit V-58. The process developers claim that the heart of the process and its unique features lie in the reactor system. No details have been released on the reactor system as of this writing. One aspect of the process that is mentioned in the literature is a low cost additive consisting of pulverized coal impregnated with iron sulfate which is purported to permit operation at reduced pressures without operability problems from coke deposition on the reactor walls.[127]

Heated heavy oil feedstock, mixed with a small proportion of the processing additive, is contacted with hydrogen and charged to an upflow tubular reactor. Residues from the additive and deposited feed metals contribute to formation of a solids bed within the reactor. Product is withdrawn overhead, depressured, cooled and separated. Light ends after gas cleanup are either compressed for recycle hydrogen or sent to a gas plant. Liquid products go to fractionation where they are further separated into naphtha, gas oil, and a residual pitch fraction. The pitch contains the bulk of the feed metals and solid residues from the additive.

Typical performance and yield data for the processing of Lloydminster atmospheric residue with the CANMET pilot plant are shown in Exhibit V-59.[128] The CANMET process developers claim the following process features:

EXHIBIT V-58

CANMET Process: Simplified Schematic

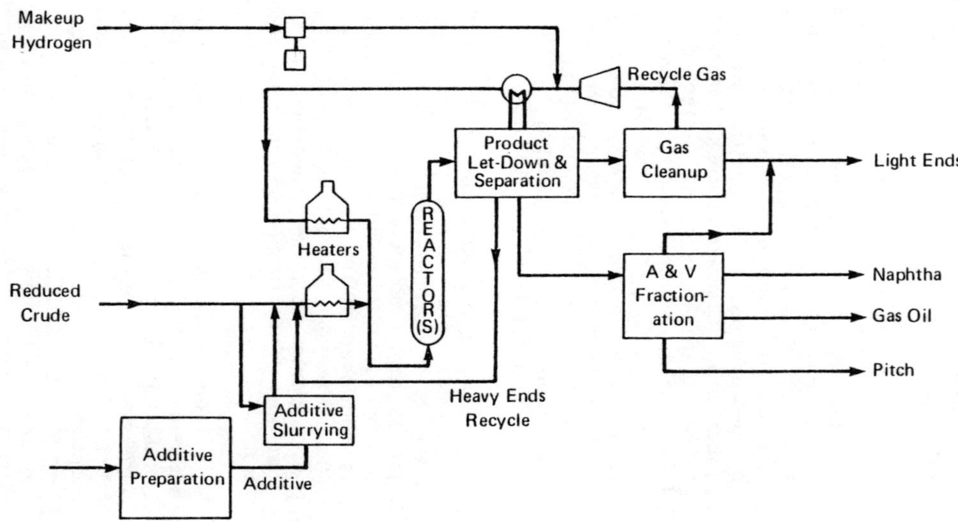

EXHIBIT V-59

CANMET Process Performance

FEEDSTOCK		LLOYDMINSTER 650+°F + RESIDUE		
Gravity °API		9.2		
Sulfur Wt %		4.2		
Metals (Ni +V) ppm		217		
PRODUCTS	**WT %**	**VOL %**	**°API**	**S WT %**
H_2S	2.65	–	–	–
CH_4	2.01	–	–	–
C_2H_6	1.53	–	–	–
C_3H_8	1.61	–	–	–
C_4H_{10}	0.85	–	–	–
C_5 - 400°F	17.07	23.1	56.4	0.61
400-650°F	38.12	43.8	28.6	1.74
650-975°F	31.39	32.6	13.2	2.13
975°F +	6.39	5.4	-10	3.81
Total	101.62	104.9	–	–
Operating Parameters				
Conversion (975°F +)		90 wt %		
Desulfurization		60 wt %		
Hydrogen Consumption		1035 SCF/bbl		

- Feedstock flexibility--insensitive to contaminants such as sulfur, nitrogen, and especially metals.
- High conversion--over 90 percent of 975+°F material.
- High overall liquid and light distillate yields.
- Reduced operating pressure relative to competitive commercial technology.
- Selective hydrogenation of the cracked distillate-minus material, therefore reducing hydrogen consumption.
- Possible application of residue pitch as a binder in the production of metallurgical coke.

Petro-Canada believes that the CANMET process is a technically and economically superior process for primary conversion of heavy oil to a synthetic crude. The product is marketable to refineries which lack bottoms-conversion capability. The firm also sees increasing potential for the process to be used as a residue conversion unit in conventional refineries.

Petro-Canada's estimates for CANMET investment utility and catalyst requirements are summarized in Exhibit V-57.[129] Data is based upon a hypothetical 50,000 bbl/d

EXHIBIT V-60

CANMET Process: Investment, Utility and Catalyst Requirements

Investment $M per bbl/d	1st Quarter 1980
Canmet with ATM & VAC Fractionation	1500
Additional Plant & Equip	1800
Indirect Cost	4000
Contingency	2000
Total Installed Cost	12600

project designed to upgrade Lloydminster crude to a high-quality hydrotreated syncrude. Additional plant and equipment includes the following units:

- Atmospheric distillation
- Naphtha HDS
- Atmospheric Gas Oil HDS
- Vacuum Gas Oil HDS
- Gas Plant
- Hydrogen/steam reformer
- Claus unit with tail gas clean-up
- Amine sweetener.

Feedstock properties and product data are provided in Exhibit V-61.[130]

4. Asphaltene Cracking - Chiyoda

Asphaltene cracking is a catalytic hydrogenation process which allows complete conversion of asphaltenes to de-asphalted oil. The process being developed by Chiyoda Chemical Engineering and Construction Company is still in its early stages of development. Chiyoda has been testing the process on a 1 liter per hour bench scale unit since May 1978. All performance reports have been based on data obtained from this laboratory unit. The process uses a proprietary heterogeneous catalyst that has also been developed by Chiyoda. The company has not announced plans for larger scale testing or commercialization.

A simplified flow schematic for the Asphaltene Cracking process in shown with possible additional downstream processing options in Exhibit V-62.[131] The process appears to be very similar to UOP's Aurabon process but differs in the use of a heterogenous catalyst. While Aurabon is described chiefly as a pretreatment for further downstream processing, Chiyoda claims that, with the addition of a solvent de-asphalting (SDA) unit, the asphaltene portion of the feed can be recycled to extinction with their process.

By using this recycle concept, vacuum resid is mixed with recycled asphaltenes from the SDA and fed to the hydrogenation reactor. The proprietary catalyst both demetallizes and hydrocracks the asphaltenes. Reactor effluent is fed to a hydrogen separator followed by the SDA unit. SDA overhead continues to conventional downstream processing units. The high asphaltene bottoms stream is recycled to extinction.

EXHIBIT V-61

CANMET Process: Feedstock Properties and Product Yields

FEEDSTOCK	Heavy Oil
Gravity °API	15.8
Sulfur wt %	3.42
Nitrogen wt %	0.30
Metals (Ni + V) ppm	155
650+°F Vol %	71.2
975+°F Vol %	43.5
Natural Gas MMSCFD	14.9
Products	**Hydrotreated Syncrude**
Gravity °API	37.6
Sulfur wt %	0.17
C_4 vol %	1.1
Naphtha vol %	25.2
Lt. Gas Oil vol %	53.7
Heavy Gas Oil vol %	19.7
Pitch vol %	3.8
Sulfur Lt/SD	229
Utilities per bbl/d	
Cooling Water gal	1525
Electric Power Import kW	.55
Steam	Balanced
Fuel FOEB	
Plant Fuel Gas	.05
Fuel Oil/HGO	.04
Total Fuel	.09
Catalysts	
Catalyst, cents	4.8
Additive, cents	8.6

EXHIBIT V-62

Asphaltene Cracking Flow Schematic and Downstream Processing Options

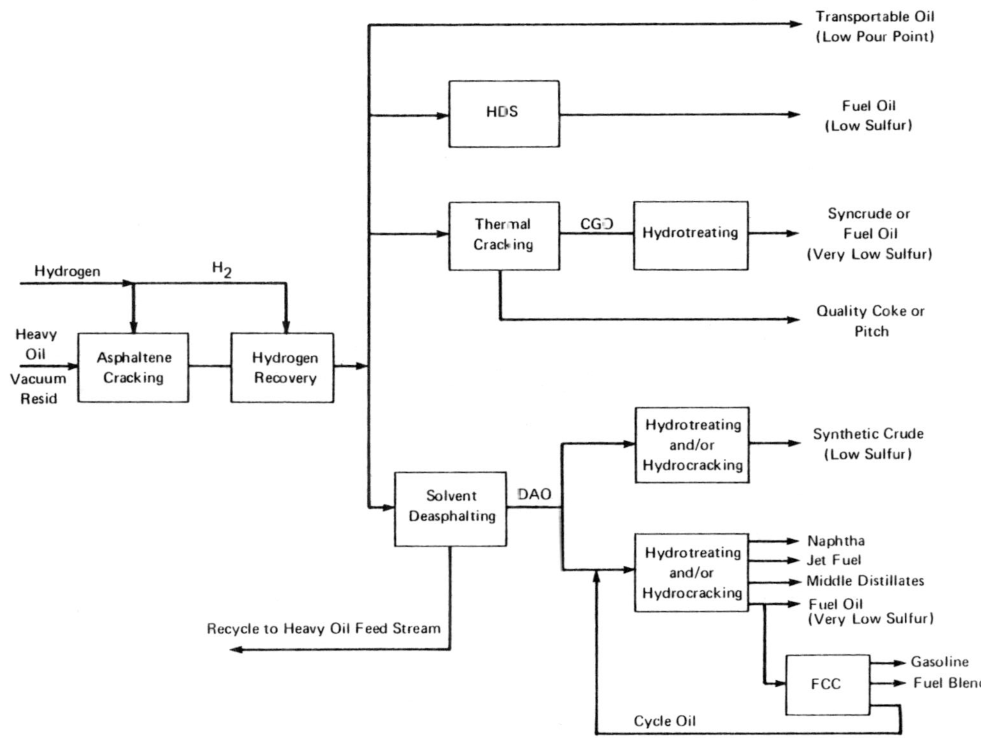

Typical process yields are listed with feedstock and recycle data in Exhibit V-63.[132] As shown, there is essentially complete removal of metals and asphaltenes with mild desulfurization and minor denitrogenation. Exhibit V-64 presents process performance data for once-through (without recycle) operation. As the data indicates, the process lends itself to substantial processing flexibility. The relative advantages and disadvantages of several possible product dispositions are listed in Exhibit V-65.[133]

Reactor conditions are listed as about 2500 psig and 800°F. These conditions correspond to a typical hydrocracking reactor which would indicate that an asphaltene cracking unit would have high capital costs and power consumption. As cracking, desulfurization and denitrogenation reactions are not a major facet of the process, hydrogen consumption is moderate in the range of 500-900 scf/bbl.

In the tested applications, asphaltene and metals contents and the pour point of each feed were substantially reduced while the Conradson carbon residue and the sulfur levels were only moderately reduced. The relative reactivities of selected feeds in the presence of the catalyst is shown in Exhibit V-66.[134] The data suggests the process' suitability to high-metal heavy feeds. Chiyoda claims the reactivity of tar sand bitumen is extremely high without the appropriate metals/asphaltene relationship. A suitable explanation for this mechanism requires additional testing.[135]

The major obstacle to the development of a process for hydrogenating high asphaltene resids has been the development of a catalyst which would not be subject to rapid deactivation. The available information on this process indicates that a catalyst life of six months is attainable. If a catalyst with these characteristics has been developed, it would have wide application beyond the present process.

5. Fixed Bed Hydrocracker

Idemitsu Kosan Company is building a 100 b/d pilot scale fixed bed catalytic hydrocracker that is scheduled for completion in 1981. The project is part of RAROP's search for a resid cracking process. Unit design is based on previous bench scale operations. The hydrocracker process flow is shown in Exhibit V-67.[136] Feed for the unit is vacuum residue and products include fuel gas, naphtha, middle distillates, and an unspecified quantity of residual oil. No meaningful yield data has been released on the process as of this writing.

EXHIBIT V-63

Asphaltene Cracking with Solvent Deasphalting: Feedstock, Product and Recycle Properties

	FEED	RECYCLE	PRODUCT
Sp. Gravity	1.0249	–	.9515
Asphaltene (wt %)	11.1	64.8	.07
Conradson Carlson (wt %)	20.1	65.6	8.2
Elemental Analysis			
C, wt %	84.04	87.59	85.52
H, wt %	10.27	6.98	11.36
S, wt %	4.90	3.81	2.54
N, wt %	.33	–	.3
V, wppm	128	137	1.2
Ni, wppm	43	131	1.8

EXHIBIT V-64

Asphaltene Cracking Process: Feedstock Properties and Product Yields

FEEDSTOCK/PRODUCT	GACH SARAN		ORINOCO		BOSCAN		TAR SAND	
Type	VACUUM RESIDUE		ATM RESIDUE		CRUDE		BITUMEN	
	FEED	PRODUCT	FEED	PRODUCT	FEED	PRODUCT	FEED	PRODUCT
Specific Gravity @60/60°F	1.031	0.989	1.022	0.975	0.998	0.950	1.006	0.957
Pour Point °F	>115	54	90	21	40	16	49	-9
Sulfur, wt %	3.83	2.61	3.92	2.28	5.18	2.53	4.4	1.8
Vanadium, wppm	282	46	478	77	1130	100	177	36
Nickel, wppm	91	28	207	30	106	27	77	16
Asphaltene, wt %	10.9	5.8	14.1	4.4	11.5	3.5	9.0	2.0
Conradson Carbon, wt %	23.0	16.5	17.9	11.2	15.9	9.3	13.5	7.3

PERFORMANCE	GACH SARAN	ORINOCO	BOSCAN	TAR SAND
Asphaltene Conversion, %	47	69	70	78
Desulfurization, %	32	42	52	60
Demetalization, %	81	82	00	80
Chemical H_2 Consumption, SCF/BBL	400-440	500-540	400-460	550-590

EXHIBIT V-65

Asphaltene Cracking
Relative Advantages and Disadvantages of Primary Product Disposition Options

ASPHALTENE CRACKING PRODUCT	PRODUCT DISPOSITION	RELATIVE ADVANTAGES	RELATIVE DISADVANTAGES
Without Recycle	• Pipelined to refinery	– Low cost	– Refinery equipment modifications may be necessary
	• Desulfurization	– Acceptable syncrude	– Hydrogen needed
	• Thermal cracking	– Marketable by product	– Handling & distribution system needed
		– Hydrotreated liquid products easily transported and refined	– Hydrogen may be needed
			– Costly
With Recycle (Deasphalted Oil)	• Hydrogen Processing	– High liquid yield quality syncrude	– Costly
		– No residue by product	– Hydrogen needed
	• Hydrogen processing w/FCC	– High liquid yield refined products	– High cost
		– No residue by product	– May duplicate downstream refinery function leaving it underutilized

EXHIBIT V-66

Asphaltene Cracking Process
Reactivity and Content of Asphaltene and Metals

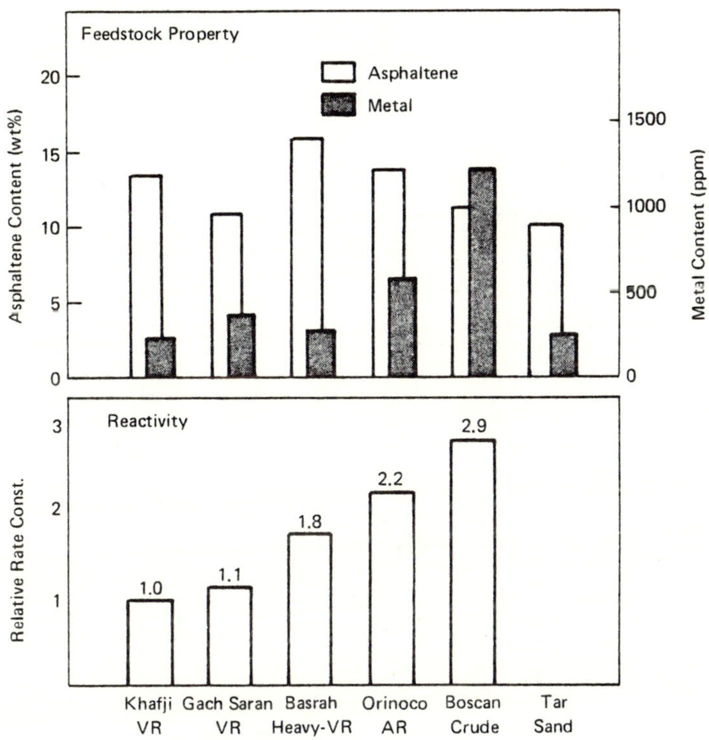

Reaction Condition
Temp. 405°C (760°F)
Press. 140 kg/cm^2

EXHIBIT V-67

Idemitsu Kosan's Residue Hydrocracker

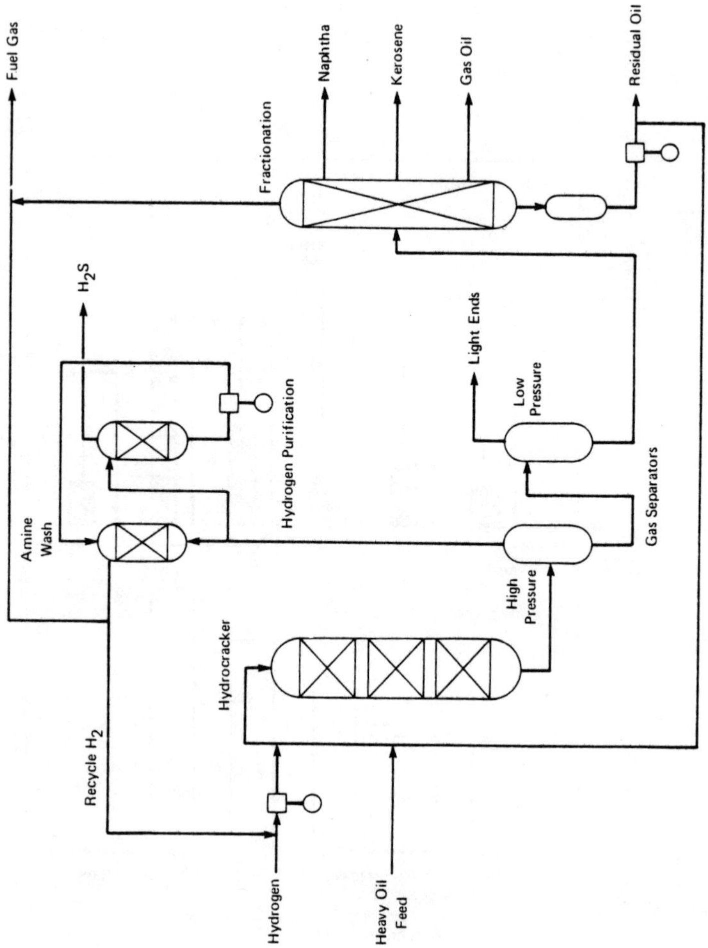

6. Kombie Kammer - Veba Oil

Veba Oil S.A. in West Germany has revived an old non catalytic hydrogen addition process that was designed for cracking coal liquids in the early 1940s. Veba officials claim that the process, called the Kombie Kammer Process, can be applied to residual oil, feeds and compares favorably with Petro Canada's CANMET Process. In addition to liquid products, the process produces a pitch residue by-product stream that would have applications similar to other hydrocracking processes. AOSTRA has just recently learned of the process and is considering it for further evaluation in their upgrading process evaluation project.[137]

REFERENCES

V. BITUMEN AND VERY HEAVY OIL UPGRADING

1. A. G. Bridge and G. D. Gould, "Chevron Hydroprocessor For Upgrading Petroleum Residua," presented at the JPI Symposium in Tokyo, October 2, 1980.

2. T. J. McCan, "Small Upgrading Plants," presented at the First UNITAR Conference on the Future of Heavy Crude and Tar Sands, June 1979.

3. A. Sloan, R. Ingham, W. L. Mann, "Pipeline Transportation of Heavy Oils, " presented at the First UNITAR Conference on the Future of Heavy Crude and Tar Sands, June 1979.

4. E. Perozo, "Orinoco Oil Belt, Exploration, Evaluation, and Exploitation," presented at the Latin American Oil Congress, Rio de Janeiro, July 1980.

5. D. Parra, "Tranport of Heavy Crudes," presented at the First UNITAR Conference on the Future of Heavy Crude and Tar Sands, June 1979.

6. D. Gurel, "How Husky Makes a Profit in Heavy Oil Production," World Oil, September 1979, p. 67.

7. The 1980 Refining Process Handbook, Hydrocarbon Processing, September 1980, p. 98.

8. "The Shell Gasification Process (SGP)," Patents and Licensing Division, Shell Oil Company.

9. E.T. Child, "Texaco: Heavy Oil Gasification," presented at University of Pittsburgh Symposium on Coal Gasification and Liquefaction: Best Prospects for Commercialization, Pittsburgh, PA, August 1974.

10. Op. cit., Reference 7.

11. "An Analysis of Potential for Upgrading Domestic Refining Capacity," prepared for AGA by Purvin & Gertz, Inc., 1980.

12. J.A. Bearhart, "More for Less From the Bottom of the Barrel," presented at the 1980 NPRA Annual Meeting, March 1980, p. 3.

13. Ford Bacon & Davis (private communication), September 1980.

14. Op. cit., Reference 7.

15. Op. cit., Reference 7.

16. A. Zanker, "Visbreak Resid? Figure Value," Hydrocarbon Processing, November 1978.

17. R. Hournac, J. Kuhn, M. Notarbartolo, "Visbreaking: More Feed for FCC," Hydrocarbon Processing, December 1979.

18. J.A. Matson, "Coke is Entering Turbulent Era Now," Oil & Gas Journal, March 24, 1980, p. 93.

19. Op. cit., Reference 9.

20. J.D. Odle, et al., "Clean Refinery Fuels by the Texaco Coal Gasification Process," 1978 Proceedings--Refining Department, 43rd Midyear Meeting API, 1978, Toronto, Ontario, Canada.

21. "Assessment of the Candidate Markets for Coal Liquid Boiler Fuels," prepared for DOE by Booz, Allen & Hamilton, December 1979.

24. Op. cit., Reference 7, p. 146.

23. D.P. Burke, "Making Cat Crackers Work on Varied Diet," Chemical Week, September 13, 1978.

24. Ibid.

25. Op. cit., Reference 7, p. 146.

26. G.H. Dale, C.L. Rogers, et al, "Commercial Experience with Passivation of Metals on FCC Catalysts," 1978 Proceedings--Refining Department, 43rd Midyear Meeting 1978, Toronto, Ontario, Canada, p. 432.

27. Oil & Gas Journal, p. 43 (March 29, 1977)

28. R.R. Edison, et al, "Metals Removal for FCC Catalyst Hydrocarbon Processing," May 1976.

29. J.R. Murphy, E.L. Whittington, C.P. Chang, "Review Ways to Upgrade Resids," Hydrocarbon Processing, September 1979.

30. W.L. Vermillion, Jr., U.S. Patent 4,025,458 assigned to UOP Inc., October 21, 1974.

31. "New Process Could Boost Gasoline Yield," Oil & Gas Journal, April 7, 1980.

32. Total Petroleum Inc. (private communication), October 1980.

33. Davidson Manufacturing (private communication), October 1980.

34. "The New Exxon RT-621 Catalyst," Exxon Research and Engineering Company publication, December 1977.

35. M.A. Christ, G.N. Shah, L.G. Sherman, "Residfining Process Upgrader Broad Range of Heavy Crudes," Exxon Researcn and Engineering Company, May 1979.

36. G.N. Shah, L.G. Sherman, M.A. Christ, "Residfining for Better Fuel Oil," Hydrocarbon Processing, May 1979.

37. Op. cit., Reference 7.

38. W.C. Van Zijll Langhout, C. Ouwerkerk, K.M.A. Prank, "Development of and Experience with the Shell Residue Hydroprocesses," a paper presented to the American Institute of Chemical Engineers, Meeting of June 8-12, 1980.

39. Report entitled, "In the Matter of an Application of GCOS, Ltd., Under Part 8 of the O&G Conservation Act," Canadian ERCB, March 1979, p. 11.

40. H.C. McIntyre, "Giant Oil Sands Plant Comes On Stream," Chemical Engineering, September 11, 1978.

41. R.A. Potkins, "Status of the Syncrude Project," presented at the First Unitar Conference on the Future of Heavy Crude and Tar Sands, June 1979.

42. Ibid.

43. Ibid.

44. The Energy Daily, September 25, 1980, p. 5.

45. "Information Statistics," Alsands Project Group publication, Calgary, Alberta.

46. "Alsands Fort McMurray Project," a report to the Lieutenant Governor in Council, Energy Resources Conservation Board, Alberta, Canada, December 1979.

47. Ibid.

48. Ibid.

49. Dynawest Projects Ltd. (a private communication), September 1980.

50. The Energy Daily, September 25, 1980, p. 2.

51. Op. cit., Reference 46, p. 41.

52. "The Cold Lake Project," a report to the Energy Resources Conservation Board by Imperial Oil Limited, May 1978.

53. Ibid.

54. "In the Matter of an Application of Esso Resources Canada, Ltd.," a report to the Lt. Governor in Council, Canadian ERCB, October 1979, p. 34.

55. Ibid.

56. Ibid, pp. 6-14.

57. Op. cit., Reference 52.

58. L.R. Aalund, "Heavy Oil Revives Venezuelan Refining," Oil & Gas Journal, September 15, 1980, p. 180.

59. Ibid.

60. Ibid.

61. Ibid.

62. Ibid.

63. Ibid.

64. "Catalogue of Aostra Technology," Issue:1, AOSTRA, December 1978.

65. "Fourth Annual Report," Alberta Oil Sands Technology and Research Authority, March 31, 1979.

66. Energy Analects, July 18, 1980.

67. AOSTRA (private communication), October 1980.

68. "Outlook of RAROP," (a brochure describing RAROP's outlook for research and development progress in Fiscal 1980), Maruzen Oil of U.S.A., Inc.

69. Ibid.

70. Ibid.

71. R.I. Berry, "A Heavy, Sour Taste for Crude-Oil Refiners," Chemical Engineering, May 19, 1980.

72. Op. cit., Reference 68.

73. H. Tominaga, S. Itoh, M. Yashiro, "Hydrovisbreaking of Athabasca Bitumen and Its Reaction Mechanism," Bulletin of the Japan Petroleum Institute, May 1977.

74. M. Nakamura. O. Togasi, T. Oho, "Hydrodenitrogenation of Residual Oils," a paper presented at the 45th Midyear Meeting, API, May 14, 1980, Houston, TX.

75. "France Launches A New Energy Plan," Chemical Week, August 6, 1980, p. 56.

76. Ibid.

77. Op. cit., Reference 71.

78. Elf France (private communication), September 1980.

79. "BASF Catalysts," Technical Leaflet, March 1980.

80. BASF Aktieng Essellschaft (private communication), October 1980.

81. BASF Catalysts, marketing brochure.

82. Lurgi Corporation (private communication), September 1980.

83. Op. cit., Reference 58, p. 182.

84. Op. cit., Reference 71.

85. O.F. Glagoleva, L.N. Sosulina, E.V. Smidovich, "Effect of Asphaltenes in Hydraulic Tars on Coker Product Yields and Coke Structure," Chemistry & Technology of Fuels & Oils, June 1975, pp. 424-427.

86. The Lurgi-Ruhrgas (L-R) Process, Tar Sand Processing, a Lurgi Corporation publication, January 1980.

87. AOSTRA, Fourth Annual Report, March 31, 1979, p. 19.

88. Ibid.

89. Great National Corporation of Dallas (private communication), October 1980.

90. AOSTRA (private communication), October 1980.

91. J.E. Filby, P.C. Flynn, J.M. Goldman, "Direct Coking of Athabasca Tar Sand," The Canadian Journal of Chemical Engineering, February 1980, p. 78.

92. R.W. Rammler, "The Production of Synthetic Crude Oil From Oil Sand by Application of the Lurgi-Ruhrgas Process," The Canadian Journal of Chemical Engineering, October 1970, p. 559.

93. Op. cit., Reference 91.

94. Op. cit., Reference 90.

95. Op. cit., Reference 91.

96. Op. cit., Reference 92.

97. G.C. Grub, M.F. Nathan, U.S. Patent 3,372,734 (1968).

98. Op. cit., Reference 92.

99. Ibid.

100. S. Gomi, R. Takahashi, Y. Kiritani, Y. Amagi, Y. Nishimura, "Thermal Cracking of Residual Oils Using Superheated Steam and Application of the Products," Kureha Chemical Industry Company, Ltd., 1975.

101. Ibid.

102. Energy Analects, July 18, 1980, p. 4.

103. F. Yamaguchi, A. Sokai, M. Yoshitake, H. Saegusa, "COSMOS Cracks Crude to Olefins," Hydrocarbon Processing, September 1979.

104. Ibid.

105. Op. cit., Reference 90.

106. R&D Trends in Heavy Oil Processing, Research Association for Residual Oil Processing - Report July 1980.

107 Shell Gasification Process (SGP), Patents & Licensing Division, Shell Development Company.

108. Ibid.

109. Ibid.

110. M. Ueda, M. Funaki, "M-Gas Process for Production of Fuel Gas From Heavy Oil," presented at First UNITAR International Conference on the Future of Heavy Crude Tar Sands.

111. Ibid.

112. Dynacracking Process Literature, published by HRI.

113. Ibid.

114. Op. cit., Reference 71.

115. Op. cit., Reference 106.

116. V.A. Gembichi, J.R. Bennett, J.G. Sikonia, "Processing Requirements and Economic Analysis of Heavy Oil and Syncrude Refining," presented at the 1980 UOP Process Division Technology Conference, Des Plains, Illinois (1980).

117. UOP (private communication), October 1980.

118. F.H. Adams, J.G. Gatsis, J.G. Sikonia, "The Aurabon Process: New Way to Upgrade Tars and Heavy Crudes," presented at the UNITAR Conference on the Future of Heavy Crude and Tar Sands, Edmonton, Alberta, Canada, June 1979.

119. Ibid.

120. S.G. Dunn, "Alternative Approaches For Bitumen Upgrading," Dynawest Projects Ltd., Calgary, Alberta, May 1980.

121. Ibid.

122. T. Sakate, T. Yagi, "Crack Resid with Spent HDS Catalyst," Hydrocarbon Processing, December 1979.

123. Ibid.

124. Ibid.

125. Ibid.

126. Ibid.

127. Menzies, M.A., T.F. Scott, J. Denis, "The CANMET Residuum Hydrocracking Process," Copyright - American Institute of Aeronautics and Astronautics, Inc., 1980.

128. Ibid.

129. Ibid.

130. Ibid.

131. C. Takeuchi, M. Nakamura, Y. Shiroto, "Upgrading of Residual Oils By Asphaltene Conversion," presented at the meeting of American Chemical Society, Honolulu, April 1979.

132. Ibid.

133. Ibid.

134. Ibid.

135. Ibid.

136. Op. cit., Reference 106.

137. AOSTRA (private communication), October 1980.

Appendix I.
Upgrading Process Descriptions
U.S. Technology

SOLVENT DEASPHALTING

Solvent deasphalting is a liquid-liquid extraction carbon removal process that was originally developed to extract heavy paraffinic lubricating oils from residual oil. The first commercial propane deasphalting unit went on steam in 1934. The high paraffinicity, reduced metals content, and low carbon residue of the deasphalted oil (DAO) have increasingly made this product a preferred feedstock for cat cracking and hydrocracking processes. A third application of solvent deasphalting has been the production of high quality specification asphalts.

Process Description

A typical solvent deasphalting unit is shown in Exhibit I-1. In the process, vacuum residue is mixed with a light paraffinic solvent such as propane, butane, or pentane and pumped to the extraction tower. A separation based on difference in solubility takes place producing a deasphalted oil solution and an asphalt solution. The solvent is then stripped from the deasphalted oil and asphalt solutions, condensed and recycled. The product is a heavy oil which is asphaltene-free and could be very low in metals depending on the solvent used.[1]

For lube oil refining, the solvent has typically been propane. Heavier solvents, such as pentane, will extract higher percentages of DAO but with respectively higher sulfur, metals and nitrogen content. The yield of DAO will depend on the quality of the feed and the required quality of the product.

The percentage contamination of DAO by various impurities is shown at different yields in Exhibit I-2.[2] The concentration of each contaminant in the DAO is plotted as a percentage of the total contaminant found in the vacuum resid feed. Product quality drops with increasing DAO yield. Downstream processing requirements and constraints will set the yield for the solvent deasphalting unit.

Solvent type, solvent to feed ratio, centrifuging time, pressure, and extractor temperature all affect the performance of the solvent deasphalting process, with temperature having the most pronounced effect. DAO yield is inversely proportional to the temperature. Typical operating temperatures range from 160°-330°F, and pressures from 400 to 600 psig.

EXHIBIT I-1
A Typical Solvent Deasphalting Process Flow Diagram

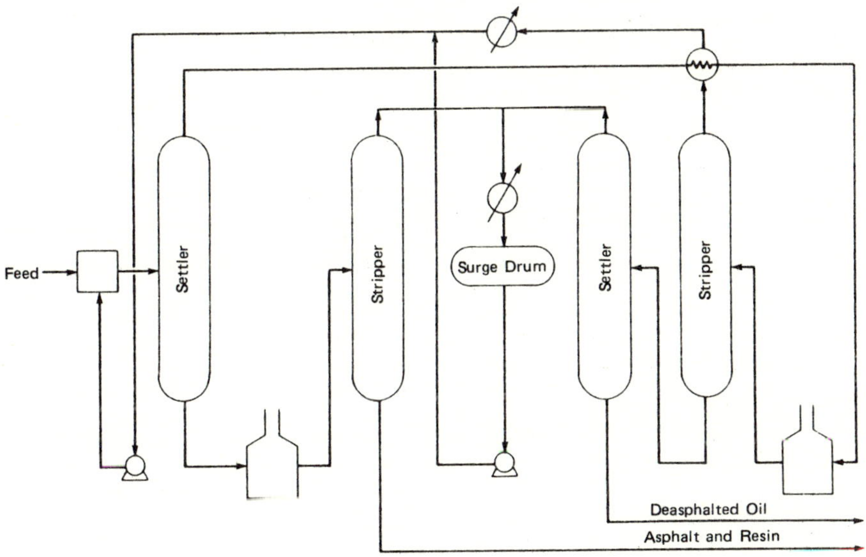

EXHIBIT I-2
Solvent Deasphalting Correlations

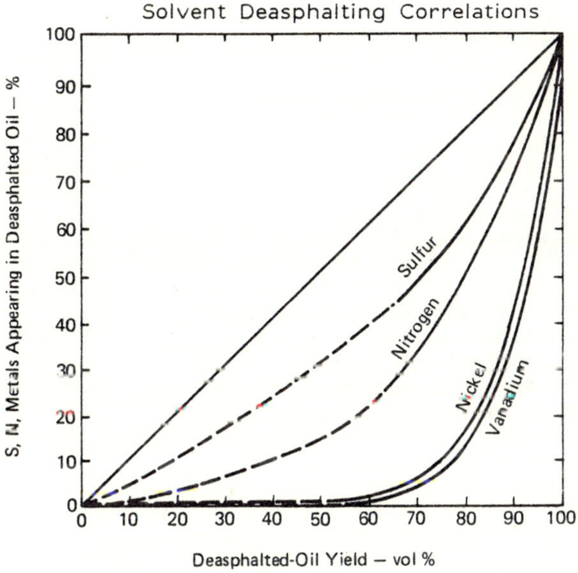

Exhibit I-3 lists the properties of deasphalted oils extracted from Boscan crude by the use of various solvents. Desulfurization (DS), denitrogenation (DN), and demetallization (DV, DNI) are all proportional to the percentage of asphaltenes removed, a function of the solvent type. For crudes with contaminants distributed more uniformly among their fractions more total sulfur could be removed in downstream hydrogenation processes.

DAO yield is not greatly affected by changes in the solvent to feed ratio nor by centrifuging time.[4] Vanadium is also relatively independent of solvent to feed ratio, but decreases linearly with increases in centrifuging time.

Several contractors offer variations of the solvent deasphalting process, including deep deasphalting which refers to the production of either DAO with viscosities greater than about 225 SSU at 210°F or asphalts with softening points above 175°F.[4] The first commercial propane deasphalting unit went on stream in 1934. Exhibit I-4 lists the various solvent diasphalting processes commercially available.

As shown in the exhibit, the utility requirements for solvent deasphalting are very high. A great deal of energy is required to recover the large quantity of recycled solvent. The process energy input is similar to that required for hydrocracking or severe hydrorefining.

The likely application of solvent deasphalting to heavy oil upgrading would be to extract from vacuum bottoms a heavy oil fraction which could be hydrotreated by more conventional processing for sulfur and metals removal. This application would be limited if no direct disposition exists for the asphalt which constitutes a substantial portion (20-30 percent) of the yield. Without an asphalt disposition, no gravity upgrade would be achieved.

Where asphalt markets exist, solvent deasphalting would be a strong contender for upgrading. Without a need for asphalt, a partial oxidation unit could be considered for production of a synthesis gas from the asphaltene fraction. For this scheme to be economical, syngas make would have to be balanced with production and upgrading fuel needs. If asphaltene content is high and system fuel requirements low, then much of the asphalt produced would be of little value as fuel and could present a liability. If stockpiling or disposing of the asphalt is acceptable and alternate processing steps cannot be justified, then solvent deasphalting could be a viable method of heavy oil upgrading. Asphaltenes or their coke could also be burned in a furnance equipped with stack gas scrubbers.

EXHIBIT I-3

Properties of Deasphalted Oils

SOLVENT	YIELD		COMPOSITION				CONTAMINANTS RETAINED BY OIL, wt. %				Gravity, °API	Con carb. wt. %	Pour pt., °F	C/H, wt./wt.
			Wppm		wt. %									
	wt. %	vol. %	V	Ni	S	N	V	Ni	S	N				
LPG	51.8	57.1	47.1	7.9	4.61	0.235	2.0	3.5	43	21	24.74	2.08	26	6.98
PENTANE	72.8	74.8	177.6	56.2	5.41	0.263	10.6	35	71	33	14.26	6.84	16	7.42
PETR. ETHER	74.5	76.7	393	55.7	5.40	0.283	24	35.5	72.5	36.3	14.26	7.51	30	7.44
HEXANE	78.0	79.7	469	68.6	5.52	0.394	30	45.7	77.6	53	13.45	9.56	30	7.61
HEPTANE	79.5	80.6	350	–	–	0.460	22.8	–	–	63	12.30	8.80	30	7.55
BOSCAN CRUDE	–	–	1,220	117	5.55	0.58	–	–	–	–	10.3	15.05	10	7.90

Source: Hydrocarbon Processing

EXHIBIT I-4

Various Solvent Deasphalting Processes
(Per Barrel of Feed)

	SOLVENT DEASPHALTING		DEEP DEASPHALTING	
PROCESS NAME	SOLVENT DECARBONIZING	SOLVENT DEASPHALTING	DEMEX PROCESS	ROSE PROCESS
Contractor	Kellogg	Foster Wheeler	UOP	Kerr McGee
Commercial Installations	9	41	3	2
Combined Feed Rate (Mb/d)	160	195	75*	27
Utilities				
Electricity (kWh)	2.2	2.1	1.44	2.7
Steam (lb)/(pressure)	30 (150 psig)	115 (150 psig)	115.2	24
Fuel (1,000 Btu)	190	86	48	143
Cooling Water (25°F Rise) (gal)	300	810	–	670

As was seen in Exhibit I-4, Kerr-McGee's ROSE (Residuum Oil Supercritical Extraction) process claims significant energy savings over the other solvent deasphalting processes. The bulk of these savings result from the substitution of a supercritical separation and heat exchange recovery system for the conventional solvent distillation step. Other savings are realized from reduced pump horsepower and cooling water requirements.

Exhibit I-5 lists product yield characteristics of the ROSE process for a light and medium Arabian and for Alaskan North Slope vacuum residues.

EXHIBIT I-5

Rose Solvent Deasphalting of Selected Vacuum Resids

	LT. ARABIAN	MEDIUM ARABIAN	ALASKAN NORTH SLOPE
Feedstock – Vacuum Residue			
Gravity, °API	8.5	5.4	5.0
Sulfur, wt %	4.45	4.94	2.2
V, ppm	66	87	178
Niwppm	25	35	77
Con Carbon Res., wt%	15	24	22
Products			
DAO			
Yield, wt % on feed	85	80	86.8
Gravity, °API	11	9.0	9.3
S, wt %	4.0	4.4	2.0
Vwppm	18	18	83
Niwppm	8	9	36
Con Carbon Res., wt %	8.8	10.2	16.1
Asphaltenes			
Yield, wt % on feed	15	20	13.2
Gravity, °API	11.0	-8.5	-8.5
S, wt %	4.0	7.2	3.5
Vwppm	18	364	803
Niwppm	8	111	347
Con Carbon Res., wt %	8.8	79.2	60.9

REFERENCES

SOLVENT DEASPHALTING

1. Refining Handbook Issue, Hydrocarbon Processing September 1980.

2. J.F. Ditman, "Solvent--Deasphalting--Versatility Preparation for Lube-Hydrotreat Feed," Oil & Gas Journal, January 14, 1974 p. 45.

3. D.A. Viloria, J.H. Krasuk, O. Rodriguez, H. Buenafama, J. Lubkowitz, "Solvent Deasphalt Boscan Crude," Hydrocarbon Processing, March 1977 p. 109.

4. Op. cit., Reference 1.

5. J.A. Gearhart, "More For Less From the Bottom of the Barrel," presented at the 1980 NPRA Annual Meeting, New Orleans, LA, March 1980.

FLUID COKING

Fluid coking is a noncatalytic thermal cracking process for converting low-grade residual oils into light liquids and coke. Fluid coking differs from delayed coking primarily in that coke is formed continuously by the cracking of a thin film of oil feed which is deposited on hot coke particles. Fluid coking is the proprietary technology of Exxon Research and Engineering Company. The process was first introduced in 1954, and a total capacity of approximately 160 Mb/sd has since been installed. The process technology is considered well demonstrated for vacuum resid and bitumen. The first fluid coker in Canada's Syncrude project started up in 1978. Another Canadian tar sands project, Alsands, which employs fluid coking, is awaiting final approval. Venezuela is considering fluid coking for its first Orinoco crude upgrading project.[1]

Process Description

A simplified process flow sheet of the Fluid Coking Process is shown in Exhibit I-6.[2] Heavy oil or vacuum residue is sprayed into the reactor vessel where it uniformly deposits on small coke particles suspended in a fluidized bed. As the feed distibutes over the surface of the coke particles, it cracks and the lighter ends vaporize, leaving layers of coke over the surface of the seed particles already present. The vapors exit the bed through a cyclone in the scrubber section of the reactor where fine coke particles are removed and recycled to the reactor as a slurry with the reactor feed. Reactor products leave the scrubber section and are sent to the fractionator for separation to liquid products. Steam introduced at the bottom of the reactor is the reactor fluidizing medium which strips the hydrocarbon vapors from the coke particles, cools the coke and is withdrawn at the bottom of the reactor, flows through a riser pipe to the burner vessel where a portion is burned with air in a fluidized bed to provide process heat.

Product coke is withdrawn from the burner while a constant inventory of solid coke particles is maintained in the system. Heated coke is recycled from the burner to the reactor, to repeat the process of feed deposition and coking.

Because of the nature of the mechanical system, more coke is produced from thermal cracking than from polymerization. The net result is a slightly higher liquid product yield and lower coke yield when compared to delayed coking. However, as with delayed coking, the primary deterent to application in a heavy oil upgrading facility is the disposition of the coke produced. A yield comparison for delayed coking versus fluid coking is presented for Alberta bitumen in Exhibit I-7.[3] Investment and utility requirements for fluid coking, as reported by Exxon, are listed in Exhibit I-8.[4]

EXHIBIT I-6

FLUID COKING PROCESS SIMPLIFIED FLOW PLAN

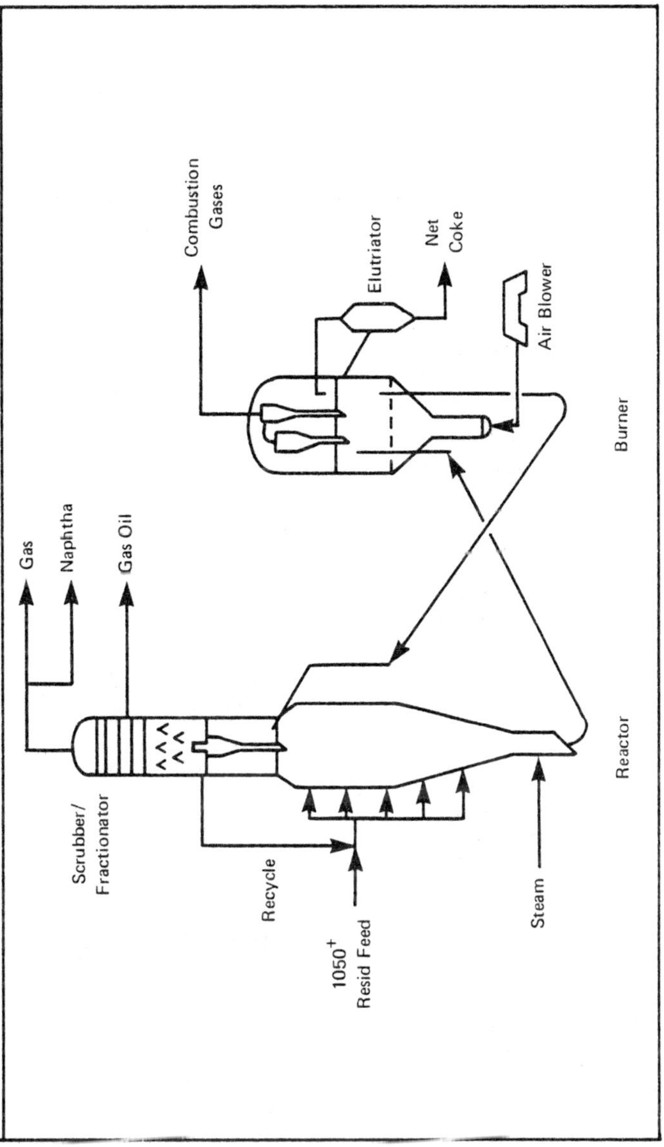

EXHIBIT I-7

Coker Yields From Bitumen
Delayed Coking Versus Fluid Coking

Feedstock				
Type	Alberta Bitumen			
Gravity, °API				
Viscosity, cs @100°F				
Distillation, °F				
IBP	350			
5%	500			
10%	600			
30%	800			
50%	1,020			
H/C Ratio, wt Atomic	1.5			
Sulfur, wt %	.125			
Nitrogen, wt %	5.0			
Metals, wppm	4.0			
V	250			
Ni	90			
Fe	100			
Asphaltenes, wt %	20			
Resins, wt %	35			
Oils, wt %	45			
Solids/Ash Content, wt %	0.75			
Carbon R esidue, wt %	14			
	Delayed Coking		Fluid Coking	
Products	Wt %	Lv %	Wt %	Lv %
C_4 and Lighter	7.9		9.8	
C_5 - 400°F	12.7	16.2	14.4	19.1
400°F - EP	57.2	60.5	60.0	63.0
Gross Coke	22.2		15.8	
Net Coke	22.2		9.4	
Total C_5 +				
Coker Liquid	69.9	76.7	74.4	82.1
Gravity, °API	24		25	
Sulphur, wt %	3-4		3-4	

EXHIBIT I-8

Fluid Coking Investment and Utility Requirements

Investment[1]	
Onsite Direct Material and Labor, $/b/sd Capacity	700-800
Utility Requirements[1]	
Electricity, kWh/bbl feed	5
Steam (125 psig), lb/bbl feed	10
Cooling Water, gal/bbl feed	60
Water, Boiler Feed Makeup, gal/bbl feed	18
Steam Generated (600 psig), lb/bbl feed	60

Notes:

[1]4th Quarter, 1977—adjusted to 4th Quarter, 1979. Based on unit capacities ranging from 15-50 Mb/sd and on processing feedstocks with Conradson carbon range of 13-23 wt %.

REFERENCES

FLUID COKING

1. The Energy Daily, September 25, 1980, p. 5.

2. Exxon Research and Engineering Company.

3. S.G. Dunn, "Alternative Approaches For Bitumen Upgrading," Dynawests Projects Ltd., May 1980.

4. Ibid.

FLEXICOKING

Flexicoking, as shown in Exhibit I-9, is Exxon's fluid coking process with a third fluidization vessel added to gasify the product coke.[1] The process produces gas oils and lighter products, and a clean, low-Btu (90-130 Btu/SCF) coke gas. Approximately 2 to 5 percent of the feed coke is discharged from the gasifier to control metals and fines build-up.

Exxon first demonstrated the commercial viability of its Flexicoker though the operation of its 750 b/d demonstration unit in Baytown, Texas. The first commercial Flexicoking unit was started up in 1976 in Japan at Toa Oil Company's Kawasaki Refinery. Rated capacity of this unit is 21,300 b/sd. A 52,000 b/sd Flexicoker is being constructed for Venezuela's Amuay Refinery. Other units totaling 150,000 b/sd are being designed, including one in Canada for the Cold Lake Project, and one in Shell's Martinez Refinery in California.[2]

Process Description

Vacuum residue at 500 to 700°F is sprayed into the coker reactor where it is thermally cracked (@ 950 to 1000°F) to lighter products and coke which is deposited on the fluidized coke particles. The cracked products are quenched in the scrubber vessel. The heavier fractions are condensed in the scrubber and can at this point be recycled back to the reactor. The lighter fractions proceed overhead to a conventional fractionator for separation into products. The sensible heat, heat of vaporization, and endothermic heat of cracking is provided by a circulating stream of hot coke from the heater.

Reactor coke is circulated to the heater where it is heated by coke and gas from the gasifier and partially devolatilized, producing a small amount of light hydrocarbon gas and residual coke. A circulating coke stream leaves the heater and enters the gasifier where it is gasified at elevated temperatures (1500-1800°F) and low pressure (25 psig) with air and steam to form a mixture of H_2, CO, N_2, CO_2, H_2O, and H_2S with a small quantity of COS. The mixture is referred to as coke gas. The primary gasification reactions are listed in Exhibit I-10. The first two reactions, the oxidation of coke and its intermediate product CO, occur rapidly in a small region surrounding the gas (air and steam) distributor. The third and fourth reactions are rate limiting and take place in the bulk of the coke bed. The final reaction is the fast water gas shift. The coke gas and entrained coke particles are returned to the heater where they provide a portion of the reactor heat requirement. The bulk of the heat requirement is supplied by a stream of coke sent from the gasifier to the heater.

EXHIBIT I-9 Simplified Flexicoking Process Flow Plan

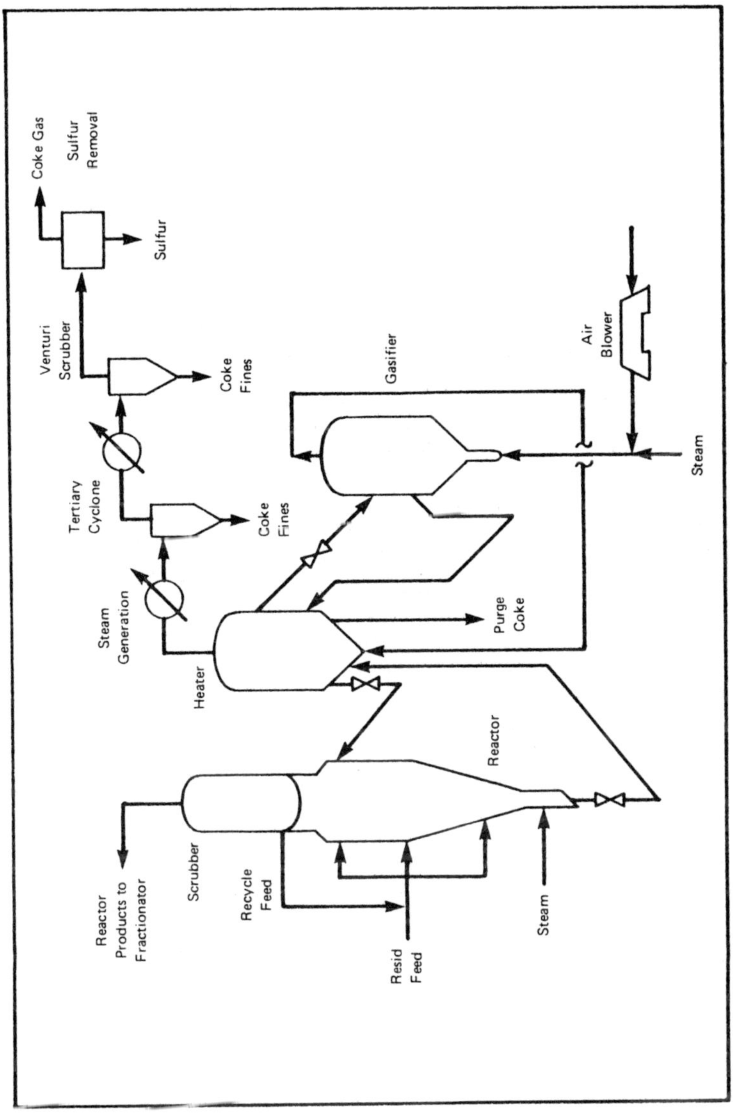

EXHIBIT I-10

Flexicoking Chemical Reactions of Gasification

FLEXICOKING CHEMICAL REACTIONS OF GASIFICATION		
LOCATION	REACTION	RATE
DILUTE PHASE	VERY LITTLE REACTIONS	
COKE BED	(5) $H_2O + CO \rightleftharpoons CO_2 + H_2$	VERY FAST
	(4) $C + H_2O \longrightarrow CO + H_2$	SLOW (RATE LIMITING)
	(3) $C + CO_2 \longrightarrow 2CO$	SLOW (RATE LIMITING)
GAS DISTRIBUTOR REGION	(2) $CO + \frac{1}{2} O_2 \longrightarrow CO_2$	VERY FAST
	(1) $C + O_2 \longrightarrow CO, CO_2$	VERY FAST

The hot coke gas leaving the heater is used to generate high pressure steam before passing through the tertiary cyclones for removal of entrained coke particles. The remaining coke fines are removed in a venturi scrubber after an additional stage of cooling. The solids free coke gas is then sent to a Stretford unit for removal of the H_2S. The low sulfur fuel gas produced can be used in process furnace or in steam/power generation after some equipment modifications are made to allow effective combustion.

Flexicoking yield data for Athabasca Bitumen is compared with Arabian Heavy Vacuum Resid in Exhibit I-11.[3] The yields for flexicoking are the same as shown for fluid coking except that all but 95-98 percent of the coke is converted to low Btu gas. The obvious advantage of flexicoking over fluid or delayed coking is the elimination of coke as a by-product for which a disposition must be found. An advantage over other processes is the availability of fuel for the production and upgrading project.

The primary disadvantages of flexicoking are:

- The high cost of sulfur removal facilities required to purify the coker gas.
- The incremental increases in investment and operating costs associated with burning coker gas as opposed to conventional fuels; Exhibit I-12 lists some of the modifications that coke gas combustion would require.
- The economic limitation on the distance the gas can be moved.
- The process' inherent limit imposed on potential liquid yield as the asphaltenes are no longer available for upgrading to transportation fuels.

Investment and utilities for flexicoking is presented in Exhibit I-13.

EXHIBIT I-11

Flexicoking Yields and Qualities

Feed	ARAB HEAVY VAC. RESID	ATHABASCA TAR SANDS BITUMEN[(1)]
Gravity, °API	3.0	8.8
Sulfur, Wt. %	6.0	4.9
Nitrogen, Wt. %	0.48	0.40
Conradson Carbon, Wt. %	27.7	12.7
Nickel, wppm	64	75
Vanadium, wppm	205	250
Yields on Feed		
H_2S, Wt. %	1.5	1.1
C_4, Wt. %	12.3	7.6
Naphtha, LV %	14.5	17.9
Distillate, LV %	18.0	23.4
Gas Oil, LV %	26.4	42.1
Coke Gas, KSCF/Bbl.	11.7	5.1
Net Coke, Wt. %	0.7	0.8
Product Qualities		
Naphtha		
Cut Range, °FVT	C_5/350	C_5/430
Gravity, °API	58.3	50.9
Sulfur[(2)], Wt. %	1.1	2.2
Nitrogen, wppm	70	60
Distillate		
Cut Range, °FVT	350/650	430/650
Gravity, °API	28.8	23.7
Sulfur[(2)], Wt. %	3.4	2.2
Nitrogen, wppm	950	800
Conradson Carbon, Wt. %	0.14	0.03
Gas Oil		
Cut Range, °FVT	650/950	650/950
Gravity, °API	10.3	12.5
Sulfur[(2)], Wt. %	5.8	4.5
Nitrogen, Wt. %	0.46	0.26
Conradson Carbon, Wt. %	4.8	1.8
Nickel, wppm	0.7	0.2
Vanadium, wppm	1.0	0.7
Coke Gas		
LHV, BTU/SCF	130	95
Sulfur[(3)], VPPM	<150	<150

Notes:

(1) For heavy feeds at remote sites, the full range feed can be fed to the FLEXICOKING unit, and the products recombined after hydrotreating to produce a high quality upgraded synthetic crude.

(2) Product sulfur contents can be reduced in downstream hydrotreating units. Sulfur content of <3 WPPM has been demonstrated for the naphtha and <500 WPPM for the distillate plus gas oil cut.

(3) Sulfur content after sulfur removal facilities.

EXHIBIT I-12 Equipment Changes Necessary for Burning Flexicoker Coke Gas

COMBUSTION CHARACTERISTIC	DIFFERENCE FROM CONVENTIONAL FUELS	REQUIRED EQUIPMENT MODIFICATIONS
• Heating Value	– 8 to 12% of conventional fuel heating values – 10 times volume of fuel fired for the same heat release	• Larger piping and values; i.e., header piping: – 8 times larger diameter than for fuel oil – 3 times larger diameter than for natural gas • Increased compressor horsepower
• Combustion Air Requirements	– 80% of conventional fuel on equal heat release basis	• Additional instrumentation and controls needed in dual-fired furnaces (coke gas/conventional gas)
• Flue Gas Rates	– 35 to 40% more flue gas than conventional for same heat release rate – 50-60% more flue gas for converted furnaces where heat absorbed is maintained the same	• More stack height or induced draft to compensate potential 140% increase pressure drop across existing convection sections and ducts
• Adiabatic Flame Temperature	– 700°-900°* decrease from conventional fuels (@20% excess air)	• Larger, more complex convection section as heat duty shifts from radiant to convection section and resultant loss in furnace efficiency • Air preheat

* Coke gas adiabatic flare temperatures between 2500°-2600°F are ample for stable self-sustaining complete combustion

EXHIBIT I-13

Flexicoking Investment and Utility Requirements

Investment	
Battery limit onsite, including product fractionator, but excluding coke gas sulfur removal	
(Basis: Direct material and labor, 4th quarter 1979, U.S. Gulf Coast), $ per bpsd capacity:	1,300-2,000
Typical Requirements, units/bbl feed	
Steam export, (600 psig), lb	200
Steam required, (125 psig), lb	100
Electricity, kWh	13
Water, cooling, gal	350
Water, boiler feed, gal	30
Instrument and utility air, scf	25

REFERENCES

FLEXICOKING

1. Exxon Research and Engineering Company, Flexicoking Information, July 1980.
2. The Energy Daily, September 25, 1980, p. 5.
3. "Refining Handbook," Hydrocarbon Processing, September 1980.

TEXACO PARTIAL OXIDATION

The Texaco partial oxidation process is a continuous noncatalytic process that reacts oxygen with heavy oils to produce a synthesis gas composed primarily of hydrogen and carbon monoxide. Over 35 commercial partial oxidation units are processing heavy oils.

Process Description

A block diagram of the overall partial oxidation process is shown in Exhibit I-14. A carefully controlled mixture of oxygen steam and residual oil is fed to the top of the synthesis gas generator. Free oxygen first reacts with the oil feed in the upper reaction zone to produce CO_2, H_2O, and heat. Hydrogen and CO are formed by a secondary reaction between the steam, combustion gas, and fuel.

Reactor effluent is charged to a shift converter with high pressure steam that is generated either in a waste heat boiler or from directquenching of the effluent gas. The water gas shift reaction takes place in this vessel producing CO_2 and H_2. The partially cooled gases leaving the shift converter are scrubbed to remove entrained particulate matter or soot resulting from incomplete combustion. The gases are then cooled to condense the water vapor. After treatment, the condensed water is returned to the boiler. Cooled gas leaving the soot removal system is sent to an acid gas removal system.

The gas cleanup system depends upon final hydrogen purity needs. For CO_2 removal, the gas can be scrubbed with cold methanol, hot potash, amine solution, caustic, or various other solvents. A liquid nitrogen can be used to remove CO and traces of other gases for a very high purity hydrogen. When nitrogen is not desired in the product, copper liquor solutions are effective for CO removal.

Generator temperatures range from 2000 to 2700°F and preheat temperatures are close to 1200°F depending on feed composition. Pressures range from atmospheric to over 1500 psig. Process oxygen requirements are from 250 to 275 cu ft per Mcf of synthesis gas produced. Feedstock properties and yield data are listed in Exhibit I-15. Hydrogen yield is near 45 percent and can be increased to 92 percent by shift conversion. With system pressures as high as 1500 psig, purified gas can be fed directly to a hydrocracking system without compression.

EXHIBIT I-14

Texaco Partial Oxidation Process Schematic

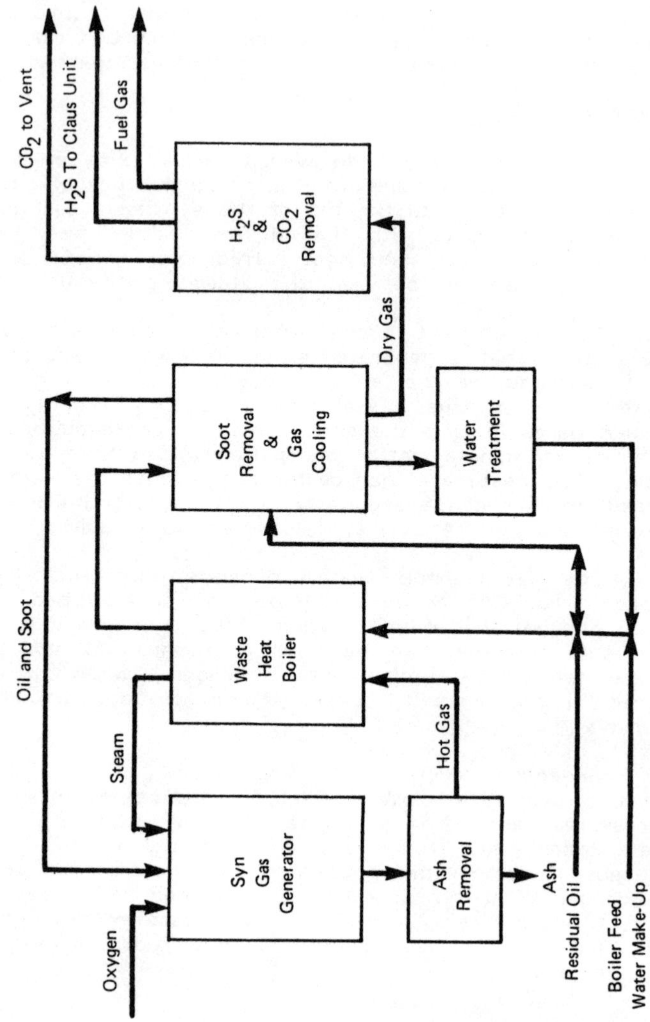

EXHIBIT I-15

Texaco Partial Oxidation Process
Feedstock Properties and Product Yields

FEEDSTOCK	KHAFJI
Type	Vacuum Resid
Gravity °API	8.2
Viscosity, SSF	
@122°F	132,000
Composition, Wt %	
Carbon	84.1
Hydrogen	10.2
Sulfur	5.0
Nitrogen	0.4
Oxygen	0.2
Ash	0.1
Metals, wppm	
Vanadium (V)	153
Nickel (Ni)	69
Iron (Fe)	24
PRODUCTS	
Composition, Vol %	
CO	16.28
H_2	42.34
CH_4	41.00
N2	.12
Ar	0.26
Gross Heating Value (HHV), Btu/scf	603

REFERENCES

TEXACO PARTIAL OXIDATION

1. Texaco process literature.

HEAVY OIL CRACKING

The Heavy Oil Cracking (HOC) process is a modified version of fluid catalytic cracking which can accept atmospheric residue directly as feed. In addition to converting heavy oils to lighter products, HOC destroys the asphaltene portion of the feed by combustion and generation of steam in the regenerator. The process is the joint development of Pullman-Kellogg and Phillips Petroleum Co. One commercial unit has been in operation since 1961; a 50,000 BPD unit which incorporates hydrodesulfurization as a feed pretreatment step will start up in Phillips' Sweeney, Texas, refinery this fall. The pretreatment involves a Unicracking/HDS atmospheric resid desulfurization unit to reduce the sulfur content of the 650+°F topped crude from 3.5 weight percent down to 0.3 weight percent. The process must incorporate this combination hydrodesulfurization step in order to process sour crudes.

Process Description

A simplified process flow diagram of Kellogg's Heavy Oil Cracking process is shown in Exhibit I-16.[1] Desulfurized atmospheric resid enters the unit at the bottom of the external vertical riser. Dispersion steam and regenerated catalyst also enter the riser at this point. Oil in the feed is vaporized by the dispersion steam. Carbon residue adheres to the catalyst as a film of coke. The steam, oil vapors, and carbon coated catalyst travel up the riser and into the converter. Vapors and steam disengage from the mixture at the top of the vessel and are sent to the fractionator. The catalyst with its carbon residue falls through the stripper and into the regenerator. The catalyst is regenerated by the combustion of the carbon in the presence of air. The carbon free catalyst flows back to the riser and the process is repeated.

It should be noted that residue feed to this unit must be low in sulfur content. Typically, a residue hydrodesulfurization unit precedes the HOC unit. The process is adversely affected by metals deposited on the catalyst. Propriety metal passivators are added to the feed to reduce the deleterious effect of metals on the catalyst. The HOC unit has all the feedstock limitations that a conventional FCC has with the exception of asphaltene content. Since the HOC unit can accommodate high asphaltene feeds, it is suited to low sulfur, low metals atmospheric resid. The major difference between the HOC unit and a typical FCC is the incorporation of steam coils in the regenerator. Heavy asphaltic crudes produce large amounts of carbon residue coke on the FCC catalyst. This coke is burned off the catalyst in the regenerator. The steam coils remove the excess heat of combustion, resulting in liquid products and steam.

Exhibit I-17 shows typical feedstock and product properties for processing Lloydminster atmospheric residues in an HOC unit.[2] Generally, feedstocks with carbon metals contents as high as Lloydminster are hydrotreated before they are fed to the Heavy Oil Converter.

EXHIBIT I-16

Heavy Oil Cracking Simplified Process Flow

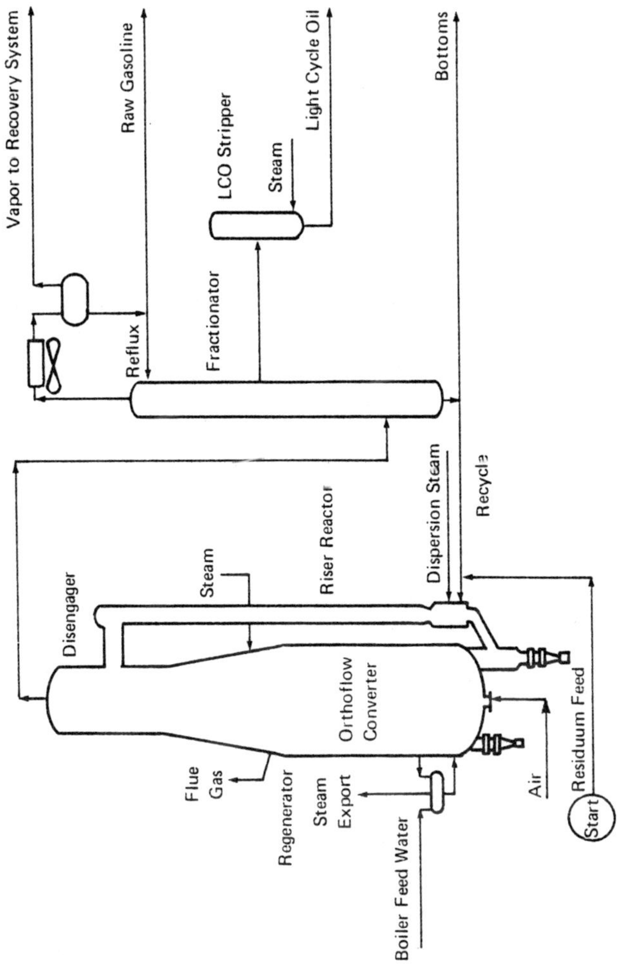

EXHIBIT I-17

Heavy Oil Cracking
Feedstock Properties and Product Yields

FEEDSTOCK	Lloydminster Atmospheric Residue
Wt % of Whole Crude	41.7
Gravity, °API	10.3
Specific Gravity @60°F	0.9982
Sulfur, Wt %	4.1
Nitrogen, Wt %	.34
Metals, wppm	
Vanadium	141
Nickel	65
Conradson Carbon Wt %	12.1

PRODUCTS	Yield Wt %	Gravity °API	Sulfur Wt %
H_2S	1.6		
C_2	2.9		
Propane-propylene	3.4		
Butane-butylene	7.0		
C_5 - 400°F, gasoline	31.1	55.5	0.27
Light Cycle Oil	24.4	12.7	2.7
Heavy Cycle Oil	10.7	3.2	5.4
Coke (gross %)	18.9	–	–

Feedstocks with vandium content ranging from 5 to 30 ppm, and carbon residue of 5 to 10 percent require use of metal passivators to minimize metal contamination penalties and the heat removal system to remove regenerator excess heat in the form of steam.[3] Utility requirements for this process vary considerably for different feedstocks.

The primary advantages of the Heavy Oil Cracking process are:

- The elimination of a vacuum unit.
- The elimination of by-product coke.

The disadvantages are:

- Desulfurization of the entire feed is necessary.
- Steam production must be integrated with other processing demands.

REFERENCES

HEAVY OIL CRACKING

1. "Heavy Oil Cracking For Greater Crude Conversion," Pullman Kellogg brochure.

2. J.R. Murphy, E.L. Whittington, C.P. Chang, "Review Ways to Upgrade Resids," Hydrocarbon Processing, September 1979.

3. Op. cit., Reference 1.

SHELL RESIDUE HYDRODESULFURIZATION, HYDRODEMETALLIZATION

The Hydrodesulfurization Process, licensed by Shell Internationale Research Maatschappij B.V. or Shell Development Company, was developed as a catalytic hydrogenation process for the removal of sulfur, nitrogen, and metal contaminants from middle distillates to vacuum gas oil. The process employs Shell's bunker reactor which contains a moving bed (as opposed to a fixed or expanded bed) trickle-flow reactor to allow the withdrawal of contaminated catalyst and the addition of fresh catalyst while the unit is on-stream. Process and catalyst development have extended application of the process to residual oils, including those that are high in metals and/or asphaltenes. A 3,000 b/d demonstration unit was started up in Shell's Gothenburg, Sweden, Refinery in 1973. The first commercial unit (45,000 BPD) came on-stream in 1976 in Seibu Sekinyu's Yamaguchi Refinery near Onada, Japan. These units have processed Kuwait long residue and Middle East residues with metal contents up to 70 ppm. In 1976 work was initiated with the close cooperation of Maraven S.A., Venezuela, to develop a hydrodemetallization process for very heavy feeds. A 400 t/d (approximately 2,500 b/d) demonstration plant in Maraven's Cardon Refinery is scheduled to come on-stream in 1981. The experimental plant will test the bunker reactor system with Tia Juana Pesado vacuum residue and asphalt. Its objective is to reduce the metals content from 800 ppm to 150 ppm.

Process Description

A simplified process flow diagram of a typical Shell Residue HDS/HDM process in shown in Exhibit I-18. Heavy oil feedstock, combined with hydrogen rich make-up and recycle gas is preheated in feed/effluent exchangers and brought up to operating temperature in the feed furnace. The combined reactor charge then trickles over the moving catalyst bed and through the reactor, exiting from the bottom.

Reactor effluent is cooled first by exchanging heat with the reactor feed, next in a condensor, and finally flashed in a high-pressure separator. The liquid product is further flashed in a low-pressure separator where final liquid product is routed to H_2S and dissolved gas removal. Low pressure off gas is sent to LPG recovery. The gas leaving the high-pressure separator is recycled as hydrogen after H_2S removal to the incoming feed.

The primary reactions occurring in the bunker reactor system are hydrodemetallization (HDM), hydrodesulfurization (HDS), and hydroconversion. Sulfur is transferred from the catalyst to the gas phase by the formation of H_2S in the HDS reaction. The HDM reaction forms metal sulfides which are deposited on the catalyst. This deposition of metal sulfides can result in pore mouth plugging when the autocatalytic HDM reaction is strongly diffusion-controlled and catalyst-core poisoning by a thin layer of metal sulfide when the

EXHIBIT I-18 Flow Scheme Demonstration Unit Shell Residue HDS Process

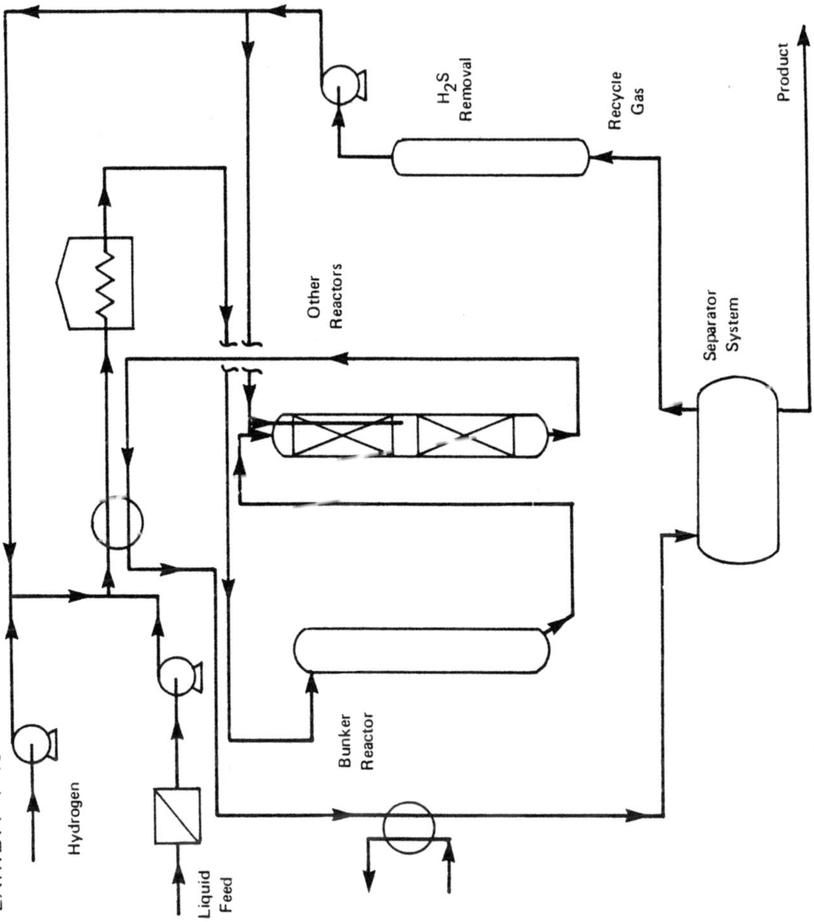

metal-containing species penetrates too deeply into the catalyst particles.[1] Feedstock properties and yield data for Venezuelan residues processed in the Shell HDS pilot plant are listed in Exhibit I-19.[2]

EXHIBIT I-19

Shell Hydrodesulfurization, Hydrodemetallization (HDS/HDM) Process:

Pilot Plant Feedstock and Product Yield Data for Venezuelan and Middle East Vacuum Residues

FEED	VENEZUELAN VAC. RESID	MIDDLE EAST VAC. RESID
Density, g/cc	1.031	1.023
Sulfur, wt %	5.9	5.1
Metals, wppm	1,500	130
C_7 – Asphaltenes, wt %	17	15
Viscosity, CS @ 212°F	4,310	2,000
Pour Point, °F	>100	–
Yield on Feed, wt %		
C_1 - C_4	2.0	2.0
C_5 +	97.4	98.0
Liquid Product (C_5+) Qualities		
Density, g/cc	0.992	–
Sulfur, wt %	3.6	1.0
Metals, wppm	320	5
C_7 – Asphaltenes, wt %	6	–
Viscosity CS @ 212°F	480	–
Pour Point, °F	17	–
Performance		
Metal Removal, wt % on feed	80	–
Chem H_2 Consumption, wt % on feed	0.9	–

REFERENCES

HYDROSULFURIZATION, HYDRODEMETALIZATION

1. W. C. Van Zijll Langhout, C. Ouwerkerk, K. M. A. Pronk, "Development of and Experience With The Shell Residue Hydroprocesses," a paper presented to the American Institute of Chemical Engineers (June 8-12, 1980).

2. Ibid.

H-OIL/LC-FINING

Both the H-Oil process and the LC-Fining process are high-pressure hydrocracking processes that use a three-phase fluidized or expanded (HRI uses the term "ebullated") bed to hydrodesulfurize or hydrocrack residual feedstock. The process and basic reactor design concepts appear to be very similar for both processes. H-Oil is licensed by Hydrocarbon Research, Inc. (HRI), while CE Lummus is the sole licensor for LC-Fining. As shown in Exhibit I-20, each process' stage of development and demonstration of commercialization is also very similar. Both processes claim an ability to maintain catalyst activity by the withdrawal and addition of catalyst while in operation --a distinct advantage over fixed bed processes. Because of the similarities of each process, the expanded bed concept is described in this study as a single concept without attention to minor differences between the two processes. Significant differences are noted, however.

Process Description

A sketch of the expanded (or ebullated) bed reactor is shown in Exhibit I-21. Feed oil and hydrogen enter at the bottom of the reactor and pass up through the bed of the catalyst, which is expanded somewhat in excess of its settled volume. The state of catalyst motion, induced by the oil and hydrogen, allows intimate contact of feed with catalyst and permits fines solids to migrate through the bed and leave the reactor system. The effluent leaves the reactor at the top and proceeds to further processing. Some effluent is recycled to help provide sufficient fluidizing volume to the bed.

Operating temperatures approximate isothermal conditions and the reaction heats are dissipated in heating the cold (100°F differential) feed. During operation, catalyst is added to and withdrawn from the reactor. As required through a system of pressurized lock pots, catalyst slurried with oil is pneumatically transferred through valves into the reactor. Catalyst loss by carryover is negligible. Further processing involves conventional separation, flashing, condensation, and fractionation. The vapor effluent is cooled, compressed, cleaned, and further separated. It is then sent back through heat exchangers and the hydrogen furnace to the feed to serve as recycled hydrogen. Some purging is necessary to control methane buildup.

The advantages, as seen by the licensors, of this processing system are summarized as follows:

- On-stream catalyst addition and withdrawal is possible which allows activity to be controlled at a desired level.

- Isothermal conditions are possible which eliminate the need for costly quench systems for temperature control.

EXHIBIT I-20

LC-Fining and H-Oil Hydrocracking Processes
Commercialization Status

LC FINING			H-OIL		
COMPANY, LOCATION	CAPACITY (b/d)	START-UP DATE	COMPANY, LOCATION	CAPACITY (b/d)	START-UP DATE
• Cities Services Refinery Lake Charles, LA		1963	• Cities Services Refinery Lake Charles, LA	6,000	1963
• Petroleos Mexicanos	18,500	1973	• KNPC, Shuaiba, Kuwait	56,000	1968
			• Humble Oil Company Bayway, NJ	16,000*	1970
			• Petroleos Mexicanos Salamanca, Mexico	30,000-40,000 18,500	1973
			• Catlettsburg, KY	(600 tpd Coal)	1980
•	(Molten Solvent Refined Coal)	In Design	• ICGG Peoria, IL	2,500	In Design

*Plant destroyed in explosion on December 5, 1970.

EXHIBIT I-21

Flow Diagram of a Typical H-Oil Process

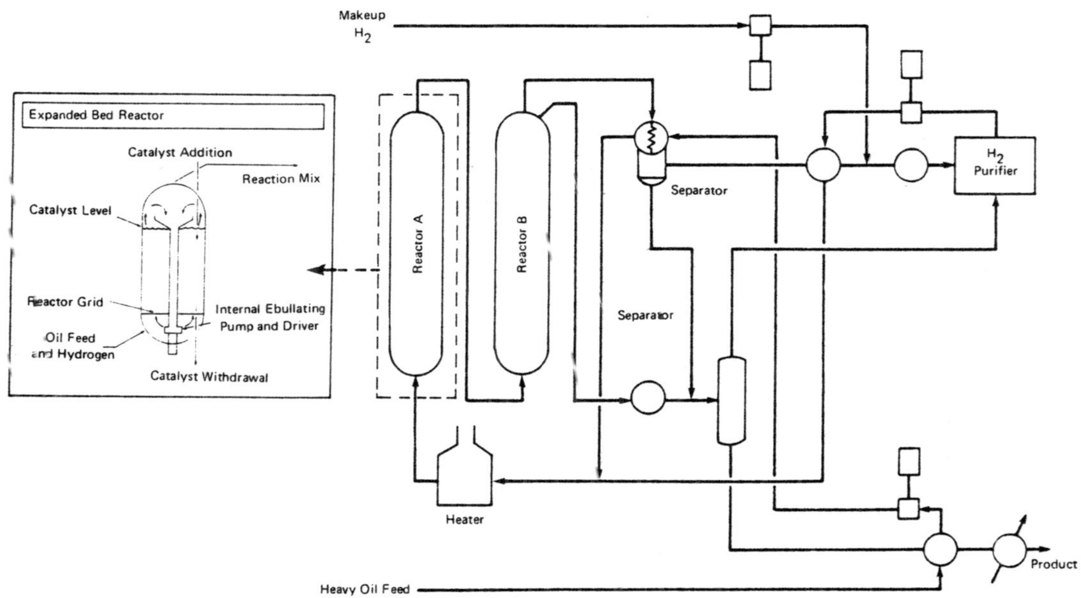

- Plugging from accumulation of inerts is avoided, thus allowing heavy feedstocks with up to 8 pounds/Mbbl entrained sediment or salts to be processed.
- A smaller diameter (1/32-inch or 0.8-mm) catalyst can be used to raise activity since it would result in only a small increase in pressure drop across the bed.

The expanded bed hydrocracking system can handle a wide variety of feedstocks ranging from gas oils to vacuum bottoms. Operation can be controlled at various severities to provide a variety of processing functions. Increasing severity can provide for:

- Residue demetallization
- Fuel oil defulfurization
- Gas oil hydrotreating for improved cat cracker performance
- Syncrude production from heavy oils and tar sands
- Resid cracking to naphtha and distillate oil products.

Temperature, space velocity, hydrogen partial pressure, and catalyst activity can each be controlled to yield desired products from specific feedstocks. As shown in Exhibit I-22, for constant pressures and catalyst activities, maximum desulfurization and conversion is obtained at high temperatures and reduced space velocities.

As with fixed bed and moving bed hydrogenation processes, additional reactors can be added to the expanded bed system for higher levels of conversion and desulfurization.

The degree of heavy oil conversion achievable for the expanded bed hydrocracker is limited by the stability of the synthetic crude product. At conversion levels in excess of approximately 80 percent of equivalent vacuum resid, the remaining asphaltenes will tend to precipitate out of solution. HRI has overcome this limitation by incorporating a vacuum bottoms recycle stream, an improvement that has evolved from their H-Coal development program.

Typical hydrocracking yields, as published by HRI and Lummus, are shown in Exhibit I-23.[1-3] The H-Oil 60 percent conversion case is for standard once-through operation, while the 90 percent conversion is for vacuum bottoms recycle. The LC-Fining 60 percent conversion case uses solids produced with the bitumen as a natural catalyst, while the 75 percent conversion case uses Lummus' synthetic catalyst that exhibits better desulfuriization characteristics than achieved with H-Oil, in spite of the lower conversion.

EXHIBIT I-22

The Effect of Space Velocity and Temperature on Hydrocracking Performance

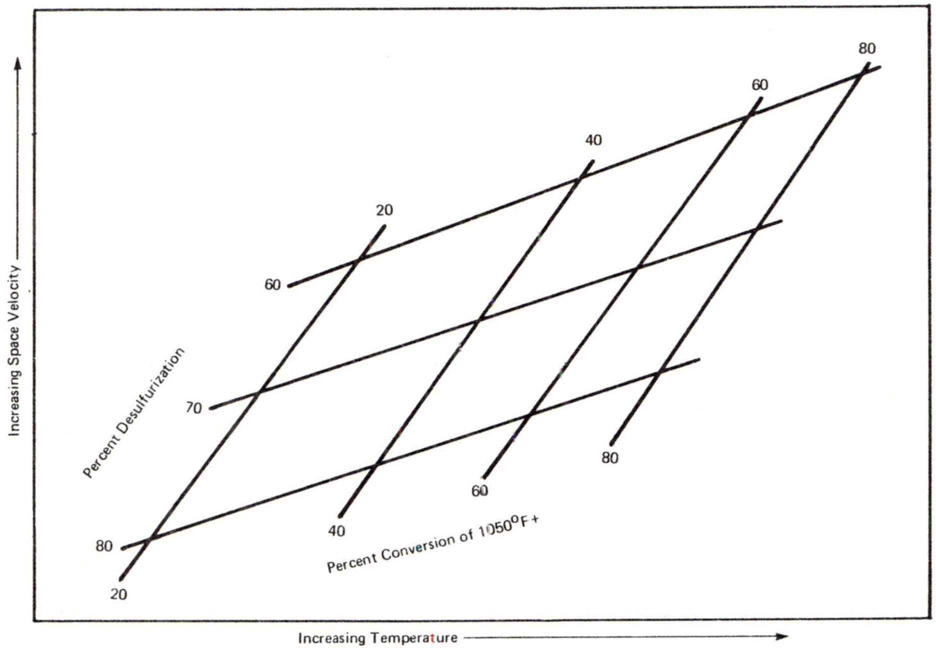

Conditions: Constant Pressure and Catalyst Activity

EXHIBIT I-23

Hydrocracking Yields from Bitumen

	H-OIL		LC-FINING	
	60% CONVERSION	90% CONVERSION	60% CONVERSION	75% CONVERSION
C_4 + Lighter, wt %	6.2	9.5	3.9	9.1
C_5 -400°F, wt %	12.3	16.7	9.8	11.9
400°F-975°F, wt %	60.1	69.9	65.7	67.7
975°F + Residue, wt %	22.2	5.2	21.0	12.9
Total C_5 -975°F				
Liquid, wt %	72.4	86.6	75.5	79.6
lv %	80.8	97.4	82.8	90.4
Gravity, °API	26	28	24	29
Sulphur, wt %	Not Reported	2	3	1
Total C_5 +				
Liquid, wt %	94.6	91.8	96.5	92.5
lv %	101.1	101.9	101.0	101.8
Gravity, °API	20	25.5	16.5	24
Sulphur, wt %	Not Reported	2.5	4	1
H_2 Consumption	540	850	300	1,100
SCF/Bbl Feed			(Natural Catalyst)	(Synthetic Catalyst)

Source: S.G. Dunn

The high degree of processing flexibility provided for by the H-Oil and LC Fining processes is achieved at the penalty of high investment and operating expense factors. Reactor pressures of 2,000 psig to 3,000 psig and temperatures of 800 to 900°F are required to achieve the high conversions. Hydrogen makeup rates of 850-1,600 scf/bbl are required. In most cases, a hydrogen manufacturing plant would be required to meet the hydrogen demand.

Additional hydrotreating of hydrocracker products would generally be required if 60 percent conversions were maintained. Hydrotreating is needed to further reduce sulfur and to improve the burning qualities of fuels. Hydrocracking naphtha and cat cracker feed that is high in nitrogen may result in losses in catalyst activity. Additional hydrotreating may be required for denitrogenation.

Economic data for the two processes is summarized in Exhibit I-24.

EXHIBIT I-24

LC-Fining, H-Oil Economic Data

	LC-FINING	H-OIL
INVESTMENT[1]		
$ Per b/d	$1,800-2,875	$2,500
UTILITIES[2] PER bbl FEED		
Fuel, MBTU		59
Power, kWh		11.9
Cooling Water, Gal (20°F Rise)		80
Total, Cents	35-45	55-75
Catalyst, Cents	45-90	42

BASIS:

1. U.S. Gulf Coast, 1980 Installed
2. Feed at 700°F, Make-up Hydrogen at 300 psig, 96% Purity

REFERENCES

H-OIL/LC-FINING

1. R. P. Van Driesen, E. Wysocki, R. A. Bell, "LC-Fining Applied to Bitumen From Oil Sands," presented at 27th Canadian Chemical Engineering Conference, Calgary, Alberta, October 1, 1977.

2. R. M. Eccles, "H-Oil Processing of Tar Sand Bitumen for Maximum Oil Yield and Cash Flow," UNITAR Conference Report No. 13, May 1979.

3. C. A. Johnson, G. Nongri, L. M. Lehnan, M. C. Chervench, "Conversion of Bitumen and Heavy Crude Oils to More Valuable Products by the H-Oil Process," 28th Annual Meeting of the Petroleum Society of CIM, Edmonton, Alberta, 1977.

Appendix II.
Abbreviations and Conversion Factors

The following abbreviations and conversion factors are used in this report.

°API	oil gravity, degrees API = $\frac{141.5}{\text{oil sp gr}} - 131.5$
AR	atmospheric resid
bbl	U.S. barrel = 42 US gallons
b/d	barrels per day
b/cd	barrels per calendar day
b/sd	barrels per stream day
Btu	British thermal unit (1 Btu = 0.252 kcal)
°C	temperature, degrees Celsius
Cp	Viscosity, centipoise
cs	Kinematic viscosity, centistokes
d	day
DAO	deasphalted oil
°F	temperature, degrees Fahrenheit
ft	foot, (1 ft = 0.304m)
gal	U.S. gallon (1 gal = 3.785 liter)
gpm	U.S. gallons per minute
GOR	gas/oil ratio, scf/b
h	hour
HHV	higher heating value
km	kilometer
KPA	Kilopascals
KWH	kilowatt hour
lb	pound
Lt	long ton = 2240 lb
m	meter
m^3	cubic meter
M	thousand, 10^3
MM	million, 10^6
md	permeability, millidarcy
mPa	millipascals

OIP	oil-in-place
OOIP	original oil-in-place
ppm	parts per million
psi	pounds per square inch
psig	pounds per square inch gage
scf	standard cubic feet, 1 atm, 60°F
SFS	kinematic viscosity, Saybolt Furol
SSU	Kinematic viscosity, Saybolt Universal
sp gr	specific gravity
t or st	short ton, 2000 lb
vol %	volume percent
VR	vacuum resid
wt %	weight percent
wppm	weight parts per million
y	year
Multiples:	
M =	thousand = 10^3
MM=	million = 10^6
	billion = 10^9
	trillion = 10^{12}
	quadrillion = 10^{15}

Appendix III.
Bibliography

Aalund, L.R., Heavy Oil Revives Venezuelan Refining, August 25, 1980, Oil and Gas Journal.

Aalund, L.R., Venezuela in Heavy Oil Push, September 15, 1980, Oil and Gas Journal.

Adams, F.H., Gatsis, J.G. and Silonia, J.G., UOP Inc., The Aurabon Process: New Way to Upgrade Tars and Heavy Crudes, UNITAR First International Conference on the Future of Heavy Crude and Tar Sands, Report No. 46.

Alberta Oil Sands and Research Authority, A Summary and Record of the First International Conference on the Future of Heavy Crude and Tar Sands.

Alberta Oil Sands Technology and Research Authority, Fourth Annual Report, March 31, 1979.

Alsands Project Group, Information Statistics, Alsands Project Group.

Andrianasolo, Colonel H., Rakoto-Andriantisilavo, M.D. and Raveloson, E., Office Militaire National Pour Les Industries, Stratesiques, Madagascar, The Bitumen and Heavy Oil Deposits in Madagascar: Feasibility and Prospects of the Bemolanga Deposit, UNITAR First International Conference on the Future of Heavy Crude and Tar Sands, Report No. 70.

Anonymous, New Catalyst Upgrades Athabasca Bitumen, July 9, 1979, Chemical and Engineering News.

Anonymous, Japanese Firm to Invest in Alberta Tar Sands, December 4, 1979, Oil and Gas Journal.

Anonymous, Steam Fireflood Delays Thwart Soviet Oil Flow, March 26, 1979, Oil and Gas Journal.

Anonymous, The Heavy Oil Industry in Venezuela, June 4, 1979, UNITAR Conference, Edmonton, Alberta, Canada.

Anonymous, Venezuela's Heavy Oil Development Prospects, March 6, 1980, Ministry of Energy and Mines.

Anonymous, The Crude from Jobo, 1971, Amoco.

Anonymous, Bechtel Group to Build Alsands Tar Sand Plant, February 25, 1980, Oil and Gas Journal.

Anonymous, Crude Oil Starts to Flow from the Sands of Athabasca, August, 1978, Process Engineering.

Anonymous, Alberta Oil Sands Technology and Research Authority, December, 1978, Catalogue of Aostra Technology, Issue: 1.

Anonymous, Alberta Poses Oil Sands Delay in Price Talks, March 24, 1980, Oil and Gas Journal.

Anonymous, Alsands Seeks More Tax, Royalty Concessions, April 2, 1979, Oil and Gas Journal.

Anonymous, Heavy Oil Project Advances in Alberta, November 12, 1979, Oil and Gas Journal.

Anonymous, New Athabasca Production Method Due Testing, January 29, 1979, Oil and Gas Journal.

Anonymous, In Situ Tar-Sand Field Test in Peace River Gets Permit, July 11, 1977, Oil and Gas Journal.

Anonymous, Alberta, Saskatchewan Heavy-Oil Plans Fused, October 16, 1978, Oil and Gas Journal.

Anonymous, Work on U.S. Oil Sands Heating Up, March, 1980, Science, Van Driesen, R.P. and Caspers, J., C.E. Lummus Company, Processing Heavier Crudes by Residual Hydrocracking, UNITAR First International Conference on the Future of Heavy Crude and Tar Sands, Report No. 50.

Anonymous, Reserves Higher for Belridge Field, August 4, 1980, Oil and Gas Journal.

Anonymous, Oil Sands, October 23, 1977, 27th Canadian Chemical Engineering Conference Oil Sands Symposium.

Anonymous, In the Matter of an Application of Esso Resources, Canada, Limited, Under Section 43 of the Oil and Gas Conservation Act and Section 27 of the Coal Conservation Act, October, 1979, Energy Resources Conservation Board, Alberta, Canada.

Anonymous, Big Alberta Carbonate Triangle Resource Seen, June 9, 1980, Oil and Gas Journal.

Anonymous, Siberian Enhanced-Recovery Tests Set, November, 1978, Oil and Gas Journal.

Anonymous, Work Still Lags on Soviet Heavy Oil Underground Mine, October 15, 1979, Oil and Gas Journal.

Anonymous, Siberian Enhanced-Recovery Tests Set; Soviet Oil Production Continues Short of Goals, November 27, 1978, Oil and Gas Journal.

Anonymous, Prospects for Enhanced Recovery from Bati Raman Field, Turkey: A Two-Billion Barrel Heavy Oil Reservoir, UNITAR First International Conference on the Future of Heavy Crude and Tar Sands, Report No. 29.

Anonymous, Russia Will Emphasize Improved Oil Recovery, February 19, 1979, Oil and Gas Journal.

Anonymous, Azerbaijan Underground Oil Mine to Start, May 7, 1979, Oil and Gas Journal.

Anonymous, Shale Oil, Tar Sands and Heavy Oils, British Petroleum.

Anonymous, Canada Sets Gas Find Record, Slows Oil Decline, March 31, 1980, Oil and Gas Journal.

Anonymous, New Processes Groomed for Bitumen, March 21, 1979, Oil and Gas Journal.

Anonymous, Synfuel from Tar, Oil Sands, Seen Promising, June 23, 1980, Oil and Gas Journal.

Anonymous, Joint Heavy Oil-Tar Sands Research Planned, June 18, 1978, Oil and Gas Journal.

Anonymous, Mining of Oil Could Multiply Recovery and Reserves at Conventional Fields, June, 1979, Engineering & Mining Journal.

Anonymous, New Canadian Agency to Monitor Oil Revenues, Costs; Alberta Unilaterally Hikes Oil and Gas Prices, August 11, 1980, Oil and Gas Journal.

Anonymous, Heavy Crudes Seen As Targets For Resid Processes, January 7, 1980, Oil and Gas Journal.

Anonymous, Politics Mires Canadian Oil Sands Work, February 11, 1980, Oil and Gas Journal.

Anonymous, Catalogue of AOSTRA Technology, Issue 1, Alberta Oil Sands Technology and Research Authority, December, 1978.

Anonymous, 1980 Refining Process Handbook, September, 1980, Hydrocarbon Processing.

Anonymous, An Analysis of Potential for Upgrading Domestic Refining Capacity, 1980, Prepared for AGA by Purvin & Gertz, Inc.

Anonymous, New Process Could Boost Gasoline Yield, April 7, 1980, Oil and Gas Journal.

Anonymous, The New Exxon RT-621 Catalyst, December, 1977, Exxon Research and Engineering Company.

Anonymous, In the Matter of an Application of GCOS, Ltd., Under Part 8 of the O&G Conservation Act, March, 1979, Canadian Energy Resources Control Board.

Anonymous, Alsands Fort McMurray Project, December, 1979, Energy Resources Conservation Board.

Anonymous, In the Matter of an Application of Esso Resources Canada, Ltd., October, 1979, Energy Resources Conservation Board.

Anonymous, Fourth Annual Report, March 31, 1979, AOSTRA.

Anonymous, R&D Trends in Heavy Oil Processing, July, 1980, Research Association for Residual Oil Processing.

Anonymous, Outlook of RAROP, Maruzen Oil of USA, Inc.

Anonymous, France Launches a New Energy Plan, August 6, 1980, Chemical Week.

Anonymous, BASF Catalysts, March, 1980, Technical Leaflet.

Araujo, J.B. and Roqueven, S.A., Experience on Direct Steam Injection in Lot 7 of the Morichal Field, 1977, The Oil Sands of Canada-Venezuela, CIM Special, Volume 17.

Balard, J.R., Thermal Recovery in the Venezuelan Heavy Oil Belt, April, 1977, The Journal of Canadian Petroleum, Vol. 16, No. 2.

Bayless, A., Alberta's Pricing Dispute with Ottawa May Stall Two Big Oil Development Projects, The Wall Street Journal.

Bearhart, J.A., More for Less from the Bottom of the Barrel, March, 1980, Presented at the 1980 NPRA Annual Meeting.

Berry, R.I., A Heavy, Sour Taste for Crude-Oil Refiners, May 19, 1980, Chemical Engineering.

Booz-Allen & Hamilton, Assessment of the Candidate Markets for Coal Liquid Boiler Fuels, December, 1979, Department of Energy.

Borresales, C.J. and Maraven, S.A., Steam Soak on the Bolivar Coast, 1977, The Oil Sands of Canada and Venezuela, CIM Special Volume 17.

Borresales, C.J., Production Characteristics and Oil Recovery in the Orinoco Oil Belt, UNITAR First International Conference on the Future of Heavy Crude and Tar Sands.

Bowan, C.W., Alberta Oil Sands Technology & Research Authority, Government Petroleum Incentives-Alberta, Canada, April 20, 1980, Paper for the First Joint SPE/DOE Symposium-Enhanced Oil Recovery.

Bridge, A.G. and Gould, G.D., Chevron Hydroprocesses for Upgrading Petroleum Residue, October 2, 1980, Presented at the JPI Symposium, Tokyo.

Bridges, J., Taflove, A. and Snow, R., IIT Research Institute, Radio-Frequency Heating to Recover Oil from Utah Tar Sands, UNITAR First International Conference on the Future of Heavy Crude and Tar Sands, Report No. 60.

Burke, D.P., Making Cat Crackers Work on Varied Diet, September 13, 1978, Chemical Week.

Burkill, G.C.C. and Maraven, S.A., Thermal Well Completion Design with Openhole Gravel Packed Liners and Methods for Selective Steam Injection, 1977, The Oil Sands of Canada-Venezuela, CIM Special Volume 17.

Burton, V.P. and Bennett, R.E., The Aurabon Process: New Way to Upgrade Tars and Heavy Oils–Refinery Economics, UNITAR First International Conference on the Future of Heavy Crude and Tar Sands, Report No. 98.

Cady, G.V., Hoffman, S.J. and Scarborough, R.M., General Crude Oil Company, Silverdale Combination Thermal Drive Project, UNITAR First International Conference on the Future of Heavy Crude and Tar Sands, Report No. 10.

Carleton, M., AOSTRA Data Base System, Alberta Oil Sands Technology and Research Authority.

Carrigy, M.A., Alberta Oil Sands Technology & Research Authority, AOSTRA International Cooperative Program for Oil Sands and Heavy Oil Technology Development, May 29, 1980, Presentation to the Seminar on Nonconventional Oil Technology, Calgary, Alberta.

Child, E.T., Texaco: Heavy Oil Gasification, August, 1974, Presented at University of Pittsburgh Symposium on Coal Gasification and Liquefaction: Best Prospects for Commercialization, Pittsburgh, Pennsylvania.

Chilingarian, G.V. and Yen, T.F., Editors, Bitumens, Asphalt and Tar Sands, 1978, Elsevier Scientific Publishing Co., Amsterdam.

Crow, P., Canada Looks to Tar Sands, Heavy Oil to Fill Conventional Oil Deficit, February 26, 1979, Oil and Gas Journal.

Cupps, C.Q., Marchant, L.D., U.S. Department of Energy, and Stosue, J.J., U.S. Department of Energy, Tapping the Vast Tar Sand Reserves of the U.S., September, 1979, World Oil.

Cyr, T.J., Alberta Oil Sands Technology and Research Authority, Status of AOSTRA International Agreements, May 29, 1980, Presentation to the Seminar on Nonconventional Oil Technology, Calgary, Alberta.

Dale, G.H. and Rogers, C.L. et al, Commercial Experience with Passivation of Metals on FCC Catalysts 1978, 1978 Proceedings–Refining Department, 43rd Midyear Meeting API.

Deroo, G., Institute Francais du Petrole, and Powell, T.G., Institute of Sedimentary and Petroleum Geology, The Oil Sands Deposits of Alberta: The Origin and Geochemical History, 1978, Verlag Chemie International, Inc.

Devaloria, S.E., Gulf Canada Resources, Ltd., Wabasca Fireflood Project, May 29, 1980, Presentation to the Seminar on Nonconventional Oil Technology, Calgary, Canada.

Ditman, J.F., Solvent Deasphalting—Versatility Preparation for Lube-Hydrotreat Feed, January 14, 1974, Oil and Gas Journal.

Doscher, T.M. and Huens, W., University of Southern California, Los Angeles, Steam-drive Performance Judged, October 22, 1979, Oil and Gas Journal.

Dunn, S.G., Alternative Approaches for Bitumen Upgrading, May, 1980, Dynawest Projects Ltd.

Eccles, R.M., H-Oil Processing of Tar Sand Bitumen for Maximum Oil Yield and Cash Flow, May, 1979, UNITAR, Johnson, C.A., Mongri, G., Lehnan, L.M. and Chervench, M.C., Conversion of Bitumen and Heavy Crude Oils to More Valuable Products by the H-Oil Process, 1977, 28th Annual Meeting of the Petroleum Society of CIM.

Edison, R.R., et al., Metals Removal for FCC Catalyst, May, 1976, Hydrocarbon Processing.

Energy Resources Conservation Board, Alsands Fort McMurray Project, December, 1979.

Energy Resources Conservation Board, In the Matter of an Application of Great Canadian Oil Sands Limited Under Part 8 of the Oil and Gas Conservation Act, March, 1979.

Energy Resources Conservation Board, In the Matter of an Application of Syncrude Canada Ltd. Under Part 8 of the Oil and Gas Conservation Act, September, 1979.

Federal Republic of Germany, Program for Energy Research and Technologies 1977-1980, Annual Report, 1977 on Efficient Uses of Energy, Fossil Sources of Primary Energy.

Filby, J.E., Flynn, P.C.,and Goldman, J.M., Direct Coking of Athabasca Tar Sand, February, 1980, The Canadian Journal of Chemical Engineering.

Finken, R.E. and Meldau, R.F., Phillips Petroleum Co., Phillips Solves Venezuelan Tar-Belt Producing Problems, July 17, 1972, Oil and Gas Journal.

Fisher, S.T., Electrical Induction Heating for the Processing of Fossil Fuels, July, 1978, Electronics & Power.

Franco, A., Petroleum Internacional, Thermal Work Humming in Venezuela, April 5, 1976, Oil and Gas Journal.

Fuchida, T., Laboratory and Field Experiments on Fire Flood Recovery Method, July, 1959, Producers Monthly.

Galavis, S. and Jose, A., Ministry of Mines and Hydrocarbons, Caracas, Velarte, H.M., Venezuelan Petroleum Corporation, Maracaibo, Geological Study—A Preliminary Evaluation of Potential Reserves of Heavy Oil of the Orinoco Tar Belt of the Eastern Basic, 1967, 7th World Petroleum Congress, Mexico City.

Gearhart, J.A., More for Less from the Bottom of the Barrel, March, 1980, Presented at the 1980 NPRA Annual Meeting.

Gembichi, V.A., Bennett, J.R., and Sikonia, J.G., Processing Requirements and Economic Analysis of Heavy Oil and Syncrude Refining, 1980, Presented at the 1980 UOP Process Division Technology Conference.

Glagoleva, O.F., Sosulina, L.M. and Smidovich, E.V., Effect of Asphaltenes in Hydraulic Tars on Coker Product Yields and Coke Structure, June, 1975, Chemistry and Technology of Fuels and Oils.

Gomi, S., Takahashi, R., Kiritani, Y., Amagi, Y. and Nishimura, Y., Thermal Cracking of Residual Oils Using Superheated Steam and Application of the Products, 1975, Kureha Chemical Industry Company, Ltd.

Greaser, G.R. and Shore, R.A., Getty Oil Co., Steamflood Performance in the Kern River Field, April 20, 1980, Presented at the First Joint SPE/DOE Symposium on Enhanced Oil Recovery at Tulsa, Oklahoma.

Grub, G.C. and Nathan, M.F., U.S. Patent 3,372,734, 1968, U.S. Government.

Gural, D., How Husky Makes a Profit in Heavy Oil Production, September, 1979, World Oil.

Guttierrez, Universidad Central de Venezuela, Facultad de Ingenieria, Escuela de Ingenieria de Petroleo, Occurrence of Heavy Crudes and Tar Sands in Latin America, UNITAR First International Conference on the Future of Heavy Crude and Tar Sands, Report No. 77.

Hall, E.S. and Tollefson, E.L., Recovery of Residual Bitumen from the Aqueous Tailings from Hot Water Extraction of Oil Sands by Air Sparging, UNITAR First International Conference on the Future of Heavy Crude and Tar Sands, Report No. 51.

Hardins, T.G., BP Exploration Canada Limited, Marguerite Lake Project Phase A, May 29, 1980, Presentation to the Seminar on Nonconventional Oil Technology, Calgary, Alberta.

Havlena, Z.G., Some Innovative Approaches Which May Facilitate Production of Heavy Crude and Tar Sands, UNITAR First International Conference on the Future of Heavy Crude and Tar Sands, Report No. 39.

Hournac, R., Kuhn, J. and Notarbartolo, M., Visbreaking: More Feed for FCC, December, 1979, Hydrocarbon Processing.

Humphreys, R.D. and Schutte, R., Syncrude Research, The Alberta Oil Sands: Energy for the Future, 1978, Verlag Chemie International, Inc.

Imperial Oil Limited, The Cold Lake Project, May, 1978, Energy Resources Conservation Board.

Kennedy, B.A., Edey, A., and Readdy, L.A., Golder Associates, Mined Oil—A Valuable Resource Opportunity, January, 1980, Mining Engineering.

Khalimov, E.M., Muslimov, R.K. and Yudin, G.T., Bitumen Deposits of the USSR and the Ways of Their Development.

Kitzan, P., Shell Canada Resources Limited, Design, Construction and Operations; Progress Report on the Peace River In-Situ Project, May 29, 1980, Presentation to the Seminar on Nonconventional Oil Technology, Calgary, Canada.

Kono, H., Central Research Institute, Ube Industries, Ltd., Chemical Engineering Approach to the Cracking and Gasification Processes of Residual Oils, Oil Sands Information Center.

Lanfranchi, E. and Amoven, S.A., Result of the Application of Alternating Steam Injection in Heavy Oil in Southeastern Venezuela, 1977, The Oil Sands of Canada-Venezuela, CIM Special Volume 17.

Lederhouse, T.P., Amoco Canada Petroleum Co., Gregoire Lake Block I Pilot; Project Description and Status, May 29, 1980, Presentation to the Seminar on Nonconventional Oil Technology, Calgary, Alberta.

Lopez, A.J. and Roqueven, S.A., Improvement in Heavy Crude Production by Squeezing Gravel and Oil Through the Perforations (Morichal), 1977, The Oil Sands of Canada-Venezuela, CIM Special Volume 17.

Lowenstein, R., The Orinoco Belt–Venezuela's Oil Bonanza, July 30, 1979, The Energy Daily.

Luhning, R.W., Alberta Oil Sands Technology and Research Authority, Availability of AOSTRA Technology, May 29, 1980, Presentation to the Seminar on Nonconventional Oil Technology, Calgary, Alberta.

Lurgi Corporation, The Lurgi-Ruhrgas (L-R) Process, January, 1980, Tar Sand Processing.

Mainland, G.G., Esso Resources Cold Lake Project, May 29, 1980, Presentation to the Seminar on Nonconventional Oil Technology, Calgary, Canada.

Matson, J.A., Coke is Entering Turbulent Era Now, March 24, 1980, Oil and Gas Journal.

McCan, T.J., Small Upgrading Plants, June, 1979, Presented at the First UNITAR Conference on the Future of Heavy Crude and Tar Sands.

McIntyre, H.C., Giant Oil Sands Plant Comes On Stream, September 11, 1978, Chemical Engineering.

Meldau, R.F. and Lumpkin, W.B., Producing Venezuelan Fireflood Wells, July, 1974, Phillips Petroleum 66.

Meldau, R.F. and Lumpkin, W.B., Phillips Petroleum Co., Phillips Testing Methods to Improve Drawdown and Producing Rates in Venezuela Fire Flood, August 12, 1974, Oil and Gas Journal.

Menzies, M.A., Scott, T.F. and Denis, J., The Canmet Residue Hydrocracking Process, 1980, American Institute of Aeronautics and Astronautics.

Meyer, R.F., The Future Supply of Nature-Made Petroleum and Gas, July 5, 1976, First UNITAR Conference on Energy and the Future.

Ministerio de Mineria Y Geologia, Oficios 154, Tar Sands in Cuba, UNITAR First International Conference on the Future of Heavy Crude and Tar Sands, Report No. 43.

Mungen, R., Amoco Production Co., and Nichols, J., Imperial Oil Ltd., Recovery of Oil from Athabasca Oil Sands and from Heavy Oil Deposits of Northern Alberta by In-Situ Methods.

Murphy, J.R., Whittington, E.L. and Chang, C.P., Review Ways to Upgrade Resids, September, 1979, Hydrocarbon Processing.

Muslimov, R.K., The Ways of Development of Heavy Oils and Bitumens in the USSR, UNITAR First International Conference on the Future of Heavy Crude and Tar Sands, Report No. 95.

Myers, R.W., Continental Oil Co., An Electrically Heated Buried Gathering System Transports High-Pour-Point Crude Oil, June, 1978, Journal of Petroleum Technology.

Myhill, N.A., A Check on Numerical Thermal Simulation, April 20, 1980, Presented at the First Joint SPE/DOE Symposium on Enhanced Oil Recovery at Tulsa, Oklahoma.

Nagel, C. and Flipping, J., Petro-Canada Explorations Inc., Project Overview: Kinsells Heavy Oil Pilot for Alberta Oilsands Technology & Research Authority Seminar on Nonconventional Oil Technology.

Makamura, M., Togasi, O. and Oho, T., Hydrodenitrogenation of Residual Oils, May 14, 1980, Paper Presented at the 45th Midyear Meeting API.

Newbold, F.R. and Perkins, T.K., Atlantic Richfield Company, Wellbore Transmission of Electrical Power, July, 1980, Technology.

Nicholls, J.H. and Luhning, R.W., Alberta Oil Sands Technology and Research Authority, Heavy Oil Sand In-Situ Pilot Plants In Alberta (Past and Present), 1977, Canada-Venezuela Oil Sand Symposium.

Nulty, P., The Tortuous Road to Synfuels, September 8, 1980.

Odle, J.D., et al., Clean Refinery Fuels by the Texaco Coal Gasification Process, 1978, 1978 Proceedings—Refining Department, 43rd Midyear Meeting API.

Ondish, G.F. and Suchanek, A.J., Gulf Science & Technology Company, Potential for Gulf Residual HDS in Upgrading Very Heavy Crudes, UNITAR First International Conference on the Future of Heavy Crude and Tar Sands, Report No. 20.

Pardo, A., Petroleos del Peru, Heavy Crudes in the Maranon Bason, Oriental Peru, UNITAR First International Conference on the Future of Heavy Crude and Tar Sands, Report No. 52.

Parra, D., Transport of Heavy Crudes, June, 1979, Presented at the First UNITAR Conference on the Future of Heavy Crude and Tar Sands.

Patents and Licensing Division, The Shell Gasification Process (SGP), Shell Oil Company.

Perozo, E., Orinoco Oil Belt Exploration, Evaluation, and Exploitation, July, 1980, Presented at the Latin American Oil Congress.

Phizackerley, Dr. P.H. and Scott, L.O., The British Petroleum Company Ltd. (Great Britian), 1967, Major Tar Sand Deposits of the World, Seventh World Petroleum Congress, Mexico City.

Potkins, R.A., Syncrude Canada Ltd., Status of the Syncrude Project, May 29, 1980, Presentation to the Seminar on Nonconventional Oil Technology, Calgary, Alberta.

Potkins, R.A., Status of the Syncrude Project, June, 1979, Presented at the First UNITAR Conference on the Future of Heavy Crude and Tar Sands.

Prats, M., Shell Development Company, A Current Appraisal of Thermal Recovery, August, 1978, Journal of Petroleum Technology.

Radvanyi, H., Oil Sands Information Centre, April 1, 1980, Alberta Oil Sands Information Centre.

Radvanyi, H., Oil Sands Information Centre and Its Relevance to Oil Sands and Heavy Oil Technology, Alberta Oil Sands Information Centre.

Rammler, R.W., Lurgi Gesellschaft fur Warme-und Chemotechnik m.b.h., The Production of Synthetic Crude Oil from Oil Land by Application of the Lurgi-Ruhrgas Process, The Canadian Journal of Chemical Engineering.

Rammler, R.W., The Production of Synthetic Crude Oil from Oil Sand by Application of the Lurgi-Ruhgas Process, October, 1970, The Canadian Journal of Chemical Engineering.

Rammler, R.W., One Process Retorts Three Feeds for Synthetic Crude, September, 1970, Mining Engineering.

Redford, D.A., Alberta Research Council, and McKay, A.S., Texaco Canada Resources Ltd., Hydrocarbon-Steam Processes for Recovery of Bitumen from Oil Sands, May 29, 1980.

Reid, T., Properties of Reservoirs and Crudes, Northern Orinoco Heavy Oil Belt, August 20, 1980, Notegra.

Reyes, A.G., Revista de la Socieda Venezolana de Ingeiards de Petroleo, (Magazine of Venezuelan Petroleum Engineers), Considerations in Producing the Orinoco Heavy Oil Belt, December, 1972.

Rosing, D.E., Heavy Oil In Western Europe: A Survey, UNITAR First International Conference on the Future of Heavy Crude and Tar Sands, Report No. 79.

Ruehl, W., Oil Mining in Germany, September, 1951, Proceedings Athabasca Oil Sands Conference, Alberta Oil Sands Information Centre.

Sakate, T and Yagi, T., Crack Resid with Spent HDS Catalyst, December, 1979, Hydrocarbon Processing.

Schultz, W.E. and Smith, H.D., Jr., Texaco Inc., Carbon/Oxygen Log Monitors Athabasca Tar Sands Recovery, December 31, 1979, Oil and Gas Journal.

Sepulveda, J.E. and Miller, J.D., Separation of Bitumen from Utah Tar Sands by a Hot Water Digestion-Flotation Technique, Mining Engineering.

Servelo, J.P. and Meneve, S.A., Filial de Petreos de Venezuela, Distrito San Tome, Performance History Improved Recovery by Gas Injection in Heavy Oil Reservoirs Oveja Field in Eastern Venezuela, UNITAR First International Conference on the Future of Heavy Crude and Tar Sands.

Shah, G.N. and Sherman, L.G., Residfining Process Upgrader Broad Range of Heavy Crudes, May, 1979, Exxon Research and Engineering Company.

Shah, G.N., Sherman, L.G. and Christ, M.A., Residfining for Better Fuel Oil, May, 1979, Hydrocarbon Processing.

Shepherd, D.W., Steam Stimulation Recovery of Cold Lake Bitumen, UNITAR First International Conference on the Future of Heavy Crude and Tar Sands, Report No. 22.

Sikonia, J.G. and Hemler, C.L., UOP, The Aurabon Process: New Way to Upgrade Tar and Heavy Crudes Product Processing, UNITAR First International Conference on the Future of Heavy Crude and Tar Sands, Report No. 69.

Sloan, A., Ingham, R and Mann, W.L., Pipeline Transportation of Heavy Oils, June, 1979, Presented at the First UNITAR Conference on the Future of Heavy Crude and Tar Sands.

Stewart-Gotfon, J., Prospects, Ability, Cash; Venezuela Has It All, January, 1980, World Oil.

Strausz, O.P. and Lown, E.M., Oil Sand and Oil Shale Chemistry, 1978, Verlag Chemie.

Takeuchi, C., Nakamura, M. and Shiroto, Y., Chiyoda Chemical Engineering and Construction Co., Ltd., Resid Upgrading, Gas Oil Cracking: Upgrading of Residual Oils by Asphaltene Conversion, April 1, 1979, American Chemical Society Honolulu Meeting.

Thurber, J.L. and Welbourn, M.E., Shell Oil Company, How Shell Attempted to Unlock Utah Tar Sands, November, 1977, Petroleum Engineer.

Tominaga, I., Itoh, S. and Yashiro, M., Hydrovisbreaking of Athabasca Bitumen and Its Reaction Mechanism, May, 1977, Bulletin of the Japan Petroleum Institute.

Ueda, M., Mitsui Engineering and Shipbuilding Co., Ltd., Fujinagata Laboratory, Funaki, M., Mitsui Kozan Coking Industry Co., Ltd., Kitakyushu Works, M-Gas Process for Production of Fuel Gas from Heavy Oil, UNITAR First International Conference on the Future of Heavy Crude and Tar Sands, Report No. 4.

Van Driesen, R.P., Wysocki, E. and Bell, R.A., LC-Fining Applied to Bitumen from Oil Sands, October 1, 1977, Presented at 27th Canadian Chemical Engineering Conference.

Van Zijill Langhout, W.C., Ouwerkerk, C. and Prank, K.M.A., Development of and Experience with the Shell Residue Hydroprocesses, June 8, 1980, Paper Presented to the American Institute of Chemical Engineers.

Vandermeer, J.G. and Presber, T.C., Union Oil Company of Canada Limited, Heavy Oil Recovery from the Grosmont Carbonates of Alberta, May 29, 1980, Presented to the Seminar on Nonconventional Oil Technology, Calgary, Alberta.

Vermillion, W.L., Jr., U.S. Patent 4,025,458, October 21, 1974, UOP Inc.

Viloria, D.A., Krasuk, J.H., Rodriguez, O., Buenafama, H. and Lubkowitz, J., Solvent Deasphalt Boscan Crude, March, 1977, Hydrocarbon Processing.

Washimi, K. and Kanbayashi, M., Iwaki-shi, Fukushima, Japan, Kanai, T. and Fujimori, A., Yokohama-shi, Kanagawa, Japan, Apparatus for the Thermal Cracking of Heavy Hydrocarbon Oils, June 18, 1974, Athabasca Oil Sands Index.

Watson, L.M., GCOS Future Looks Bright With Improved Mining Methods, Rising Oil Prices, June, 1979, Canadian Mining Journal.

Wels, B.B., Alberta Energy Company, Alberta Technology and Research Authority, Kaiser Oil, Ltd., Westcoast Petroleum Ltd., Suffield Heavy Oil Pilot(shop) Project, Alberta Energy Company, March, 1980.

Yamaguchi, F., Sokai, A., Yoshitake, M. and Saegusa, H., COSMOS Cracks Crude to Olefins, September, 1979, Hydrocarbon Processing.

Zanker, A., Visbreak Resid: Figure Value, November, 1978, Hydrocarbon Processing.

Zheltov, Y.P., Boxerman, A.A., Rakovsky, N.L., Egurtsov, A.N., Sakhalinneft, O., Maltsev, M.B., Sivous, M.B. and Kleshev, V.A., Sakhalin Research and Design Petroleum Institute, Studies of Oil Displacement Efficiency with Heat Carrier Slug, 1974, Oilsand Symposium, Moscow.

Other Noyes Publications

MATERIALS OF CONSTRUCTION FOR ADVANCED COAL CONVERSION SYSTEMS

by Vinod K. Nangia

The Engineering Societies Commission on Energy, Inc. (ESCOE)

Energy Technology Review No. 75

This book describes materials of construction, and materials problems for equipment used in advanced coal conversion systems. The need for cost effective industrial operation is always a prime concern, particularly in this age of energy consciousness. Industry is continually seeking improved materials for more efficient systems. The information presented here is intended to be of use in the design and planning of these systems.

Coal conversion and utilization impose severe demands on construction materials because of high temperature, high pressure, corrosive/erosive, and other hostile environmental factors. Successful economic development of these processes can be achieved only to the extent that working materials can withstand increasingly more aggressive operating conditions.

The book, which reviews present and past work on the behavior of materials in the environments of advanced coal conversion systems, is divided into three parts: atmospheric fluidized bed combustion, coal gasification and liquefaction, and advanced power systems. A **condensed table of contents** is given below.

ISBN 0-8155-0884-0 (1982) **511 pages**

Other Noyes Publications

HYDROPROCESSING CATALYSTS FOR HEAVY OIL AND COAL

Edited by M.J. Satriana

Chemical Technology Review No. 202
Energy Technology Review No. 74

Hydroprocessing catalysts, as described in over 230 processes covered in this book, are hydrogenation catalysts used in the upgrading of heavy crudes and coal to products expected to be in great demand as the world's primary oil supplies gradually dwindle. The techniques employed in hydroprocessing result in the removal of contaminants, the transformation of lower grade materials such as heavy crudes to valuable fuels, or the conversion of hydrocarbonaceous solids into gaseous or liquid fuel products. All of these techniques are, of course, carried out in the presence of hydrogen.

Some of the brightest energy prospects for the future lie in heavy oil reservoirs and coal reserves. Heavy oils, defined in this book as having gravities of $<20°$API, are crudes so thick that they are not readily extracted from their reservoirs. However, processing of these crudes is of great importance, because the U.S. resource alone is enormous.

The main types of processing catalysts covered in the book are hydrorefining catalysts for the removal of contaminants such as metals, asphaltenes, sulfur, and nitrogen from heavy oils; hydrocracking catalysts, for thermal decomposition to lower boiling, more valuable fractions; plus some combinations of the two.

Catalysts for the conversion of hydrocarbonaceous materials to gaseous or liquid fuels are also covered. The primary starting material for these conversions is coal, but wood, lignin, oil shale, tar sands, and peat are other possibilities. The final chapter describes the preparation of various catalyst support systems.

Listed below is a condensed table of contents giving **chapter titles and selected subtitles.** Parenthetic numbers indicate the number of processes per topic.

ISBN 0-8155-0883-2 (1982)

308 pages

Other Noyes Publications

COAL-OIL MIXTURE COMBUSTION TECHNOLOGY

Edited by M.M. Schumacher

Energy Technology Review No. 73

This book reflects the current status, on an international basis, of coal-oil mixtures, their preparation and stabilization, combustion, and commercialization. The book is based on selected papers from the Third International Symposium on Coal-Oil Mixture Combustion sponsored by the U.S. Department of Energy.

The objective of any coal-oil mixture (COM) combustion program is generally to reduce domestic dependence on imported petroleum products, by substituting coal for an appreciable portion (an estimated 300,000 barrels/day in the case of the U.S.) of the oil used in existing boilers and combustors. COM has been demonstrated to be a viable and economic alternate fuel for potential utility and industrial applications. It is anticipated that the complete commercialization of COM will be accomplished in the near-term time frame (1982-1985) with private sector impetus. The extent to which COM will be used to replace oil in the mid to late 1980s will depend on the relative prices of coal and oil, and the site-specific economics of COM vs oil.

The chapters in the book cover the various aspects of COM technology. After an overview of research and development in the U.S. and abroad, specific chapters cover demonstration projects, pilot and bench scale research, COM preparation and stabilization, equipment, COM rheology, commercialization and economics, and combustion and emissions. The condensed table of contents listed below gives **chapter titles and selected subtitles.**

ISBN 0-8155-0878-6 (1981) **480 pages**

Other Noyes Publications

HOW TO DISPOSE OF OIL AND HAZARDOUS CHEMICAL SPILL DEBRIS

Edited by A. Breuel

Pollution Technology Review No. 87

This book describes various techniques which can be used to dispose of the collected debris from oil and hazardous chemical spills. It is based on research prepared by *SCS Engineers* and *CONCAWE* (the European oil companies' international study group for conservation of clean air and water).

Engineering constraints and equipment requirements for handling and disposal are evaluated. Debris management aspects considered are storage, transport, treatment, reprocessing, and disposal. Hardware and processing systems are identified and conceptual transport/disposal plans are developed. A literature review and several case studies are included. U.S. and European technology are covered.

Disposal is a complex problem. Spill incidents which have occurred in recent years have highlighted the difficulties in dealing with the large quantities of emulsions and debris which are collected during control and cleanup operations. Close cooperation between authorities and industry is a necessity at every stage, to ensure disposal in an environmentally acceptable, cost-effective, and energy-conserving manner.

Each case is different and no specific rules can be formulated, but guidance is given on the choice of options and their state of development. The condensed table of contents listed below gives **chapter titles and selected subtitles.**

ISBN 0-8155-0876-X (1981) **420 pages**

Other Noyes Publications

COAL GASIFICATION PROCESSES

Edited by Perry Nowacki

Energy Technology Review No. 70

This book presents, in concise form, the important processes, including latest available technology and economy, for converting coal into gaseous fuels.

The availability of coal in the U.S. exceeds that of all forms of petroleum and natural gas combined; thus it seems only natural to pursue the preparation of clean fuels from this source.

Processes are discussed based on gasifier categories—fixed bed, fluidized bed, entrained bed or molten bath bed. Project histories, process descriptions, operating conditions, reactants, utilities, products, efficiency, test results, and other process considerations are included as available.

Below is a **condensed table of contents, including selected subtitles.**

ISBN 0-8155-0864-0 (1981) **378 pages**

Other Noyes Publications

POLLUTION CONTROL TECHNOLOGY FOR INDUSTRIAL WASTEWATER

Edited by D.J. De Renzo

Pollution Technology Review No. 80

This book provides an extensive survey of the reliability and effectiveness of 56 unit operations in industrial water pollution control. These operations include 32 generic wastewater treatment technologies, classified as preliminary, primary, secondary or tertiary, and 24 sludge treatment and disposal technologies.

Each process is briefly described by the type of control equipment required, the major variations of design, flow diagrams, and information on the following: design criteria, common modifications, typical performance, applications and limitations of the process, reliability, chemicals required for operation, residuals generated, and environmental impacts.

Summary tables for most technologies are provided showing the concentrations of various pollutants in the effluents, the minimum, maximum, median and mean removal efficiencies, and the number of data points used to generate this information. Conventional pollutants as well as EPA-categorized "priority pollutants" and "hazardous substances" are covered.

The information reviewed here should be extremely useful to engineers, management personnel, and others involved with regulatory requirements, guidelines, and decisions. **Contents** are:

ISBN 0-8155-0855-7 (1981)

712 pages

Other Noyes Publications

OIL SHALE TECHNICAL DATA HANDBOOK

Edited by Perry Nowacki

Energy Technology Review No. 63
Chemical Technology Review No. 182

A ready reference for the evaluation of appropriate technology for oil shale development, this book will be extremely useful to engineering and managerial personnel. The book defines the oil resource and lists the properties of shale and the oil and gas derived from it. Recovery technologies potentially compatible with the resource are described. Associated processes, such as mining, materials handling, beneficiation, upgrading, wastewater treatment, and spent shale disposal, have also been included. Emphasis has been placed on documenting available design information such as maximum module size, operating conditions, yields, utility requirements, etc.

The following is a partial table of contents, illustrating chapter headings and some important subtitles.

ISBN 0-8155-0835-2 (1981) **309 pages**

Other Noyes Publications

FUELS FROM BIOMASS 1980

Edited by J.S. Robinson

Energy Technology Review No. 61
Chemical Technology Review No. 176

A study of biomass resources and the various processing options are provided in this book, ranging in size from laboratory scale to commercially proven, which can be applied to convert biomass feedstock to gaseous or liquid fuels.

Consideration is given to sources of biomass, such as municipal wastes, manures, crop and forestry residues and energy farms yielding crops such as trees and plants. Thermal and biochemical conversions are discussed in detail, with emphasis on case study technology including processing data and economic considerations.

This book should be of particular interest to engineers, researchers and managers concerned with reducing fuel consumption from conventional sources and increasing the use of so-called "synthetic" fuels. The chapter titles and examples of subtitles are given below.

ISBN 0-8155-0824-7

377 pages

Other Noyes Publications

SMALL AND MICRO HYDROELECTRIC POWER PLANTS 1980

Technology and Feasibility

Edited by Robert Noyes

Energy Technology Review No. 60

The need for alternative sources of energy, as well as new capacity, is becoming more urgent as fossil fuel becomes less available and more expensive; thus there is increasing interest in the development of small scale (<15 MW) or low head (<20 m) hydroelectric power plants.

This book provides a general survey of the feasibility and technology for small scale and micro (<100 kW) hydroelectric power production and is intended for use by public and private agencies and utilities, private investors, engineers, researchers, and educational institutions. The material presented is based on research conducted by the U.S. Army Corps of Engineers, Stone & Webster, and the National Center for Appropriate Technology. Part 1 is a brief discussion of the resource potential for 6 regions comprising the U.S. Part 2 details the feasibility of a small scale facility, including discussions on economic and financial analysis, hydrologic studies, facility integrity, electromechanical systems, and civil features. Part 3 gives data on available equipment, and Part 4 presents an information package for the individual investor or property owner who is considering micro hydropower—from the initial idea to the final production of power.

Today, with fuel costs escalating, it would seem that future economic conditions will favor small scale hydroelectric power generation for at least a part of the national power supply. The fundamental concepts in the design and construction of a small scale plant are that the impacts of implementation are modest; the project is likely to be noncontroversial; and, therefore, license and permit granting should be possible within a relatively short time frame. A condensed table of contents is given below.

ISBN 0-8155-0819-0 **457 pages**

Other Noyes Publications

LIGNITE TECHNOLOGY 1980

Edited by Perry Nowacki

Energy Technology Review No. 53
Chemical Technology Review No. 146

This book discusses the present and future use of lignite for electric power generation, and the problems associated with the use of low-rank coals.

Lignite and other subbituminous coals represent a large domestic, yet relatively unused source of low-cost energy. The abundant deposits are located largely in the Northern Great Plains and Rock Mountain regions, as well as in Texas, and are near the surface.

Lignite is a low-sulfur coal, but produces copious amounts of alkaline fly ash, and ash fouling of boiler tubes is the most serious operating problem when burning it. Yet lignite has high permeability and reactivity which properties make it a superior feedstock for gasification and liquefaction.

A partial and condensed table of contents with **chapter headings and examples of some subtitles** follows here.

1. LIGNITE RESOURCES
Reserves of U.S. Low-Rank Coals
Ash Composition

2. MINING & RECOVERY
Drainage Control
Cleaning and Blasting Exposed Lignite
Magnitude of Proven Reserves
Sodium Content of Lignite Ash
Mining of Texas Lignite
Mining of Estevan Lignite (Canada)
Opencast Mining in Germany
Reclamation of Lands

3. TRANSPORTATION & STORAGE

4. DRYING & PULVERIZATION
Commercial Scale Drying
Handling and Shipment Characteristics
Density of Lignite in Railroad Car
Lignite Pulverization
Status of Pulverized Firing
Grindability
Pellet-Forming Process

5. COMBUSTION
Modern Boiler Design for Lignite Firing
Ash Fouling Research
Coal Creek Station Steam Generators
Lignite Handling System
Furnace Designs
Soot Blowers
Electrostatic Precipitators
Spray Tower Scrubbers
Fluidized-Bed Combustion
Sulfur Retention
Nitric Oxide Emission
Combustion Efficienty
Experience in Latrobe Valley (Australia)

6. GASIFICATION
CO_2 Acceptor Process
Rapid City Pilot Plant
Mathematical Model
Intermediate Fines Buildup
Process Variables
Alternate Acceptor
Slagging Fixed-Bed Gasification
Operational Considerations
Slag Characteristics
Oxygen-Steam Ratio
Taphole Erosion
Stability of Operation
Effect of Pressure on Gasification Parameters
HYGAS Process
Catalytic Methanation
High-Temperature Winkler Gasification
Hydrogasification of Brown Coal Using Process Heat from Nuclear Reactors
Hydrogasification Processes in Germany

7. LIQUEFACTION
CO-Steam Process
Batch Reaction Kinetics
Time-Sampled Autoclave
Conversion and Coking
Rates of Reaction for Carbon Monoxide and Hydrogen
Mechanism and the Future of the CO-Steam Process
Process Development of Lignite Liquefaction
Studies on Liquefaction
Typical Process Development Unit Operation
Flash Hydropyrolysis Process for Conversion of Lignite
U.T. Process for Carbonization of Texas Lignite

ISBN 0-8155-0783-6 **228 pages**

Other Noyes Publications

COAL LIQUEFACTION PROCESSES 1979

by Perry Nowacki

Chemical Technology Review No. 131
Energy Technology Review No. 45

With the precarious position of petroleum supplies in the world market and the unfavorable balance of payments caused by overdependence on foreign sources, it may be expedient to exploit coal further as a source of liquid fuels. In the United States coal is the most abundant natural energy resource. Through its conversion to liquid fuels, it could supply a portion of the energy required for transportation, heating, etc.

Coal can furnish a complete range of liquid fuels, as well as chemical feedstocks. To achieve maximum yields, however, judicious adjustment of temperature, pressure, time and catalyst, as elaborated in this book, is essential. The text reviews the histories of important domestic and foreign coal liquefaction processes—pyrolysis, solvent extraction, and catalytic and indirect liquefaction. It details the chemistry of the processes and their inherent problems, also discussing economics, developmental status, and environmental impacts.

Chapter headings and **examples of some** important subtitles follow below:

1. **INTRODUCTION**
2. **COAL RESOURCES**
3. **COAL LIQUEFACTION CHEMISTRY**
4. **OVERVIEW OF PROCESSES**
5. **PYROLYTIC PROCESSES**
   Char Oil Energy Development (COED)
   Clean Coke
   TOSCOAL
   Occidental Flash Pyrolysis
   Coalcon Hydrocarbonization
   Oak Ridge Hydrocarbonization
   Flash Hydropyrolysis of Lignite
   Flash Hydrogenation
   Lurgi-Ruhrgas Flash Carbonization
6. **SOLVENT EXTRACTION**
   Solvent-Refined Coal (SRC) Process
   Pilot Plants—Ft. Lewis & Wilsonville
   Consol Synthetic Fuel Process
   Cresap Pilot Plant
   Exxon Donor Solvent Process
   CO-Steam Process
   Supercritical Gas Solvent Extraction
   UOP Solvent Extraction
7. **CATALYTIC LIQUEFACTION**
   H-Coal Process
   Trenton, N.J. Pilot Plant
   Synthoil Process
   Clean Fuel from Coal Process
   Zinc Halide Coal Liquefaction
   Dow Coal Liquefaction
8. **INDIRECT LIQUEFACTION**
   Fischer-Tropsch Process
   SASOL Commercial Plant, South Africa
   Mobil M Gasoline Process
9. **COAL LIQUIDS FROM H-COAL**
   Analysis and Properties of Products
   Distillate Separation & Characterization
10. **SOLVENT-REFINED EXTRACT—COMMERCIAL, EXPANDED-BED HYDROPROCESS**
11. **ECONOMICS OF SRC PROCESS**
   Reclaiming, Preparing, Liquefying Coal
   Hydrogenation of Fuel Oil and Naphtha
   Stack Gas Desulfurization
   Hydrogen Manufacture
   Capital Investment and Costs
12. **ENGINEERING ASSESSMENT OF SYNTHOIL PROCESS**
   Feed Slurry Pumping and Heating
   Reactor Design
   Removal of Spent Catalyst
   Low-Temperature Carbonization
13. **FISCHER-TROPSCH DESIGN & COST**
   Recovery of Heat, Liquids, Sulfur
   Gasification and Methanation
   Water Systems
14. **EQUIPMENT AND MATERIALS**
   Pumps, Heat Exchangers, Fired Heaters
   Vessels and Pipe Systems
   Instrumentation and Metallurgy
15. **ENVIRONMENTAL, HEALTH AND SAFETY ISSUES**
   Solid Waste Management
   Hazards in Product Transfer and Storage
   Federal Legislation
16. **COMPARISON OF PROCESSES**
   Technical and Economic Aspects
   Probability of Commercial Success

ISBN 0-8155-0756-9 339 pages